Farming in Miniature

STEERING WHEEL CONTROL

METTOYS *for the Farmer*
– fitted with powerful motors

STEERING WHEEL
CONTROL

CLICK CLICK CLICK

3262/1
FARM TRACTOR

Powered by strong clockwork contained in rear driving wheel. Designed to pull farm implements, includes detachable driver and steering wheel control. Packed in decorated box as illustrated.

7½" x 5" x 5¼"	15 ozs.
190 x 125 x 135 mm.	425 gms.

HEAVY STEEL MECHANICAL
FARM TRACTOR
MADE IN GREAT BRITAIN
PERFECT MECHANICAL TOYS
METTOY PLAYTHINGS

FITTED WITH
RUBBER TYRES

3264/1
GIANT TRACTOR

Heavy model fitted with rubber tyres. Driven by strong clockwork in rear wheel. Cranking "Starting Handle" causes realistic noise. Designed to pull farm implements. Equipped with steering wheel control and driver. Packed in strong, illustrated box.

9¼" x 6½" x 5¾"	23 ozs.
235 x 160 x 145 mm.	644 gms.

METTOY
PLAYTHINGS

Farming in Miniature

A Review of British-Made Toy Farm Vehicles up to 1980

Volume 2: Dinky to Wend-al

Robert Newson – Peter Wade-Martins – Adrian Little

With contributions by Brian Bastiman and Mervyn Spokes

Old Pond
PUBLISHING

ISBN 978-1-908397-56-0

A catalogue record for this book is available from the British Library

Published by

Old Pond Publishing
5M Publishing Ltd
Benchmark House, Smithy Wood Drive
Sheffield, South Yorkshire
S35 1QN
United Kingdom

www.oldpond.com

Cover design and book layout by Liz Whatling
Printed by Beamreach Printing (www.beamreachuk.co.uk)

ACKNOWLEDGEMENTS

The authors are grateful to the following who allowed their collections to be photographed, provided us with photographs or helped with information: Peter Baron, David Bastiman, John Begg, Colin Boor, Philip Bowdidge, Alwyn Brice, Barney Brown, Nigel Cooper, Ray Crilley, Philip Dean, Clive Drakes, Steve Ebbage, Keith Elphick, Stuart Gibbard, Peter Gilpin, Richard Gray of Grays International, Warner Hall, Eric Hewes, Lucie Hobbs, Hotchkiss Group, Norman Joplin, Robin Ketley, John King, Jim Lindsay, Dave and Terry Lucas, Hans and Marco Ludwig, Shaun Magee, Graham Miller, Anthony Molay of ASAM Models, Rod Moore, Paul Morehead of Plastic Warrior Magazine, Alfred Plath, Bas Poolen, Martyn Reeve, Roy and Heather Rowswell, Jim Russell, Ken Simmons, Derek Smith, Richard Smith, Special Auction Services, Mervyn Spokes, Patrick Trench, Alan Turner, V. & A. Museum of Childhood, Vectis Auctions, Hugh Walters, West Wales Museum of Childhood.

In addition, the main authors would like to express a particular debt of gratitude to Brian Bastiman for acting as copy editor. His support and encouragement has helped us see this demanding project through to completion.

Contents

Introduction to Volume 2

So much material has been assembled for the book that we have divided it into two volumes. Each manufacturer has been given its own chapter, identified by the brand name rather than the manufacturer's name (e.g. Dinky rather than Meccano Ltd). Volume 1 covers manufacturers alphabetically from Airfix to Denzil Skinner, while this volume, Volume 2, covers Dinky to Wend-al. Also in this volume are a Guide to Harnessed Horses, allowing the easy identification of horses that may have become separated from their carts or wagons, and two indexes (covering both volumes): a Model Index for particular tractors and types of model, and a General Index for personalities, companies etc.

The Introduction to Volume 1 explained the scope of the book, as well as giving a historical survey of British farm toys in the twentieth century, highlighting the changes in materials and the development of toys as scale models. Toy collecting was also discussed. In this volume are the Conclusions, which reiterate and pull together some of these themes.

A Glossary of technical terms appeared in Volume 1. We stated in the Introduction in Volume 1 that the Glossary would be included in both volumes, but rather than repeat information, we thought it better to use the space for an Addenda with details of additional items for the manufacturers in Volume 1 that have come to our attention since the book was finished. One of the fascinating aspects of toy collecting is that previously unrecorded variations seem to turn up quite regularly, so that in the time since Volume 1 was completed (less than a year) we have filled several pages with new finds!

Company History

'Dinky' is synonymous with British diecast toys in the same way that 'Biro' is with ballpoint pens. It is a household word, and millions of Dinky Toys were sold in Britain and throughout the world before and after the last world war. The story of the Meccano company, which manufactured the toys, was a remarkable one of rapid pre-war and post-war growth and slow decline from the mid-1950s. It has been told in an excellent book *Factory of Dreams – A History of Meccano Ltd, 1901-1979* by Kenneth Brown (2007). The following summary of the company's history draws heavily on that book.

Meccano Ltd. of Liverpool, which produced Meccano sets, Hornby trains and Dinky toys, was the brainchild of Frank Hornby (1863-1936). In a quarter of a century he transformed his hobby of toy making into an international company with annual sales of almost £5 million. In 1901 he finalised his patent for his sets of 'Mechanics Made Easy' construction kits. His business grew slowly to start with, but by 1906 he was making a profit. He named his construction kits 'Meccano' in 1907. With expanding domestic and overseas sales he moved into a purpose-built factory at Binns Road in Liverpool in 1914. By 1915 Meccano was the largest toy company in the world.

The first clockwork train appeared in 1920, and the first electric train in 1925. By 1930 Hornby train sets made up more than 40 per cent of the Binns Road output. The move into a new range of diecast toys, called 'Modelled Miniatures', came in 1933. To begin with, the accuracy and quality of the toys was poor. Many models were generic approximations, and unsatisfactory metal alloys led to metal fatigue. The remarkably successful name 'Dinky' was given to the toys in 1934, but it was only in 1939/40 that Dinky raised their standard to the level of excellence which ensured they dominated the post-war market for over ten years. By 1938 there were some 300 Dinky items in the range which, by 1939, represented 16 per cent of the company's sales.

During the war, production was switched to munitions, but when toy making resumed in 1946, Dinky production was given higher priority, with 60,000 items sold in that year. With the high quality they were now achieving, sales rose to 1,281,000 a year by 1954. Dinky Toy output doubled between 1951 and 1956. They were usually designed at a very popular scale of a quarter of an inch to a foot, or 1:48. Farm models, however, were normally at 1:38, and the Halesowen Harvest Trailer was, for some reason, at 1:45. A photograph of an original Meccano list of scales, dated 30 October 1949 was published in Richardson, M. & S. (1981), pp.293-4. Models to match the 00 ('Dublo') gauge railways were at 1:77.

The company's profitability peaked in 1956, but declined sharply from 1958. This was largely due to the failure of senior managers to modernise methods of production and develop new products. Changes, when they came, were forced upon the company by competition, rather than being the result of innovative review and development. Complacent managers, deeply entrenched production methods and uncompetitive pricing caused the inexorable decline of the company during the 1960s. It was seriously in debt by 1963. In 1964 the Meccano shareholders accepted a cut-price take-over by Lines Brothers, the makers of Tri-ang toys. Roland Hornby, who had run the company since his father's death in 1936, resigned.

Strong competition came first from Matchbox toys from 1953 and Mettoy's remarkable new range of Corgi Toys from 1956; these raised standards of realism, detailing and attractive packaging. Then came Tri-ang's equally competitive 'Spot-On' range from 1959. During the 1960s these three companies quickly took a large share of the quality diecast market.

In response to declining profits, Lines Brothers soon replaced senior management at Meccano and re-organised production and marketing. The Dinky models were modernised and the range of toys was extended. Nevertheless, conservative and inefficient working practices remained. Trading profits continued to fall during the 1960s, made worse by strong American competition. In 1971 Lines Brothers went into voluntary liquidation, partly because Meccano had placed such a severe strain on the group's resources.

The next company brave enough to take on Meccano was Airfix Industries in 1971. Meccano remained a self-contained trading unit with its own identifiable products but still did little to revive product development. In the first year, profits were made, but then in 1972/3 these changed to heavy losses. Senior managers were still resistant to more modern working practices, and the sales director even rejected the idea that their products should be sold through supermarkets. Year after year there were heavy losses, and for 1975/6 there was a deficit of almost £545,000.

This could not last, and in November 1977 the staff were all given termination notices with a view to starting afresh on a green-field site at Huyton. This proved unworkable, and Airfix laid off the entire workforce at the end of November 1979, announcing that Meccano was for sale. The Airfix group went into liquidation in early 1981 with debts of £15 million. Meccano Ltd had dragged down both Lines Brothers and Airfix. Dinky products had actually been uncompetitive since the 1950s, and it was ironic that they had managed to hang on almost as long as their superior competitors, Lesney and Corgi. No doubt it was the name Dinky which had helped them to survive for a while.

Models

All Dinky farm models were of tractors and their implements. No horse-drawn items were ever made, even though at the time Meccano started making diecast toys, horses were still plentiful in the countryside. This was, no doubt, a reflection of Frank Hornby's personal enthusiasm for technological innovation.

The main sources used here are Mike and Sue Richardson's *The Great Book of Dinky Toys* (2000), Cecil Gibson's *A History of British Dinky Toys 1934-1964* (1966), extracts from the *Meccano Magazine* in Peter Randall's *The Products of Binns Road: A General Survey* (1977; revised 1981) and contemporary Dinky catalogues. These catalogues do, however, have to be used with great care, because changes were made to the models which were not always reflected in the catalogue illustrations or even on the boxes. The boxes are sometimes difficult to date, because books and articles tend to focus on the history of the models, and boxes are usually ignored.

No.22e Tractor (1933–41)

In 1933 Hornby produced their first tractor, modelled on the Fordson Model N with low mudguards, as one of six vehicles in the Modelled Miniatures no.22 Motor Vehicles boxed set illustrated in Richardson, M. & S. (2000), p.25. All six were available for sale by October 1933, and the tractor was numbered 22e. The vehicles in the set were the Sports Car, Sports Coupé, Motor Truck, Delivery Van, Farm Tractor and Army Tank. Most were withdrawn in 1935, but the Army Tank lasted until 1939 and the Farm Tractor until 1941.

The first version of the tractor had *HORNBY SERIES MECCANO LTD LIVERPOOL* cast underneath, and was without a tow hook. It was sold in the boxed set or separately for 9d. (**1** to **3**). In April 1934 Modelled Miniatures were re-named Dinky Toys, and the lettering underneath the tractor was changed to *MECCANO DINKY TOYS MADE IN ENGLAND* and it was given a tow hook (**5**). The first 1933 version without a hook was made entirely of lead except for the steel axles, and was usually blue and yellow on red wheels (**1** and **2**), although we have seen one in yellow and green (**3**). The underneath could be bare metal or coloured the same as the engine block and rear mudguards. The 1934-41 tractors with tow hooks were also made of lead, except that some had the front wheels cast in zinc alloy, and these are now often cracked or broken with metal fatigue (**7** and **12**). The second version was painted in a wide range of two-colour combinations (**4** to **13**), but a genuine all-yellow tractor on red wheels is known (**14**). The underneath was usually bare metal. The tractor when sold separately was delivered to shops in cream (**15**) or yellow (**16**) trade boxes of six.

No.27a/300 Massey-Harris/Massey-Ferguson Tractor (1948–71)

Toy production ceased in 1941 for the rest of the war and was not resumed until 1946. The first post-war tractor was the no.27a Massey-Harris, which appeared in June 1948 (**17** to **21**). *MASSEY-HARRIS* was on transfers on both sides of the engine block. The boxes bought in by Meccano had the serial numbers followed by capital letters (as in no.27A), but the toys were actually catalogued and advertised with lower-case letters (as in no.27a). The tractor was launched with a flourish in the pages of the *Meccano Magazine* in June 1948 as follows:

'Massey-Harris is a famous name in the agricultural world. The firm was founded just over 100 years ago, and throughout its life it has been in the forefront with the design and manufacture of farming machinery and requisites of all kinds. Its range of tractors includes five of these power units, ranging from the small "Pony" specially suitable for the small farm or the market garden, to model No. 55, a giant machine for the large scale farm. The one that has been chosen for reproduction in miniature in the Dinky Toys series is model No. 44, the second largest in the range. The model is true in form to its original in every respect, from the radiator grille at the front to the large rear wheels with their giant ribbed tyres that take a firm grip on the ground. Every possible detail of the tractor is accurately reproduced in correct colours, red for the chassis, bonnet and mudguards over the rear wheels, yellow for the centres of the wheels, and black for the steering column and wheel, and for the silencer and "breather" that emerge from the top of the bonnet. The name of the makers of the prototype is reproduced on the sides of the bonnet, which are completely closed in the model. At the back of the new Dinky Toy is a hook, which may be used for hauling trailers or implements, and a finishing touch to a really fine model is provided by the driver, seated correctly at the wheel.'

One unique piece to surface recently is a trial green colour Massey-Harris casting found originally in a rubbish skip at Binns Road (**22**). No finished examples in this colour are known.

This section should be read in conjunction with a discussion of the boxes for Massey-Harris, Massey-Ferguson and Field-Marshall tractors and their implements which follows.

The Massey-Harris was re-numbered 300 in 1954 when a simpler three-digit system was introduced (**21** and **23** to **37**). Dinky catalogues show that the tractor was given rubber tyres in 1962.

The Massey-Harris and Ferguson tractor companies were merged in 1953, but the Dinky Massey-Harris was not re-named in the catalogues until 1966. The change in lettering on the models from *MASSEY-HARRIS* to

MASSEY-FERGUSON must have happened earlier, for the Massey-Ferguson (**26** to **37**) was advertised in July 1963 in the Meccano Magazine as being in the no.398 Gift Set (**77**). This advertisement could have been ahead of model production, because the Massey-Harris was certainly used in the set to start with, and the Massey-Ferguson then followed. The change was therefore in about 1964. This coincided with the introduction of a blue plastic driver on a cast seat, shown in the 1964 catalogue, to replace the metal driver on a tinplate seat. There are transitional versions with the *MASSEY-HARRIS* printing and a black tinplate seat but with a plastic driver (**24** and **25**) and also with the *MASSEY-FERGUSON* printing and a cast seat but with a metal driver (**26**); these must surely date from 1963/1964.

A plastic exhaust, first yellow and then red, was introduced to the Massey-Ferguson (**33** to **37**), presumably soon after it came into use on the David Brown tractor in 1966. However, the main casting remained unchanged from 1948 until the tractor was discontinued in 1971. This was an extraordinarily long period of 23 years reflecting the company's reluctance, unlike Britains, to be innovative in the 1960s.

There are very rare examples where the wheels intended for the Field-Marshall were put on the Massey-Harris (**27** to **30**), but we have not seen any convincing examples of the reverse.

All the tractors had the same wording stamped in the baseplate: *DINKY TOYS MASSEY-HARRIS MADE IN ENGLAND MECCANO LTD.* No model number was shown on the baseplate.

Implements for the Massey-Harris and Massey-Ferguson Tractors

Five implements for the tractor followed in the 27 Series over the five years from 1949, followed by a tipping trailer introduced in 1961. The details cast underneath the implements reflected the change in model numbering. The earliest implements were without any model number. Then the 27 Series numbering was sometimes, but not always, added, and subsequently these numbers were replaced by the 300 Series. These changes seem to have happened at different times for different models. In any case, the amount of writing cast underneath the models depended on the space available.

No.27b/320 Halesowen Harvest Trailer (July 1949–70)

The early trailers were tan with red trim and had bare metal tyres and red or tan raves (**38** to **41**). Yellow plastic hubs and rubber tyres were first shown in the catalogue for 1963 (**42**) and then from 1966 the trailers were red with yellow raves (**43**). The change from tan to red trailer body corresponded to an alteration in the casting of the towbar with the addition of a metal plate joining the two arms of the towbar together (compare **42** and **43**).

The earliest tan examples with tan raves were without numbers in the casting. On others in trade boxes and some in the individual dark yellow boxes was cast the number *27B* (with a capital *B*). All the rest were numbered *320*. Cast underneath was *DINKY TOYS HALESOWEN FARM TRAILER MADE IN ENGLAND MECCANO LTD.*

No.27c/321 Massey-Harris Manure Spreader (November 1949–71)

The manure spreaders retained the name Massey-Harris right up to the time they were discontinued, although many examples had no name printed on them at all (**46**). When the name is found on a model the words Massey and Harris are usually transposed compared with the illustration on the boxes, for no obvious reason. An exception, with the word *MASSEY* above *HARRIS* is shown in (**53**). The red trailer body remained consistent, with two sets of silver-painted rotating arms, each with three sets of six spikes and a screw behind, all turned by a belt of fine twisted wire driven off the rear axle. The main changes were from metal wheels (**44** and **45**) to metal hubs with rubber tyres in 1963 (**46** and **47**) and then to plastic hubs with rubber tyres (**48** to 52). The metal hubs for the rubber tyres were either yellow (**46**) or red (**47**). The plastic hubs were also either yellow (**48** to **52**) or red (not illustrated).

Cast under the trailer floor was *DINKY TOYS MASSEY HARRIS MANURE SPREADER MADE IN ENGLAND MECCANO LTD*, with no number on the earliest models, then with *27C* (with a capital *C*) and then *321*.

No.27ac Tracteur "Massey Harris" et Remorque Épandeur D'Engrais (June-December 1950)

In 1950 an attempt was made in France to bring out the tractor and manure spreader, both with metal wheels, as a set in a plain red front-opening box with a lid, but it was withdrawn within six months (**53**).

No.27h/322 Disc Harrow (May 1951–63)

The disc harrow with a red-and-yellow frame with two sets of six silver-painted discs and a rear black tinplate tow hook was not changed at all before it was withdrawn in 1963 (**54** to **57**). It was re-issued in 1967 as no.322 in white, or white and red, for the David Brown tractor (**105**, **106**, **113** and **114**). All models in trade boxes appear not to have had numbers cast in them, just *DINKY TOYS DISC HARROW MADE IN ENGLAND MECCANO LTD.* The models in individual yellow boxes then had *322* added.

No.27j/323 Triple Gang Mower (October 1952–59)

The triple gang mower remained unaltered during its relatively short life until it ceased to appear in the catalogues after 1959 (**58** to **60**). There were three sets of yellow blades with green wheels set in a red frame. Cast underneath the first models was *MECCANO LTD MADE IN ENGLAND* without a number and *323* was added later.

No.27k/324 Hay Rake (April 1953–70)

The only variation with this popular item was the lever to raise and lower the tines, which was sometimes black and sometimes bare metal (**61** to **66**). The bare metal tines were held in a red frame on yellow spoked wheels. Cast under the examples in the earliest trade boxes (as in (**61**)) was *MECCANO LTD MADE IN ENGLAND DINKY TOYS* without a number, but *324* was added on (**62**), released when the 300 Series was being introduced.

No.27ak/310 Farm Tractor and Hay Rake set (1953–65)

This set was the only tractor and implement combination sold by Dinky over this period. It was released in April 1953, and first of all sold in a blue box with a blue and white striped lid with the Massey-Harris tractor (**67**), then in a blue box with a yellow lid for the Massey-Harris tractor (**68**) and finally in a yellow box with a yellow lid for the Massey-Ferguson (**69**).

No.319 Weeks Farm Trailer (Tipping) (1961–70)

Added late to the group, this red tipping trailer with a yellow chassis only came with rubber tyres. The first hubs were metal (**70**, **71**, **73** and **74**) and then plastic (**72**). The metal hubs were usually red, but occasionally yellow (**73**). One rare colour variation was the blue trailer (**71**). Cast under the yellow chassis was *DINKY TOYS MECCANO LTD MADE IN ENGLAND 319 WEEKS TIPPING FARM TRAILER.*

Gift Set No.1 Farm Gear (1952–54)

Two farm sets with a range of implements were produced. The first, in 1952, was the *GIFT SET No.1: FARM GEAR*, containing the Massey-Harris Tractor, Halesowen Harvest Trailer, Massey-Harris Manure Spreader, Disc Harrow and the Motocart, although the latter was hardly a farm item (**75** and **76**). The hay rake, which could have made up the set, was not released for another year.

Gift Set No.398 Farm Equipment (1963–65)

After a gap of ten years the *Gift Set no.398 "FARM EQUIPMENT"* followed in 1963, first with the Massey-Harris tractor and then with the Massey-Ferguson version, but again only for a short while. This had the tractor, trailer and manure spreader, all by then on rubber tyres, and the disc harrow and the hay rake (**77**).

At 17s 3d and 21s 0d respectively the two gift sets were good value, but neither was available for long.

Gift Set No.399 Farm Tractor and Trailer (1969–72)

In 1969 a combination set in a window box with the Massey-Ferguson tractor and a red four-wheeled lorry trailer came out for a short while as the no.399 Gift Set, probably to use up ageing stock (**78**).

No.27n/301 Field-Marshall Tractor (1953–65)

In 1953 came the no.27n (later no. 301) Field-Marshall tractor (**79** to **85**), which was presumably intended to use the Massey-Harris implements. This had changes of hub colour from silver (**79** to **81**) to green (**82** and **83**), and changes from metal wheels to metal hubs with rubber tyres (**84** and **85**). Drivers changed from metal (**79** to **83**) to plastic (**84** and **85**) and exhausts from orange (**79** to **82**) to black (**83** to **85**). However, the catalogue illustrations show no change to the tractor throughout its life except for the switch to rubber tyres in 1962. Other changes are therefore very difficult to date. The Field-Marshall, which never had quite the same popularity as the Massey-Harris, was discontinued in 1965.

Stamped on the baseplate was *301 FIELD MARSHALL DINKY TOYS MECCANO MADE IN ENGLAND.* We have not seen any which carry the 27n number on the baseplate.

Boxes for Massey-Harris/Massey-Ferguson and Field-Marshall Tractors and their implements

The study of these boxes is an interesting and rather complex subject. After the war, models were sold in cream, yellow or brown card trade boxes until the end of 1953 (**17** to **21**, **38**, **39**, **44**, **54**, **58**, **61**, **62** and **79**), when lightweight individual dark yellow boxes were introduced. When the new numbering began trade boxes displayed both the old and new numbers on the ends of the lids (**21**, **39**, **62** and **79**). On the end flaps of the new boxes the old number was sometimes shown in red to either side of the new number in white on a black oval (**45** and **46**, **81** and **82**). The boxes were usually illustrated with a picture of the item they contained, although a few rare early un-illustrated boxes with all-red lettering, with or without dual numbering, can be found (**55**, **63** and **80**). In 1955, with most of the re-numbering process complete, fresh illustrated boxes were introduced bearing only the new numbers (**23** to **34**, **40** to **43**, **47**, **48**, **56**, **59**, **60**, **64**, **65**, **83** and **84**). Dark yellow boxes were replaced with light yellow ones in 1959. Boxes were then printed for a short while with red side panels without illustrations and from 1965 with white side panels with illustrations and with the name changed to *MASSEY-FERGUSON TRACTOR* on the appropriate tractor boxes, but again for a very limited period (**36**, **37** and **50**). The date of 1965 for the introduction of the white panels is derived from the fact that the Field-Marshall tractor was discontinued before the white panels were introduced. From mid-1966 into 1968, yellow and then gold window boxes were introduced for the export market, mainly for the U.S.A. (**51**, **52**, **73** and **74**).

It is interesting that the box illustrations and the wording on the boxes did not keep up with the change from Massey-Harris to Massey-Ferguson or with the change to rubber tyres and blue plastic drivers on either the Massey-Harris or the Field-Marshall tractors.

The sequence of boxes for later Dinky models is simpler and will be covered under the appropriate models.

No.563 and No.963 Supertoys Heavy Tractor (1948–59)

Alongside the wheeled tractors were the crawlers, described first as no.563 Heavy Tractor (**86** to **92**) and then from 1955 as the no.963 Blaw Knox Heavy Tractor (**93** and **94**). There was also a slight alteration to the casting, most apparent in the introduction of a vertical ridge down the back of the driver's seat (**95**). In the picture sequence this occurred between (**91**) and (**92**), so it was shortly before the model name and number were changed. With their rubber tracks, these crawlers were relatively unwieldy and were never as popular as their wheeled counterparts. They came in dark blue (**87**, **88** and **90**), orange (**86**, **89** and **91**), and later red (**92** and **93**) and pale yellow (**94**). None were manufactured beyond 1959.

Stamped on the baseplate was *DINKY SUPERTOYS HEAVY TRACTOR MADE IN ENGLAND BY MECCANO LTD.*

The boxes were first all of plain brown card with illustrated labels (**86** to **89**). These labels were white (**86** and **87**) and then reddish-orange and white (**88** and **89**). The card was then covered in blue paper with an orange-and-white label (**90** to **92**). Finally the boxes were covered in a fetching blue-and-white striped paper with a coloured illustration, usually reflecting the colour of the model inside (**93** and **94**).

Dublo No.069 Massey-Harris-Ferguson Tractor (1959–65)

A small blue tractor, no.069, with grey plastic wheels, was produced to suit 00 scale model railways, at a scale of 1:77, for six years from 1959. There was a hole in the seat, but no driver, and a black tinplate tow hook, but no implements were made for this small tractor. The rear wheels were light (**97**) or a darker grey (**96**). *MECCANO ENGLAND DUBLO DINKY TOYS* was cast inside. It came in a simple red-printed yellow box. We have also seen a rare unboxed factory-painted version in orange (**97**).

No.305 David Brown 990 Tractor (Red and yellow) (1965–67)

The David Brown tractor company launched its 990 tractor in 1961, and Dinky followed with their version in 1965, numbered 305. This had a distinctive red bonnet with yellow hubs and a detachable yellow cab, and there were paper stick-on labels for the grille and bonnet side lettering. Although the tractor was steerable, it was not possible to reach the steering wheel without removing the cab! The engine block was usually black (**98**), but there was also a very rare version with a red engine block (**99**). This was the first Dinky tractor to carry the three-point linkage. There was no lettering under the tractor, but *DINKY TOYS* was cast on the front of the cab on the right-hand side and *MECCANO LTD* on the left.

The catalogue illustrators were inadequately briefed for the 1965 catalogues; the first edition showed an extraordinary red/white hybrid with a red cab (**112**), and the second edition an all-red tractor, but by 1966 the mistakes had been rectified. The tractor was sold in an attractive colourfully illustrated card box showing a pair of tractors at a field gate (**98**), and for the American market it was sold in a yellow window box from mid-1966 (**100**).

No.305 David Brown Selectamatic 990 Tractor (White) (1968–73)

In 1965 David Brown produced their white 990 tractor with a chocolate-brown engine block, and Dinky followed this with their model in 1968 in the same colours, still numbered 305. It had a new bonnet casting with headlights above the grille, represented by paper stick-on labels, and had *DAVID BROWN SELECTAMATIC 990* on labels on the bonnet sides. At first, both *DINKY TOYS* and *MECCANO LTD* were cast on the front of the cab (**103** to **105**), but later *MECCANO LTD* was replaced by a blank panel.

To start with, it was sold in the old box used for the red-and-yellow tractor (**101**). Later, it had its own box (**102**), then a yellow window box for the American market (**104**) and then a plastic lift-off case (**107**). From late 1972 it was given a red engine block and red hubs. Rare examples had a brown three-point linkage (**108**). There was also a very rare all-red box which simply had *David Brown 990 Selectamatic Tractor* on the side, presumably a promotional model to sell the real thing (**103**).

No.325 David Brown Tractor and Disc Harrow Set (1967–72)

The no.305 David Brown was also produced in a set with the disc harrow in a box with an illustration showing the tractor and harrow in a farming scene (**105**) or against a plain background (**106**).

No.305 Case David Brown 995 Tractor (White) (1974–75)

The white no.305 David Brown in the rigid plastic lift-off case became the no.305 Case David Brown 995 in 1974, after the David Brown and Case tractor companies had merged in 1973 (**109** to **111**). This was the same as the previous version of the tractor except for the *Case DAVID BROWN 995* bonnet labels. There were three colour variations of the exhaust, black, red and yellow.

No.322 Disc Harrow for the David Brown Tractors (1967–72)

The only implement manufactured for the white David Brown was the no.322 Disc Harrow, which was a re-issue of the old 1951-63 no.27h/322 red-and-yellow model. It was sold separately or in a combination pack with the tractor. It came with or without a red top to the frame (**105**, **106**, **113** and **114**).

No.308 Leyland 384 Tractor (1971–80)

The no.308 Leyland 384 was the last Dinky tractor Meccano produced. It was usually blue (**116**, **119**, **120** and **123**), metallic red (**115**, **117** and **121**) or from 1978 was orange (**118**, **122** and **124**) with white hubs. One pair of metallic bronze unfinished prototype models with white hubs is known (**125** and **126**). There was also a metallic red version with yellow hubs (**115**), as used on the no.305 David Brown 990, and a rare blue variation with red hubs using wheels left over from the no.305 Case David Brown 995 (**119**). Initially the model was released in card boxes (**116** to **118**), then plastic lift-off cases (**119** to **122**), and finally in hanging window boxes (**123** and **124**) before sales ceased in 1979.

Cast into the bright metal baseplate was *308 DINKY TOYS MADE IN ENGLAND LEYLAND.*

No.3265 Mogul Steel Toys Tractor & Trailer (1974–77)

In 1974, Meccano introduced the Mogul range, 'A new and exciting range of Big Steel Toys'. These were large toys, mainly in tinplate, and included a tractor and trailer set. The red steerable tractor had a roll bar and a large red *M* for Mogul on the black grille (**127**). The trailer could only be tipped up by hand. This backward-looking use of tinplate, reminiscent of the 1950s, could not have been a great success, because this set is very seldom seen now.

The pleasure they gave

Of the thousand or so different models produced by Meccano between 1933 and 1979, the range of farm items was actually quite small, yet their impact on the play scene, particularly with the no.27a (300) Massey-Harris tractor in the 1950s, was enormous. It was surely because they were so popular that they remained unchanged for so long. Few small boys could have grown up in the 1950s without getting immense pleasure from these small red tractors.

The authors are grateful to John King for his comments on the dating of Dinky boxes.

Model List

Serial No.	Model	Dates
22e	Tractor (Fordson)	1933-41
27a or 300	Massey-Harris or Massey-Ferguson Tractor	1948-71
27b or 320	Halesowen Harvest Trailer	1949-70
27c or 321	Massey-Harris Manure Spreader	1949-71
27ac	Tracteur "Massey Harris" et Remorque Épandeur D'Engrais (France)	June-December only 1950
27h or 322	Disc Harrow (red and yellow)	1951-63
27j or 323	Triple Gang Mower	1952-59
27k or 324	Hay Rake	1953-70
27ak or 310	Massey-Harris Tractor and Hay Rake Set	1953-65
319	Weeks Farm Trailer (Tipping)	1961-70
Gift Set No.1	Farm Gear	1952-54
Gift Set No.398	Farm Equipment	1963-65
Gift Set No.399	Farm Tractor and Trailer Set	1969-72
27n or 301	Field-Marshall Tractor	1953-65
563	Supertoys Heavy Tractor	1948-54
963	Supertoys Blaw Knox Heavy Tractor	1955-59
305	David Brown 990 Tractor (red and yellow)	1965-67
305	David Brown Selectamatic 990 Tractor (white)	1968-73
305	Case David Brown 995 Tractor (white)	1974-75
325	David Brown Tractor and Disc Harrow set	1967-72
322	Disc Harrow (white or white and red)	1967-72
308	Leyland 384 Tractor	1971-80
Dublo 069	Massey-Harris-Ferguson Tractor	1959-65
3265	Mogul Steel Toys Tractor & Trailer	1974-77

Further Reading

Bell, Brian, 1999, *Fifty Years of Farm Tractors*. Farming Press.

Brown, Kenneth D., 2007. *Factory of Dreams – A History of Meccano Ltd, 1901–1979*. Crucible Books.

Cooke, David, 1999, *Dinky Toys*. Shire Publications.

Farnworth, John, 1999, *The Advertising of Massey-Harris, Ferguson and Massey Ferguson*. Farming Press.

Force, Edward, 1996, *Dinky Toys*. Schiffer.

Gibson, Cecil, 1966, *A History of British Dinky Toys 1934-1964*. Model Aeronautical Press.

Harrington, Peter, and Sutherland, Hugh, 1984, *A Concise Detailed and Illustrated Listing of Dinky Model Cars, 1933-1980*. Collecta Books Series privately published.

McReavy, Anthony, 2002, *The Toy Story: The Life and Times of Inventor Frank Hornby*. Ebury Press.

Randall, Peter, 1977, *The Products of Binns Road – A General Survey*. New Cavendish Books.

Richardson, Mike, 2001, *Collecting Dinky Toys*. Francis Joseph.

Richardson, Mike and Sue, 1981, *Dinky Toys & Modelled Miniatures*. New Cavendish Books.

Richardson, Mike and Sue, 2000, *The Great Book of Dinky Toys*. New Cavendish Books.

Stanford, Tony, undated, *A Meccano Magazine Digest Review* in four parts. Cranbourn Press.

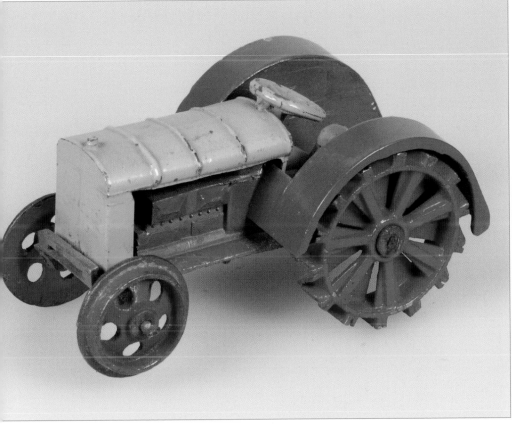

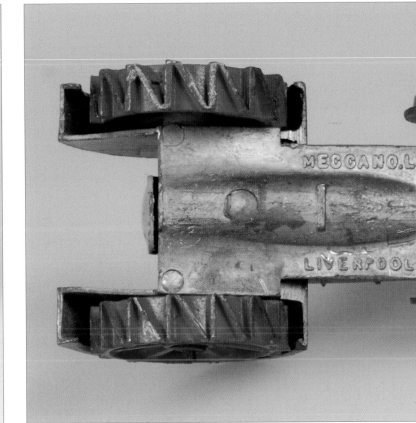

1. *Modelled Miniatures no.22e Tractor without a tow hook in blue and yellow with red wheels (70 mm).*

2. *Underside of (1) showing the HORNBY SERIES casting.*

3. *Modelled Miniatures no.22e Tractor without a hook in yellow and green with red wheels.*

4. *Dinky no.22e Tractor with tow hook, dark blue and yellow with red wheels (70 mm).*

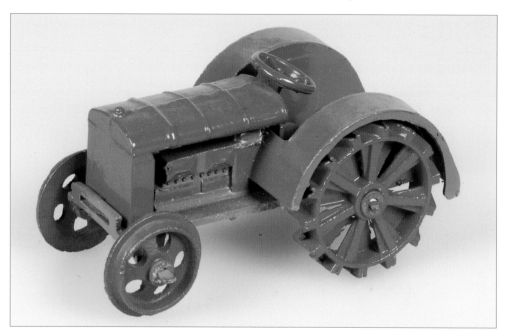

5. Underside of (**4**) showing DINKY TOYS casting.

6. Dinky no.22e Tractor as (**4**) but blue and creamy-white.

7. Dinky no.22e Tractor as (**4**) but yellow and green.

8. Dinky no.22e Tractor as (**4**) but light blue and red.

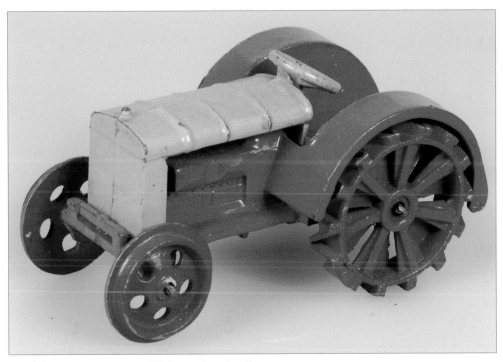

9. *Dinky no.22e Tractor as (**4**) but red and yellow.*

10. *Dinky no.22e Tractor as (**4**) but light blue and yellow.*

11. *Dinky no.22e Tractor as (**4**) but cream and light blue.*

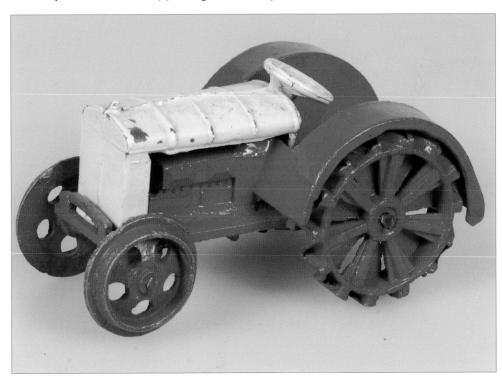

12. *Dinky no.22e Tractor as (**4**) but red and cream.*

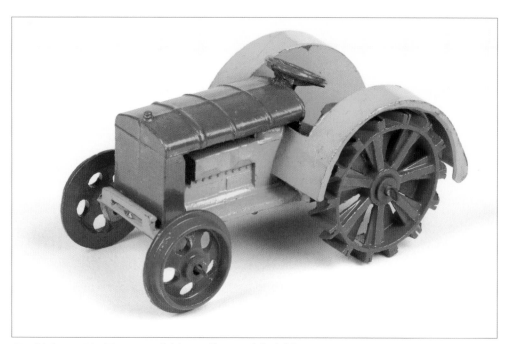

13. *Dinky no.22e Tractor as (4) but yellow and dark blue.*

14. *Dinky no.22e Tractor as (4) but all-yellow.*

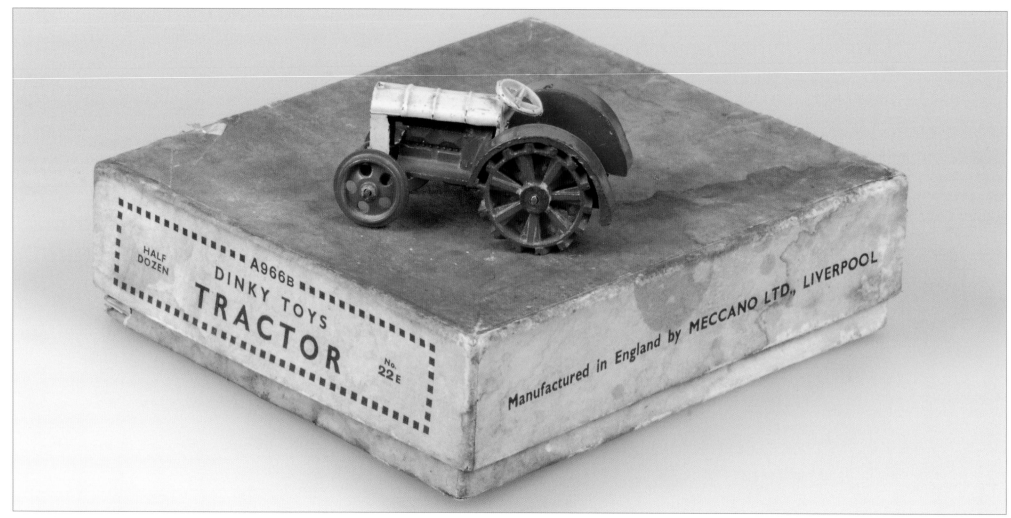

15. *Dinky no.22e Tractor as (6) on a faded and stained rare cream trade box for holding six no.22e tractors. The label reads A966B HALF DOZEN DINKY TOYS TRACTOR No.22E. (Note the 'E' in capitals.) The tractor is added for scale and did not come with the box.*

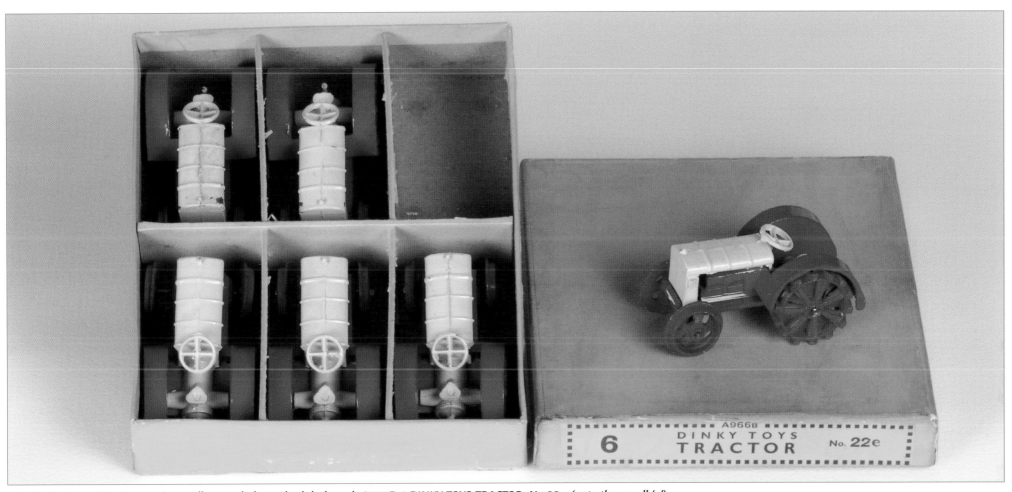

16. *Dinky six no.22e Tractors in a yellow trade box. The label reads* A966B 6 DINKY TOYS TRACTOR No.22e *(note the small 'e').*

17. *Dinky no.27a Massey-Harris Tractors, three in a cream trade box. The label reads* DINKY TOYS 3 MASSEY-HARRIS TRACTOR 27A THE FRONT WHEELS OF THIS MODEL ARE ADJUSTABLE FOR RUNNING TO LEFT OR RIGHT. *Tractors with red body, light brown metal driver on a black tinplate seat, yellow hubs and bare metal tyres with black metal exhaust and air filter (88 mm).*

18. *Dinky no.27a Massey-Harris Tractors, three in a cream trade box as (**17**) but with* 50029 *over* DINKY TOYS. *It also has an* H. HUDSON DOBSON *label on the lid. Tractors as (**17**).*

19. *Dinky no.27a Massey-Harris Tractors, three in a brown trade box with a stick-on yellow end label and another label stuck diagonally across the lid reading* Dinky Toys No. 27a Massey-Harris Tractor. The Front Wheels of this Model are adjustable for running to left or right. *Tractors as (17).*

20. *Dinky no.27a Massey-Harris Tractors, three in a yellow trade box with end label printed as on (18). Tractors as (17).*

21. *Dinky no.300/27a Massey-Harris Tractors, three in a yellow trade box. Tractors as (17).*

22. *Dinky no.27a or no.300 Massey-Harris Tractor, a probably unique pre-production unused tractor body in green, found in a skip at the Binns Road factory (85 mm).*

23. *Dinky no.300 Massey-Harris Tractor in an illustrated dark yellow box. Tractor as (**17**).*

24. *Dinky no.300 Massey-Harris Tractor in a light yellow illustrated box. The tractor with a blue plastic driver on a black tin seat. The driver had a projection under the left foot to fit into a hole in the tractor floor, as normally found on metal drivers, but rare on later plastic drivers. Yellow metal rear hubs with a wavy detail around the perimeter and plastic front hubs, both with rubber tyres.*

25. *Dinky no.300 Massey-Harris Tractor rear view of (**24**) to illustrate the plastic driver on the tin seat.*

26. *Dinky no.300 Massey-Ferguson Tractor in a light yellow illustrated Massey-Harris box. The tractor with a metal driver on a solid cast seat, yellow metal rear hubs with wavy detailing around the perimeter and plastic front hubs.*

27. *Dinky no.300 Massey-Ferguson Tractor in a light yellow illustrated Massey-Harris box. The tractor with a blue plastic driver on a cast seat, green metal rear and green plastic front hubs, rare on the Massey-Ferguson but usual on the Field-Marshall tractor.*

28. *Dinky no.300 Massey-Ferguson Tractor as (27) showing the Field-Marshall green plastic front hubs as on (85).*

29. *Dinky no.300 Massey-Ferguson Tractor as (27) but with different green plastic front hubs.*

30. *Dinky no.300 Massey-Ferguson Tractor as (29) showing the variation of the Field-Marshall green plastic front hubs seen in (84).*

31. *Dinky no.300 Massey-Ferguson Tractor in a light yellow illustrated Massey-Harris box. Tractor as (27) but with yellow metal rear hubs and yellow plastic front hubs.*

32. *Dinky no.300 Massey-Ferguson Tractor in a light yellow illustrated Massey-Harris box. Tractor as (31) but with a silver steering wheel.*

33. *Dinky no.300 Massey-Ferguson Tractor in a light yellow illustrated Massey-Harris box. Tractor as (31) but with a yellow plastic exhaust.*

34. *Dinky no.300 Massey-Ferguson Tractor in a light yellow illustrated Massey-Harris box. Tractor as (33) but without transfers.*

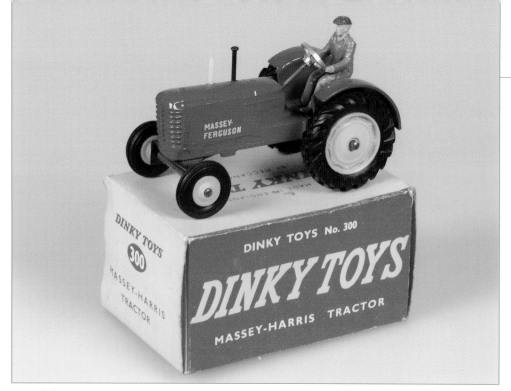

35. *Dinky no.300 Massey-Ferguson Tractor in a light yellow un-illustrated Massey-Harris box with red side panels. Tractor as (**33**) but with a silver steering wheel.*

36. *Dinky no.300 Massey-Ferguson Tractor in a light yellow illustrated Massey-Ferguson box with white side panels. Tractor as (**35**).*

38. *Dinky no.27b Halesowen Harvest Trailers, three in a yellow trade box. The label reads DINKY TOYS 3 HARVEST TRAILER 27B. Varying shades of tan trailers with red trim and chassis, red raves, yellow metal hubs and bare metal tyres (135 mm).*

37. *Dinky no.300 Massey-Ferguson Tractor in a light yellow illustrated Massey-Ferguson box. Tractor as (**35**) but with a red plastic exhaust.*

39. *Dinky no.27b/320 Halesowen Harvest Trailers, three in a yellow trade box. The label reads DINKY TOYS 320 THREE HARVEST TRAILER 27B. Trailers as (**38**) but with tan raves.*

40. *Dinky no.320 Halesowen Harvest Trailer in a dark yellow illustrated box. Dark tan trailer with red trim and chassis, dark tan raves, yellow metal hubs and bare metal tyres.*

41. *Dinky no.320 Halesowen Harvest Trailer in a dark yellow illustrated box. Light tan trailer with red trim and chassis, red raves, yellow metal hubs and bare metal tyres.*

42. *Dinky no.320 Halesowen Harvest Trailer in a light yellow illustrated box. Trailer as (41) but with yellow plastic hubs and rubber tyres.*

43. *Dinky no.320 Halesowen Harvest Trailer in a light yellow illustrated box. Red trailer with yellow raves, yellow plastic hubs and rubber tyres.*

44. *Dinky no.27c Massey-Harris Manure Spreaders, three in a yellow trade box. The label reads DINKY TOYS 3 MANURE SPREADER 27C. Manure spreaders with yellow metal hubs and bare metal tyres (118 mm).*

45. *Dinky no.27c/321 Massey-Harris Manure Spreader in a dark yellow illustrated box. Manure spreader as (**44**).*

46. *Dinky no.27c/321 Massey-Harris Manure Spreader in a dark yellow illustrated box. Manure spreader without transfers, and with yellow metal hubs and rubber tyres.*

47. *Dinky no.321 Massey-Harris Manure Spreader in a light yellow illustrated box. Manure spreader as (**46**) but with yellow transfers and red metal hubs.*

48. Dinky no.321 Massey-Harris Manure Spreader in a light yellow illustrated box. Manure spreader as (**47**) but with yellow plastic hubs.

49. Dinky no.321 Massey-Harris Manure Spreader in a light yellow un-illustrated box with red side panels. Manure spreader as (**46**) but with lemon-yellow plastic hubs.

50. Dinky no.321 Massey-Harris Manure Spreader, in a light yellow illustrated box with white side panels. Manure spreader as (**48**).

51. Dinky no.321 Massey-Harris Manure Spreader in a yellow window box for export to the U.S.A. Manure spreader as (**50**).

52. Dinky no.321 Massey-Harris Manure Spreader in a gold window box for export to the U.S.A. Manure spreader as (**51**).

53. *Dinky no.27ac labelled* DINKY TOYS 27 AC TRACTEUR "MASSEY HARRIS" et REMORQUE ÉPANDEUR D'ENGRAIS C'est une fabrication MECCANO *in a plain red box with a front flap and lid. A British-made set for the French market, consisting of a Massey-Harris tractor and manure spreader, both with bare metal wheels. Unusually, the word MASSEY is above HARRIS on the manure spreader.*

54. *Dinky no.27h Disc Harrow four in a yellow trade box. The label reads* DINKY TOYS 27H DISC HARROW 4. *Disc harrow with red-and-yellow frames with silver discs (85 mm)*

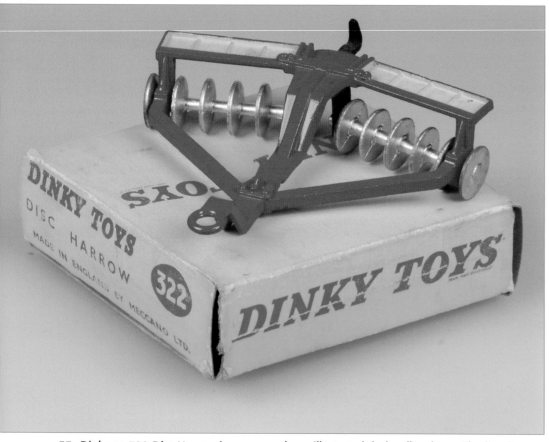

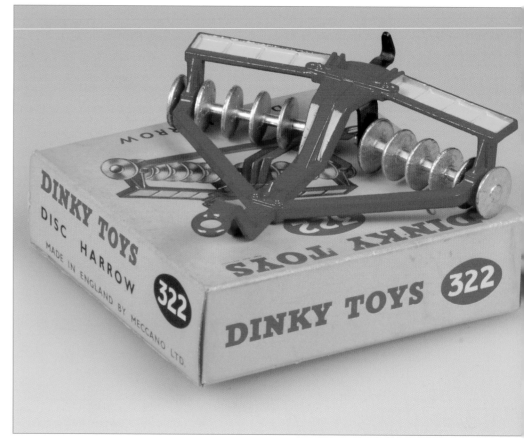

55. *Dinky no.322 Disc Harrow in a rare early un-illustrated dark yellow box with all-red printing. Disc harrow as (54).*

56. *Dinky no.322 Disc Harrow in a dark yellow illustrated box. Disc harrow as (54).*

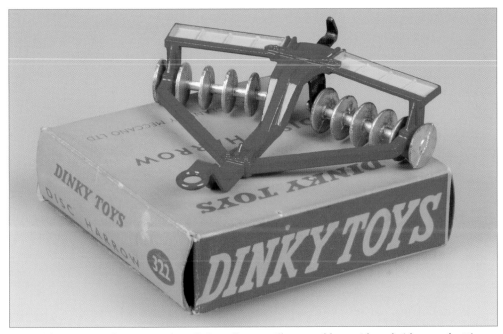

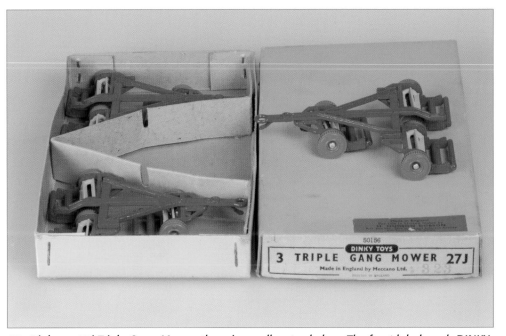

57. Dinky no.322 Disc Harrow in a light yellow un-illustrated box with red side panels. Disc harrow as (**54**).

58. Dinky no.27j Triple Gang Mower, three in a yellow trade box. The front label reads DINKY TOYS 3 TRIPLE GANG MOWER 27J. The rear illustrated label described the contents in French, Spanish and German. Mowers with red frames, yellow cutter blades and green wheels (110 mm).

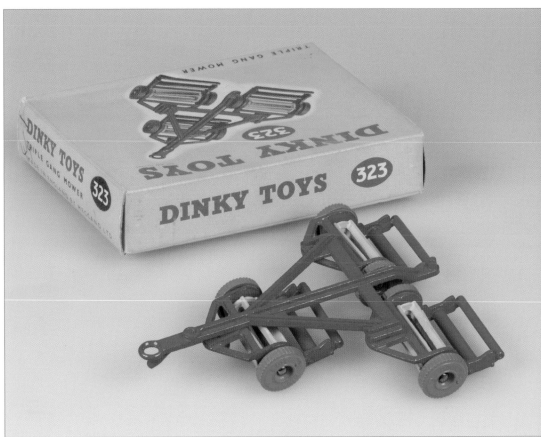

59. Dinky no.323 Triple Gang Mower in a dark yellow illustrated box. Mower as (**58**).

60. Dinky no.323 Triple Gang Mower in a light yellow illustrated box. Mower as (**58**).

61. *Dinky no.27k Hay Rake in a yellow trade box. The label reads* DINKY TOYS 1 HAY RAKE 27K. *Rake with black lever, red frame, yellow wheels and bare metal tines (75 mm).*

62. *Dinky no.27k/324 Hay Rake in a yellow trade box. The label reads* DINKY TOYS ONE 324 HAY RAKE 27K. *Rake as (61).*

63. *Dinky no.324 Hay Rake in a dark yellow rare early un-illustrated box with all-red printing. Rake as (61).*

64. *Dinky no.324 Hay Rake in a dark yellow illustrated box. Rake as (61) but with a bare metal lever.*

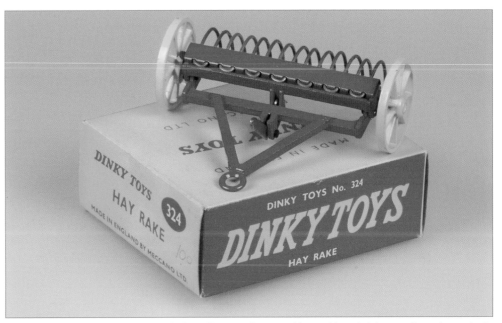

65. *Dinky no.324 Hay Rake in a light yellow illustrated box. Rake as (**61**).*

66. *Dinky no.324 Hay Rake in a light yellow un-illustrated box with red side panels. Rake as (**64**).*

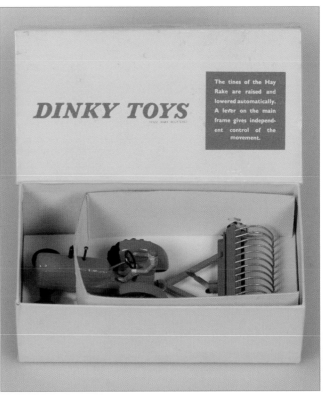

67. *Dinky no.27ak Farm Tractor & Hay Rake set in a box with a blue-striped illustrated lid, Massey-Harris tractor with bare metal tyres and black metal exhaust as (**17**) and hay rake as (**61**).*

68. *Dinky no.310 Farm Tractor & Hay Rake set in a box with a yellow un-illustrated lid. Tractor and hay rake as (**67**).*

69. *Dinky no.310 Farm Tractor & Hay Rake set in a box with a yellow un-illustrated lid. Massey-Ferguson tractor with blue plastic driver and rubber tyres as (**31**) and hay rake with bare metal lever as (**64**).*

70. *Dinky no.319 Weeks Farm Trailer (Tipping) in a light yellow illustrated box. Trailer with red body, yellow chassis, red metal hubs and rubber tyres (105 mm).*

71. *Dinky no.319 Weeks Farm Trailer (Tipping) in a light yellow illustrated box. Trailer as (**70**) but with scarce blue body.*

73. *Dinky no.319 Weeks Tipping Farm Trailer in a yellow window box for export to the U.S.A. Trailer as (**70**) but with yellow metal hubs.*

72. *Dinky no.319 Weeks Farm Trailer (Tipping) in a light yellow illustrated box. Trailer as (**70**) but with red plastic hubs.*

74. *Dinky no.319 Weeks Tipping Farm Trailer in a gold window box for export to the U.S.A. Trailer as (**70**).*

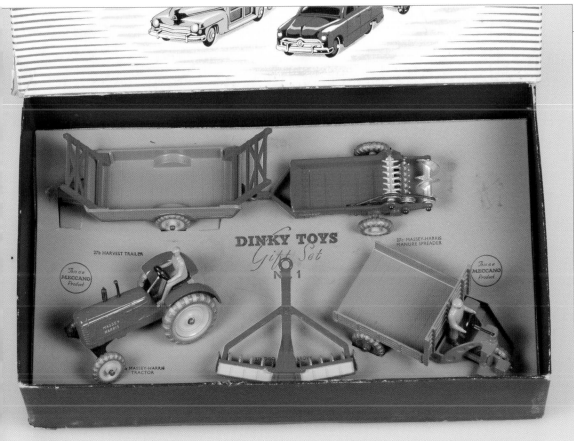

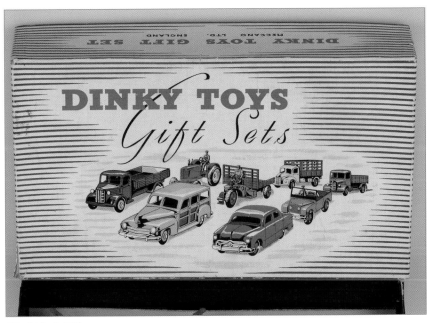

75. *Dinky* GIFT SET No.1: FARM GEAR *in a box with an illustrated blue-striped lift-off lid with blue insert labelled* 27a MASSEY-HARRIS TRACTOR, 27b HARVEST TRAILER, 27c MASSEY-HARRIS MANURE SPREADER, 27h DISC HARROW *and* 27g MOTOCART. *All items have bare metal tyres.*

76. *Lid of* (**75**).

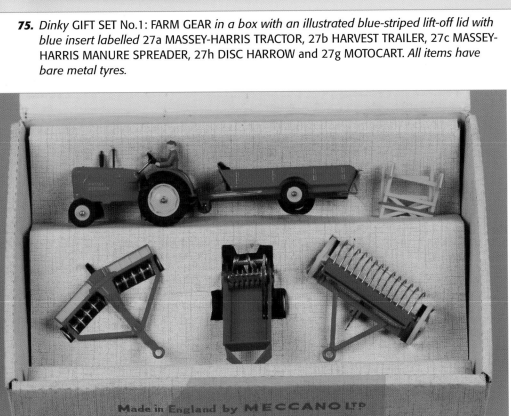

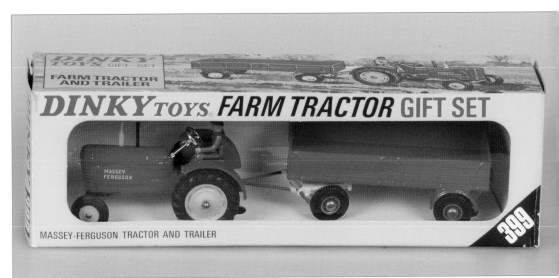

78. *Dinky no.399 Farm Tractor Gift Set: Massey-Ferguson Tractor and Trailer in a yellow window box; tractor with blue plastic driver, silver steering wheel, yellow plastic exhaust and rubber tyres, and a red four-wheeled trailer with rubber tyres.*

77. *Dinky Gift Set No.398 "FARM EQUIPMENT" in a white textured box with a hinged lid containing a Massey-Ferguson Tractor with rubber tyres, blue plastic driver and black metal exhaust, a red Harvest Trailer with yellow raves and rubber tyres, a red-and-yellow Disc Harrow, a red Manure Spreader with rubber tyres and a red-and-yellow Hay Rake.*

79. Dinky no.27n/301 Field-Marshall Tractor in a rare, rough brown trade box for three tractors. Printed on the box was DINKY TOYS 27N 301 27N FIELD-MARSHALL TRACTOR THREE. Tractor with dark orange body with silver highlights, light brown metal driver, silver hubs and bare metal tyres (75 mm).

80. Dinky no.27n/301 Field-Marshall Tractor in a dark yellow very rare un-illustrated box with all-red printing. Tractor as (**79**).

81. Dinky no.27n/301 Field-Marshall Tractor in a dark yellow illustrated box. Tractor as (**79**).

82. *Dinky no.27n/301 Field-Marshall Tractor in a dark yellow illustrated box. Tractor as (**79**) but with dark green hubs, not shown on box.*

84. *Dinky no.301 Field-Marshall Tractor in a light yellow box with illustration which reverts to silver hubs. Tractor with green metal rear hubs with wavy detailing around the perimeter, green plastic front hubs as (**30**), rubber tyres, black exhaust and blue plastic driver.*

83. *Dinky no.301 Field-Marshall Tractor in a dark yellow illustrated box. Tractor as (**79**) but with light green hubs and black exhaust, shown on box.*

85. *Dinky no.301 Field-Marshall Tractor in a light yellow un-illustrated box with red side panels. Tractor as (**84**) but with a different style of front hubs as (**28**).*

86. *Dinky Supertoys no.563 Heavy Tractor in a plain card box with an illustrated white label. Orange crawler with light green wheels and dark green rubber tracks, light brown metal driver, black exhaust, air cleaner and levers (113 mm)*

87. *Dinky Supertoys no.563 Heavy Tractor in a plain card box as (86). Tractor as (86) but dark blue with light blue wheels.*

88. *Dinky Supertoys no.563 Heavy Tractor in a plain card box with an illustrated reddish-orange and white label. Tractor as (87).*

89. *Dinky Supertoys no.563 Heavy Tractor in a plain card box as (88). Tractor as (86).*

90. *Dinky Supertoys no.563 Heavy Tractor in a blue box with an illustrated orange-and-white label. Tractor as (87).*

91. *Dinky Supertoys no.563 Heavy Tractor in a blue box with an illustrated orange-and-white label. Tractor as (86).*

92. *Dinky Supertoys no.563 Heavy Tractor in a blue box as (91). Tractor as (86) but red with black wheels.*

93. *Dinky Supertoys no.963 Blaw Knox Heavy Tractor in a blue-and-white striped box with an illustration of a red crawler. Tractor as (92).*

94. *Dinky Supertoys no.963 Blaw Knox Heavy Tractor in a blue-and-white striped box with an illustration of a yellow crawler. Tractor as (92) but with yellow body.*

95. *Dinky Supertoys no.563 Heavy Tractor showing the change in casting on the back of the driver's seat between (91) on the left and (92).*

96. *Dinky Dublo no.069 Massey-Harris-Ferguson Tractor in an un-illustrated yellow box with all-red printing and a trade box to hold six boxed tractors. A blue tractor with a silver grille and grey hard plastic wheels; a variation of this has light grey rear wheels (36 mm).*

97. *Dinky Dublo no.069 Massey-Harris-Ferguson Tractor as (**97**) but re-sprayed in the factory in orange, with dark grey front and light grey rear wheels.*

99. *Dinky no.305 David Brown 990 Tractor as (**98**) but with a rare red engine block.*

98. *Dinky no.305 David Brown 990 Tractor in a yellow colourfully illustrated box showing a farming scene. Red tractor with black engine block, yellow detachable cab with DINKY TOYS and MECCANO LTD cast on the front, yellow plastic exhaust, yellow metal rear and yellow plastic front hubs and red three-point linkage (83 mm).*

101. *Dinky no.305 David Brown Selectamatic 990 Tractor which first came out in this yellow illustrated box designed for the previous red version of the David Brown. White tractor with a white detachable cab, chocolate-brown engine block, red plastic exhaust, white metal rear and white plastic front hubs, black stick-on label with white lettering and a red surround on both sides of the bonnet,* DINKY TOYS *cast on the front of the cab but a blank panel cast in place of* MECCANO LTD *(83 mm).*

100. *Dinky no.305 David Brown 990 Tractor in a yellow window box for export to the U.S.A. with a protective tear-off card over the top with a stick-on label reading* Mr DEALER – please remove this strip. *The tractor as (98) stands on a black card plinth with white road markings.*

102. *Dinky no.305 David Brown Selectamatic 990 Tractor in a correctly illustrated box. Tractor as (101).*

104. *Dinky no.305 David Brown Selectamatic 990 Tractor in the same yellow export window box used for the red David Brown tractor in (100). Tractor as (103).*

103. *Dinky no.305 David Brown Selectamatic 990 Tractor in a very rare promotional red box with white lettering reading* David Brown 990 Selectamatic Tractor. *Tractor as (101) but with both* DINKY TOYS *and* MECCANO LTD *on the front of the cab.*

105. *Dinky no.325 David Brown Selectamatic Tractor with Disc Harrow in an illustrated yellow box showing a farming scene. Tractor as (**103**), disc harrow with white-and-red frame and silver discs.*

106. *Dinky no.325 David Brown Tractor with Disc Harrow in a yellow box with an illustrated white front. Tractor as (**101**), disc harrow as (**105**) but with an all-white frame.*

107. *Dinky no.305 David Brown Selectamatic 990 Tractor in a plastic lift-off case. Tractor as (**101**) but with a black plastic exhaust.*

108. *Dinky no.305 David Brown Selectamatic 990 Tractor. Tractor as (**107**) but with red hubs, red engine block and the rear three-point linkage in an unusual brown.*

109. *Dinky no.305 Case David Brown 995 Tractor as (107) but with a new-style stick-on label with white lettering on a black background without the red surround, red hubs and red engine block.*

110. *Dinky no. 305 Case David Brown 995 Tractor as (109) but with a red plastic exhaust.*

111. *Dinky no.305 Case David Brown 995 Tractor as (109) but with a yellow plastic exhaust.*

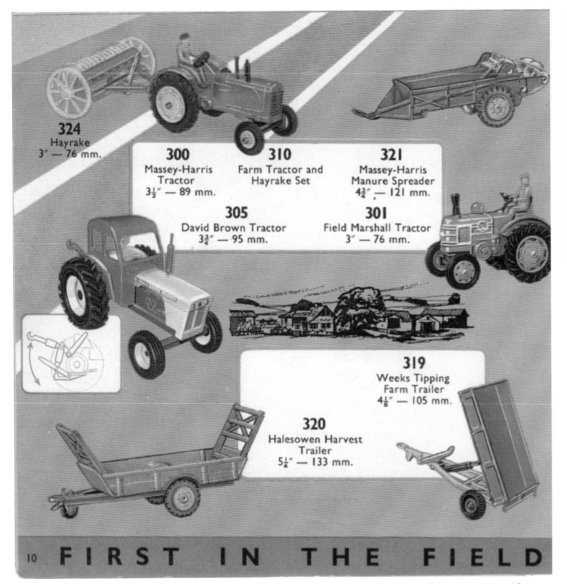

112. *Dinky first edition of the 1965 catalogue showing a strange hybrid version of the no.305 David Brown tractor with a white bonnet, red cab and white hubs. The second edition showed the tractor all in red with yellow hubs, and then the 1966 edition showed the model in its correct red and yellow colours.*

113. *Dinky no.322 Disc Harrow in a yellow illustrated box. Disc harrow as (105) (77 mm).*

114. *Dinky no.322 Disc Harrow in a yellow illustrated box. Disc harrow as (106).*

115. *Dinky no.308 Leyland 384 Tractor. Metallic red tractor with blue stick-on strip with white lettering, silver steering wheel, white plastic exhaust, yellow metal rear and yellow plastic front hubs with rubber tyres (85 mm).*

116. *Dinky no.308 Leyland 384 Tractor in a yellow illustrated box. Blue tractor with black steering wheel, white plastic exhaust, white metal rear and white plastic front hubs.*

117. *Dinky no.308 Leyland 384 Tractor in a yellow illustrated box as (**116**). Tractor as (**116**) but metallic red with silver steering wheel.*

118. *Dinky no.308 Leyland 384 Tractor in a yellow illustrated box as (**116**). Tractor as (**117**) but orange.*

119. *Dinky no.308 Leyland 384 Tractor in a rigid plastic lift-off case. Tractor as (**116**) but with silver steering wheel and red metal rear and red plastic front hubs.*

120. *Dinky no.308 Leyland 384 Tractor in a rigid plastic lift-off case. Tractor as (**116**) but with silver steering wheel.*

121. *Dinky no.308 Leyland 384 Tractor in a rigid plastic lift-off case. Tractor as (**117**).*

122. *Dinky no.308 Leyland 384 Tractor in a rigid plastic lift-off case. Tractor as (**118**).*

123. *Dinky no.308 Leyland 384 Tractor in a hanging window box. Tractor as (**116**).*

124. *Dinky no.308 Leyland 384 Tractor in a hanging window box. Tractor as (**118**).*

125. *Dinky no.308 Leyland 384 Tractors on an AEC Hoynor car transporter. Two very rare and unfinished prototype tractors in metallic bronze with white hubs without drivers and exhausts and with small black plastic steering wheels. The tractor unit of the transporter was bright yellow with SC on the doors and with a white chassis. The single-deck transporter was in metallic lime with yellow rear drop-down ramps (320 mm). Sold at Wallis and Wallis Auctioneers, Lewes, Sussex on 5 October 2009, it is understood that only two of these sets were produced as salesmen's samples for the Leyland tractor.*

127. *Dinky no.3265 Mogul Steel Toy Tractor and Trailer. Red, black and silver tinplate tractor with black plastic wheels with white plastic hubs, black plastic air cleaner and towing hook. The red-and-silver steel trailer with wheels similar to those on the front of the tractor and with a black plastic towing loop (465 mm).*

126. *Dinky no.308 Leyland 384 Tractor. A close-up view of one of the tractors in (125). While the other had front tyres as in (115) to (124), this had tyres with different treads and HEAVY DUTY moulded on both sides.*

Dragon Toys

Company History

Very little is known about the manufacturer of the toy tractor illustrated (**1**). In addition to the tractor, we have found a boxed toy described as a 'constructional landscape toy', which had the Dragon Toys logo, as well as the brand name *SCENOVA* on its label. The toy consisted of approximately thirty hardwood bases, each 127 mm square, on which were set all the elements to form an English village. Individual bases were modelled with farm and village buildings, plain fields and ploughed fields, hills, meadows and road sections, and the bases could be arranged in multiple ways.

Although the company logo resembled a Welsh dragon, the boxed toy was marked *Made in England.* As both the boxed village and the tractor were made mainly of wood, the company was probably one of many that took advantage of the expansion of the U.K. economy immediately after the Second World War, when all metals were in short supply but wood was generally available. The toys mentioned are the only examples of this company's products that we know of. We assume that it was a very small company that did not survive the mid-1950s when wooden toys were falling out of favour as diecast metal toys became increasingly popular.

Model

Wooden Tractor

From the materials used, it would be easy to assume that the tractor was home-made were it not for the professional-looking logo (**1**). This was a red 'Welsh' dragon, set in a gold roundel with black outline and *Dragon Toys* in white letters. The dragon logo on the boxed English village toy was black and white. The tractor was made from seven separate pieces of wood fixed together with small nails. The rear wheels were also made of wood and were fixed separately to each side of the tractor with screws. The front wheels of the tractor and the steering wheel were formed from tinplate, very similar to wheels used contemporaneously by other small toy companies such as Kayron, A.V.H. and Olson (see the Kayron, A.V.H. and Olson chapter), so were probably bought-in. They were fixed to the tractor with small nails. A small wooden dowel was used as an exhaust but it is not known what the similarly coloured flat piece of plywood nailed to the top of the engine block is supposed to represent. It is not known whether the screw set into the rear wheel or the small hole in the side of the tractor are original.

1. *Dragon Toys wooden Tractor in red with yellow steering wheel and light blue wheels (200 mm).*

Edith Reynolds

Company History

Edith Reynolds specialised in making equestrian models for sale through prestige London shops such as Harrods, Hamleys, Liberty and Fortnum and Mason from the 1930s to the 1960s. The business was run by a 'Miss Reynolds' from her house at 194 Bellingham Road, Catford, London S.E.6, and she is recorded at this address in telephone directories in 1939 and from 1946 to 1961. Her models were mainly leather horses with removable riding harness and their riders (**5**). Her advertising claimed that her sets of huntsmen and hounds were popular gifts for boys and girls. She provided a model sectional stable building and mangers with standing for two Shetland ponies, harness rack and hay rack (**2** and **4**). There was also a very fine wooden tipping cart (**1**). From the advertised prices one can see that her range of models was expensive.

The Reynolds horses were made from leather stretched over a wire frame filled with rubber. Accessories included riders, tack, rugs and hounds. The horses had her labels sewn underneath, and almost every piece of her wooden sectional buildings was stamped with her name.

Models

Tipping cart

It is only the lovely two-wheeled farm tipping cart or tumbrel (**1**) which need concern us here. Her other products were purely equestrian (**2** to **5**). A tumbrel seems a strange choice, since a gig would surely have been more appropriate for her clientele. This fine wooden tumbrel was cream, with red trim on the wavy front edges, along the spokes and rims of the large wheels, and on the cart sides. Assembling such realistic tumbrel wheels must have been a work of art in itself. The ends of the shafts were black, as were the rims of the wheels, which were made of thin strips of metal nailed onto the wheels to represent the iron tyres. The tumbrel tipped up by releasing a wire loop on the front right-hand corner visible in (**1**), and the tailboard was removable. Stamped under the floor of the cart was *EDITH REYNOLDS CONSTRUCTIONAL WOODWORK MADE IN ENGLAND.*

The horse had a cart harness and reins made of leather strips, and on the saddle was a curved sheet of brass covered with two leather strips, one of which is missing in (**1**). The figure of the driver was covered with sewn brown leather for the trousers and green leather for the rather fine coat. The head and cap were hand-painted plaster of Paris. Sewn on the belly of all her horses was a label reading *EDITH REYNOLDS HAND-MADE REAL SKIN AND LEATHER ANIMALS SERIES MADE IN ENGLAND* (**3**).

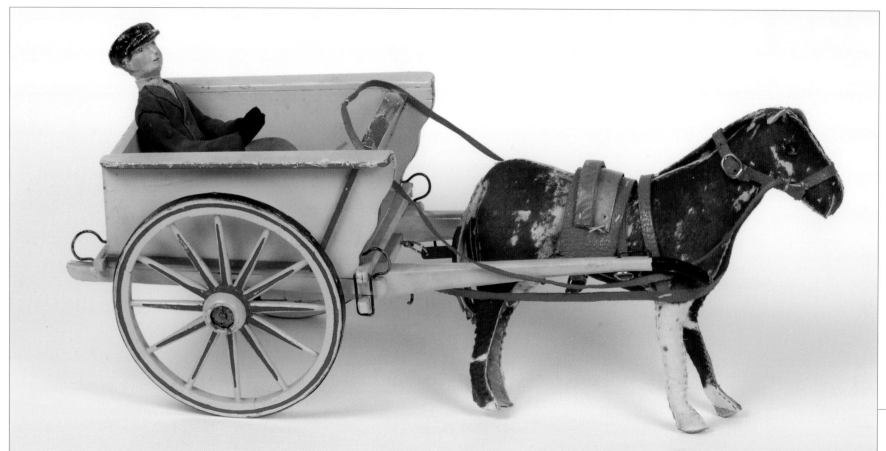

1. Edith Reynolds wooden tipping cart with horse and driver. The cart and shafts were cream with red trim on the cart and on the wheels. The ends of the shafts and the metal tyres were black. The driver had brown trousers, green jacket and black peaked cap (420 mm).

3. *Edith Reynolds label as sewn on the belly of her horses and ponies.*

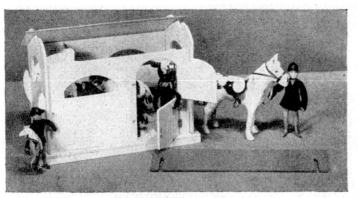

The Complete Stable Outfit

These are the **EDITH REYNOLDS** leather and real-skin horses.

The STABLE has standing for two Shetland ponies; with manger, hay-rack and harness-rack.

Ponies and horses range from six to eleven inches in height: mounted by Huntsman, Lady, Military and Police Officers, and young riders.

Please order as early as possible, as Miss Reynolds finds it difficult to keep pace as Christmas approaches.

Obtainable at HARRODS

HAMLEYS : FORTNUM & MASON, London.

2. *Edith Reynolds cream sectional stables for two ponies with hay rack, dividing wall and half-doors. Outside is an Edith Reynolds pony and rider.*

4. *Edith Reynolds advertisement in* Riding *Vol.11, No.11 November 1951, p.455, showing a sectional stable building for two Shetland ponies as in (***2***) described as* The Complete Stable Outfit.

The **Edith Reynolds** real-skin and leather animals, of which the horses have taken first place, are filled with rubber, which gives them a quite unique, life-like, "feel"; and everlasting qualities. They are entirely hand-made; the harness takes off, and each horse has its individual rider, dressed in habit, or in brightly coloured, and tan, suedes.

Order in good time for Christmas, as supply is limited.

Obtainable at HARRODS; HAMLEYS; FORTNUM & MASON, London.

5. *Edith Reynolds advertisement in* Riding and Driving *Vol.9, No.103 November 1949, showing a hunt meet with hounds.*

Escor Toys

Company History

Escor Toys, formed in 1938 by E.S. Corner, who had previously worked as a tea planter in India, became a limited company in 1946. Based in Christchurch, Hampshire, Escor made a wide range of traditional wooden children's toys, predominantly aimed at the pre-school and younger children's market. Upon E.S. Corner's retirement, his nephew Peter Corner took over the running of the company, and in the early 1970s it was sold to the Earl of Ronaldshay.

Peter Thorne joined the company as woodworking foreman in 1972, and was soon promoted, at first to factory manager and then in 1974 to managing director. Thorne began to acquire shares in the company, and when the Earl of Ronaldshay withdrew from the business in 1985, Thorne took over responsibility for running Escor, eventually becoming owner in 1995.

In the late 1980s the business had entered into partnership with the local council and merged with Dorset Enterprises, an organisation set up in 1914 by a Bournemouth councillor to provide work for injured servicemen returning from the First World War. Escor began to specialise in providing work for people with disabilities, and when Thorne left the business in 2000, Bournemouth Council took over full control.

The range produced by Escor Toys included push-along vehicles, carpet railway engines, wooden animals, roundabouts, constructional toys and educational toys such as abacuses, as well as dolls' house accessories. Escor products were of high quality, and were distinguished by their hygienic, hard-wearing finish and the primary colours of the painted toys. Most examples will be found bearing the *ESCOR* logo. As well as export sales to the U.S.A., Australia, Norway and Japan, Escor made toys for U.K. brands including Galt Toys and Kiddicraft, as well as Boots and John Lewis own-label products. Despite winning a 'Best Buy' award in November 2006 from the Ethical Consumer organisation, Escor was unable to compete with imported products. The company ceased trading in 2012, and Dorset Enterprises was closed during 2013.

Model

Pull-along Horse and Cart

We have found only one example of a farm vehicle made by Escor (**1**). This was a naïve cart in red with red shafts, on solid yellow wheels that were held on by square wooden wheel nuts screwed onto threaded axle ends. The cart was drawn by a horse with four wooden dowel legs, which stood on a light green wooden platform with four yellow, turned-wood wheels. The horse was painted white with black mane, hooves and tail, and had red detailing to the eyes, nostrils and mouth, but no harness. The horse was attached to the shafts of the cart by means of a wooden dowel through the horse's body, which fitted through holes in the shafts either side. The toy could be pulled along by means of a piece of multi-coloured string which went through a hole in the horse's neck and tied at the throat. It had a wooden ball attached to the end to help small hands get a good grip!

The horse and cart each had a wooden figure, one in blue and one in green, and both horse and cart were marked *ESCOR* in yellow letters on a red oval paper sticker with yellow border (**2**).

Further Reading

V. & A. Museum of Childhood website: www.museumofchildhood.org.uk.

1. *Escor wooden Pull-along Horse and Cart. Red cart pulled by a white horse on green platform, with naïve green and blue figures (209 mm).*

2. *Escor wooden Pull-along Horse and Cart in (1) showing detail of the ESCOR logo.*

Company History

The 1964 telephone directory listed Fairchild Plastics Ltd, Toys, at Walmar House, Regent Street, London W1, with a warehouse at Avon Trading Estate, London W14. In 1965, Fairchild Plastics Ltd was listed as Toy Manufacturers, at Waterside Mill, Todmorden, Yorkshire. Then, in 1966, the company was again listed at Avon Trading Estate, implying that the office at Regent Street was no longer in use. Because a number of the products that we have seen in original packaging bore the brand name *SELCOL FAIRCHILD*, we know that for at least some of its history, Fairchild was part of Selcol Ltd. Selcol was the plastic products division of the Selmer Company, a maker of musical instruments which had been started in London in 1928. Although it had a similar name, it was not directly related to either the French- or American-based Selmer companies.

We do not know if Fairchild Plastics was a separate company that was already in existence before Selcol bought it, or whether it was itself a subsidiary brand of Selcol. Selcol's head-office operations were run from the Selmer premises in Charing Cross Road, London WC2, which does imply that Fairchild and Selcol were totally separate companies before they combined their products.

The Selmer Company started Selcol after the Second World War as a means of diversifying from musical instruments and amplification equipment - their main products, for which they were the U.K. industry leader prior to the Second World War. We know that Selcol was in existence by 1953 as an associate company of Selmer, because in that year the two became part of the Musical and Plastics Industries holding company.

Based at a factory in Woolpack Lane, Braintree, Essex, Selcol made a range of garden furniture and ornaments as well as household goods such as cups and saucers, in addition to toys. Although the toys were made of plastic, many of them were music-related, such as record players and record racks, and a range of guitars featuring artists of the day, including The Beatles, Elvis Presley and Sooty. After a few years of poor performance, in 1968 the Selcol plastics operation was shut down and the Selmer Company moved its operations into the Braintree factory. Selmer was itself shut down in the early 1980s.

It is possible that products with only the *FAIRCHILD* name were produced in the early 1960s, and that the toys found in packaging marked *SELCOL FAIRCHILD* were later products, from when the plastics operations of the Selmer Company were already experiencing problems and they were therefore consolidating factories and brands.

The tractors shown are unmarked, and consequently difficult to identify outside their packaging.

Model

Heavy Duty Farm Tractor

We have found only one farm vehicle made by Fairchild, the *HEAVY DUTY FARM TRACTOR* (**1** to **3**). All the versions we have found were packaged in polythene bags with a header card. The header card had the same illustration on both sides, of a tractor pulling a manure spreader. Apart from the description of the contents, it was marked *FAIRCHILD* and *MADE IN ENGLAND.* When found in original packaging, Selcol Fairchild products have an item number, and this is another means of differentiating between Fairchild and Selcol Fairchild items.

The tractor body and mudguards were moulded in one piece, which would have required substantial investment in a large mould. The grille was a separate moulding which carried the headlights, and snap-fitted into a hole in the front of the tractor. All the tyres were made of the same black plastic as the grille and had separate white plastic hubs which also formed a separate back-piece to each tyre. Both axles were made of metal rod. The headlight lenses, seat, exhaust, gear lever, steering wheel and belt pulley were all made from the same white plastic as the wheel hubs.

1. *Fairchild Heavy Duty Farm Tractor. Light blue tractor with black grille and tyres and white accessories, showing the header card packaging (235 mm).*

2. *Fairchild Heavy Duty Farm Tractor, as (1) but view of left side showing the belt pulley.*

3. *Fairchild Heavy Duty Farm Tractor, as (2) but in red.*

Fairylite

Company History

Fairylite was a brand name of Graham Brothers, a firm of London-based toy wholesalers that had been in business since 1887. Telephone directories listed the firm of Graham Bros, Far Eastern Merchants, at 73 Endell Street, London WC2 before 1959, and Graham Bros (Fairylite) Ltd, Far Eastern Merchants, at the same address between 1961 and 1968. We have seen a wholesale catalogue for 1949 which listed a Fairylite plastic Steam Roller, so obviously the brand name was being used by Graham Bros before it was incorporated into the company name. Between 1968 and 1969 the company of Fairylite Moulded Plastics Ltd was listed at Tingewick Road, Buckingham, under the Toys & Games Manufacturers & Suppliers section of the *Yellow Pages* telephone directory.

The various names and addresses of the company imply that there were two parts to the Graham Bros business, one involved with the importation of toys and novelties, and a second that acted as a wholesaler for toys sourced in the U.K., perhaps with its own manufacturing facility.

This theory is reinforced by the various brand markings found on Fairylite packaging and products. Products at the cheaper, or novelty, end of the plastic toy market can be found with the Fairylite name and marked either *Empire Made* or *Made in Hong Kong*, and we assume that these were the toys imported by Graham Bros as part of their Far Eastern Merchants business. Products marked *MADE IN ENGLAND* have the abbreviation *REGD.* beneath the *Fairylite* name (**1**), and obviously this was used in an attempt by Graham Bros to protect the Fairylite brand within the toy markets of the British Empire after the Second World War. We are aware of similar plastic toys sold under the 'Fairylight' brand, and we assume this variation was also used by the company; additionally we have seen a book with the *Fairylite REGD.* brand that was also marked *G.B. LONDON*. Despite all the information we have, it is not possible to determine accurately which toys were of pre- or post-Second World War production. Graham Bros stood at the 1947 British Industries Fair in the Toys and Games section, and an advertisement in the fair guide showed that their range encompassed most of the products for children in post-Second World War Britain. They listed themselves as 'Producers and Publishers of Rubber Toys, Metal Toys, Fairylite Series, Toy Books, Rubber Teats, Soothers, Plastic Toys and Games, Gaeltarra Eirann Dolls and Animals, Rubber Playballs'. (Gaeltarra Eirann referred to items made from Donegal yarn.)

Unfortunately we have been unable to find any information about Graham Bros after 1969, and we assume the company closed at some point during the late 1960s or early 1970s.

Model

Plastic Rowcrop Tractor

We have found only one farm vehicle made in the U.K. by Fairylite (**2** and **3**). This was a rowcrop tractor, similar to products made by Tudor Rose in the U.K. (see the Tudor Rose chapter) and Thomas Manufacturing of the U.S.A., and based on a tractor used in North America. The tractor body, steering wheel, axles and driver were moulded in one piece from a brittle, cellulose acetate-type of plastic. The four wheels were separate mouldings, attached to the tractor by being pushed onto stub axles before heat was applied to the axle ends to form crude hubs to each wheel. The models were marked underneath with *Fairylite, REGD.*, similar to (**1**), and *MADE IN ENGLAND.*

Research indicates that the Fairylite brand was used between 1949 and 1969, and the tractor illustrated was obviously one of the products made in the U.K. Given the nature of the company's main business, it is possible that the tractor was made from moulds obtained directly from an American-based plastic toy maker, or from one of the U.K. plastic toy companies.

Further reading

Fawdry, Marguerite, 1990, *British Tin Toys*. New Cavendish Books.

Grace's Guide to British Industrial History website: www.gracesguide.co.uk.

1. Fairylite REGD. *brand mark on children's book printed in England.*

2. *Fairylite Rowcrop Tractor in green with dark blue wheels (75 mm).*

3. *Fairylite Rowcrop Tractor, as (2) but pale blue with green wheels.*

Forest Toys

Company History

Forest Toys was started at the end of the First World War by Frank Whittington at Brockenhurst in the New Forest, Hampshire, hence the company name. Initially working from Whittington's home, the company made wooden toys from deal, the relatively soft and easily worked sawn wood of pine or fir. The figures and animals were constructed from three sections of wood stuck together. Using a jig-saw, all three pieces would be cut to the required shape from large flat sheets and the detail was then hand carved into the two outer sections. The larger pieces might be made with separate limbs which were then attached to the body. All of the products were produced to a constant scale of 1:16 or, as a 1920s catalogue stated, 'The scale of size has been carefully considered, ¾ in = 1 ft, as this makes the animals of great value in the schools, and they are used more and more for educational purposes.'

Forest Toys regularly attended the British Industries Fairs held in London between the First and Second World Wars and enjoyed some success. A small factory was opened in Brockenhurst in 1922, using local women to paint the toys, and at its peak the company employed 16 people. During the 1920s, a visit to the factory at Brockenhurst became a regular organised day out for passengers of cruise liners docked at Southampton, and despite the toys being relatively expensive, this led to Forest Toys' products being popular in the U.S.A. A mail order business was developed to service that demand.

The company produced a broad range of standard products. The toys were available through Hamleys of London, and direct through the Medici Society Ltd, Forest Toys' London agent. The late 1920s catalogue, published by the Medici Society, listed over thirty wild animals, including three poses of wolf, a Noah's Ark 'as supplied to Her Majesty the Queen' with a complement of fifteen pairs of animals, a dove, Noah and Sarah, an Arab scene with Arabs mounted on camels near a well, and a scene from the *Jungle Book* with Mowgli, Kaa the Python, a cave and trees.

Additionally the company produced a whole range of toys reflecting the life of the Hampshire countryside. There were two sets of fox-hunting figures – 'Full Cry' and 'The Meet'– and a stable was produced for the horses. Also available were a Gypsy caravan with male and female Gypsies, a farmyard with animals, farmer and dairymaid, ponds, trees, a pigsty, a timber wagon and a tipping farm cart.

Commissions were taken for almost any subject. The catalogue stated that dogs were available 'of various kinds, made to order, from 7/6 each' – a considerable sum for a single item in the late 1920s. The non-standard range included figures of army personnel in dress uniform, a large fairground carousel, and stagecoaches in motion and being stopped by a highwayman. Nativity scenes were created for local churches, a model was made of the Bournemouth Symphony Orchestra and a coronation coach was made, presumably for the 1937 coronation. Forest Toys made a model of Little Red Riding Hood and the Wolf, and even made a public house, complete with publican and customers.

Demand seemed to outstrip supply, but the Second World War brought an end to production, a shortage of raw materials and labour forcing the company to close in 1939. Although it is known that the majority of Forest Toys' tools and working drawings were still in existence in the late 1980s, for reasons unknown, production was never resumed after the Second World War. Mr Whittington died in 1973, aged 97.

Models

Timber Waggon, with 2 horses

This was described in the catalogue as 'TIMBER WAGGON, with 2 Horses, in Box' and sold for 9 shillings. Sold individually boxed, this carried a large log 'tree'. The wagon and horses were made of wood, and the wagon wheels were made of cast iron. The wagon was supplied with two horses which could be modelled as standing or walking, and were coloured either white or brown with docked, dark brown tails. All the horses were fitted with blinkers which were made of small pieces of leather, the horses' harnesses were made of thin strips of leather and the horses were attached to the shafts by small link chains. Nails were used as pins to keep the logs on the wagon. The front turntable was made from a flat piece of wood, as were the shafts, and the rest of the wagon was cleverly made from wooden rod with a square cross-section.

There were two methods of attaching the shafts to the turntable. On earlier versions, the turntable had a piece cut out at the front, which left a protrusion on each side. The rear of the shaft had cut-outs each side which allowed it to fit between the protrusions on the front of the turntable (**2**), and a wire pin hinge ran through the assembly. The later version just had a piece of leather pinned on top of the joint (**3**), and used less material and was simpler to assemble. Both versions allowed the joint to hinge in the vertical plane. The bolsters were joined by a square-section wooden rod which slid through the rear bolster to allow adjustment of the length of the wagon. The front bolster was permanently fixed to the wooden rod, and the turntable turned by means of a nail through the front bolster and the wooden rod into the turntable. A substantial amount of the original paint is missing from the examples we have found, but it seems as though the wagon was painted one colour throughout (red in the example illustrated), and the wheels were painted the same colour (**1**).

Farm Cart to tip up, with 1 horse

This was described in the catalogue as 'FARM CART to tip up, with 1 Horse, in Box', and sold for 7 shillings. Originally sold boxed, the cart and horse were made of wood, and the cart wheels were made of cast iron. The horse which was supplied with the cart was similar to those supplied with the wagon and could be modelled as standing (**4**) or walking (**6**). The horses were fitted with blinkers which were made of small pieces of leather; the horses' harnesses were made of thin strips of leather and the horses were attached to the shafts by small link chains. Wire clips, two on the body of the cart and two on the shafts, were used to hold a small wood

dowel which would stop the cart from tipping. The cart had a removable flat wooden tailboard at the rear (**5**), which was painted the same colour as the body of the cart. In the examples we have seen the base and the shafts of the cart were painted red, as were the wheels, and the body of the cart was painted blue. We have found one example of the Farm Cart in an original box (**7**). The box was made of plain cardboard with a lift-off lid. The box had an original label which showed the Forest Toys logo of a leopard drinking and bore the words *THE FOREST TOYS. Farm Cart. MADE IN THE NEW FOREST BY F.H. WHITTINGTON, BROCKENHURST.*

Further Reading

Hampshire Museums Service, Childhood collection, *The Forest Toys.*

www.hants.gov.uk/childhood-collections/toys/forest-toys

1. Forest Toys Timber Waggon, made of wood with cast iron wheels (630 mm.)

2. *Forest Toys Timber Waggon, as (1) but first version, showing detail of the early method of attaching shafts to turntable, involving cut-outs to turntable and shaft piece.*

3. *Forest Toys Timber Waggon, detail of (1), showing the later method of attaching the shafts to the turntable using a piece of leather.*

4. *Forest Toys Farm Cart, made of wood with cast iron wheels. The two cork 'buckets' came with the cart and are believed to be original (320 mm.)*

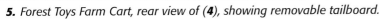

5. *Forest Toys Farm Cart, rear view of (**4**), showing removable tailboard.*

6. *Forest Toys Farm Cart, as (**4**), but with a walking white horse (missing the tailboard).*

7. *Forest Toys Farm Cart, as (**4**), in original box.*

Company History

Very little is known about the manufacturer of Glyntoys. A boxed Gate Unit has been found, and printed on the box was *GLYNTOYS FARMYARD SERIES and the address 241 DERBY ROAD LOUGHBOROUGH.* However we could not find Glyntoys in any telephone directories to establish their dates of operation. We suspect they were active in the immediate post-war years, when even crudely-made toys would find a ready market, but it is also possible that their production dates from the 1930s.

The farm Gate Unit consisted of lead posts soldered to a tinplate base, a lead gate and wire hinges and fence rails (**1**). We have also found the same lead fence posts joined by wire fence rails in 2'6" lengths, without a baseplate, to make field fencing (**2**). The horse-drawn vehicles were solid-cast heavy lead models. No doubt there were also some solid cast lead figures and animals making up the *FARMYARD SERIES*, but we are not aware of any boxed or carded sets which might identify these items, and all Glyntoys models are rare.

Models

Horse-Drawn Tumbrel with Plain Sides

The cart body was hinged on a single axle which also carried the wheels, and there were bent wire catches on each side which could be moved to allow the body to tip backwards. The body had slots for a separate tailboard, which presumably was provided although we have never actually seen one. The model was painted all over in a medium or dark shade of green, or was dark blue with red shafts and wheels (**3** to **5**). The examples pictured had minor differences between the castings, notably in the lettering cast underneath, which on (**6**) was absent. On (**7**) *GLYNTOYS LOBRO* was cast transversely underneath, *LOBRO* obviously being short for Loughborough, and on (**8**) the same wording was cast at an angle. The horse, in cart harness, was a solid lead casting painted brown or dappled white, with trim in white, black and gold (**9** and **10**) or black, brown and gold (**11**). Some of the horses pictured had metal rings attached to string traces, and these could be used to harness the horses to the carts, by slipping the front rings over the ends of the shafts and attaching the rear rings to hooks provided on each side of many of the models, although the tumbrels with plain sides did not have the hooks.

Horse-Drawn Tumbrel with Panelled Sides

This tumbrel was loosely copied from the Britains model, with panelling on the sides and projections which might have retained a pair of raves. There appears to have been a tailboard (missing from the example pictured), and *GLYNTOYS LOBRO* was cast at an angle underneath.

There were hooks at the rear of the shafts for the metal rings mentioned under the plain-sided tumbrel; however, the only example we have seen of the panelled-sided tumbrel had a different horse without the rings and string traces. The model was painted red with light brown shafts and blue wheels. The horse, without a collar, was painted dappled white, with grey, black, brown, gold and red trim (**12** and **13**).

Horse-Drawn Water Cart

Despite being so crudely made, this is an attractive toy. The shafts and wheels were similar to the tumbrels, attached by the axle to a very heavy solid lead barrel, and the shafts had hooks for harness rings. To prevent the body from tipping backwards, the body and shafts were held together by a piece of bent wire. The model had a brown body and shafts with black trim and red wheels, or red-brown body with green shafts and wheels. No identification was cast on it (**14** to **16**).

Horse-Drawn Hay Rake

The lead cross-piece of the rake was moulded onto the wire tines, and also had a wire lever so that the tines could be raised and lowered. The casting of the shafts included a driver's seat, and had hooks for harness rings. The components were held together by two stub axles carrying the wheels. The model had a blue rake with green shafts and wheels, or an orange rake with blue shafts and wheels. No identification was cast on it (**17** and **18**).

Flat Wagon

This was probably a horse-drawn wagon, but the shafts are missing on the only example we have seen. We cannot rule out the possibility that it had a drawbar for a tractor or lorry instead, but since the other Glyntoys models were all horse-drawn, this seems the most likely means of power for the wagon as well. There was a loose turntable which carried the front axle, and a similar but larger support for the rear axle was attached underneath the wagon body by soldering. The model was painted light brown with red edges to the body, a light blue floor and red wheels. No identification was cast on it (**19**).

Model List

Model	Dates
Horse-Drawn Tumbrel with Plain Sides	?
Horse-Drawn Tumbrel with Panelled Sides	?
Horse-Drawn Water Cart	?
Horse-Drawn Hay Rake	?
Flat Wagon	?

1. *Glyntoys Gate Unit, green fence sections, white gate, green and brown base (223 mm).*

3. *Glyntoys Tumbrel with Plain Sides, dark green, part of the shafts broken off, missing the horse.*

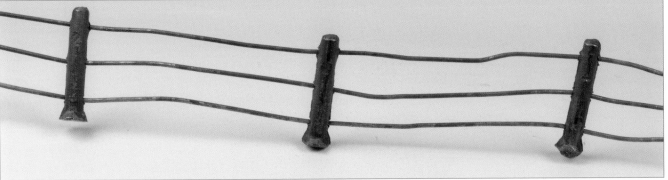

2. *Glyntoys Field Fencing, green, using fence posts as (1) wired together in 2'6" lengths.*

4. *Glyntoys Tumbrel with Plain Sides, as (3) but medium green, brown horse (145 mm).*

5. *Glyntoys Tumbrel with Plain Sides, as (4) but dark blue with red shafts and wheels.*

6. *Glyntoys Tumbrel with Plain Sides, underneath view of the model in (3), without any cast lettering.*

7. *Glyntoys Tumbrel with Plain Sides, underneath view of the model in (4) showing GLYNTOYS LOBRO cast transversely, and other minor casting differences from (6).*

8. *Glyntoys Tumbrel with Plain Sides, underneath view of the model in (5) showing GLYNTOYS LOBRO cast at an angle, and other minor casting differences from (6) and (7).*

9. *Glyntoys horses, brown with white, black and gold trim, the example on the right having string traces with metal rings which could be used to harness the horse to the carts and implements.*

10. *Glyntoys horse, dappled white with white, black and gold trim, string traces with metal rings as (9).*

11. *Glyntoys horse, dappled white with black, brown and gold trim, string traces with metal rings as (9).*

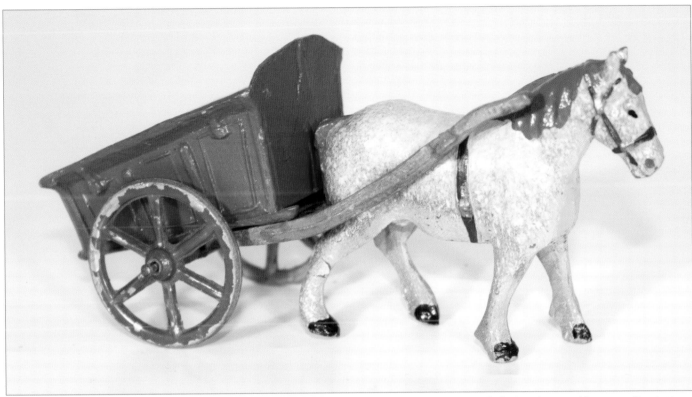

12. *Glyntoys Tumbrel with Panelled Sides, red with light brown shafts and blue wheels, dappled white horse without a collar (112 mm excluding the horse).*

13. *Glyntoys Tumbrel with Panelled Sides, underneath view of the model in (12) showing* GLYNTOYS LOBRO *cast.*

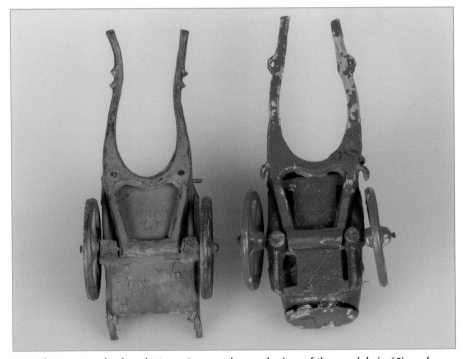

14. *Glyntoys Water Cart, brown with black trim and red wheels, missing the horse (116 mm).*

15. *Glyntoys Tumbrel and Water Cart, underneath view of the models in (4) and (14). Note the hooks on the Water Cart at the rear of the shafts, to which the metal rings of the harness could be attached.*

16. *Glyntoys Water Cart, as (**14**) but red-brown with green shafts and wheels. The barrel has brown paint on top of the red-brown colour, but we are not sure whether this is original.*

17. *Glyntoys Hay Rake, blue with green shafts and wheels, missing the horse (128 mm).*

19. *Glyntoys Flat Wagon, light brown with red edges to the body, light blue floor, red wheels, missing the shafts and horse (110 mm).*

18. *Glyntoys Hay Rake, as (**17**) but orange with blue shafts and wheels.*

Grace Toys

Company History

The earliest mention we have of E. & H. Grace is from telephone directories, which listed the company as 'Cabinet Makers' based at Melbourne Yard in Upper Norwood, South-East London, between 1939 and 1941.

The company seems to have decided to make wooden toys to take advantage of the war-time shortage of metal. An editorial in the edition of *Games & Toys* magazine for April 1941 stated that the company's range 'contains such topical lines as camouflaged army trucks, aircraft and carriers, fire engines, ambulances, streamlined camouflaged cars, as well as railway engines, timber lorries, tradesmen's vans etc.' Additionally the editorial mentioned a model London Transport bus which was already popular with overseas buyers, so perhaps the company was in existence before 1939, although we have been unable to find any information to suggest that E. & H. Grace started earlier than that year.

The Manchester Fortnight trade fair took place between 9 and 13 June 1941, and Grace Toys were exhibited by Strome & Co., a full-page advertisement (1) in *Games & Toys* proclaiming, 'Because of their individuality, the appeal of Grace Wooden Toys is instant. There's nothing "mass-production" about them. The exceptional sturdiness, brilliant finish, pleasing design and lasting qualities of each toy are the result of individual craftsmanship and skill. The range is large, both in variety and price.'

The vehicles shown – a London Transport bus, a Fire Engine and a Camouflaged Army Truck – had catalogue numbers from G20 to G31 and we can only assume from their sturdiness that they were relatively expensive and individually made as would befit toys from a firm of cabinet makers.

E. & H. Grace moved between various addresses in the Norwood area of South-East London between 1943 and 1977. From 1943 to 1949 the company's business was described in directories as 'Models', but they seem to have been less successful with their wooden toys as they entered the 1950s, when E. & H. Grace were then described as 'Manufacturers in Wood'. From 1975 until 1977 the company was advertised as 'Furniture Manufacturers' but we have been unable to find any information about the business after 1977.

Model

Wooden Horse and Cart

We have found only one farm vehicle made by E. & H. Grace Ltd (2). A wooden horse and cart, the toy was attractively finished in multiple colours. The cart was made from sheets of wood, predominantly painted red with red shafts, but was cleverly decorated with blue and yellow panels to represent the slatted sides of full-size farm carts. The wheels were also made from wood

1. *Advertisement for Grace Toys from* Games & Toys *magazine, June 1941.*

sheet as opposed to plywood, and had black panels painted to represent the areas between the spokes on a full-size wheel. The cart had a transfer with the Grace Toys trade-mark of *A PRODUCT OF E. & H. GRACE LTD. LONDON* set in a scroll beneath a line drawing of a deer set against a black background, similar to the logo used in the *Games & Toys* advertisement.

The horse was white with black tail, mane, hooves and face detail, blue collar and cart saddle, and red bridle, belly band and breeching. The horse stood on a white wooden base with red wheels, and was attached to the cart by means of a piece of string strung between the shafts which lay across the horse's back in the groove on the pad.

What at first appeared a simple wooden toy, was in fact a sophisticated representation of the real thing.

Further Reading

Games & Toys, April and June 1941.

2. *Grace Toys Wooden Horse and Cart (320 mm).*

Company History

Hercules was the brand name for a range of wooden toys made by Critchley Bros Ltd, a company based in Brimscombe, near Stroud in Gloucestershire. The range included wooden building bricks, pull-along carpet toys, lorries and train sets, as well as the tractor and timber wagon illustrated. In addition to the wooden toys, the company also made handicraft outfits, such as knitting, embroidery and sewing sets, some of which were sold under the Wimberdar brand.

It is not known how long the company was in business, but the illustrations are taken from the Critchley Bros catalogue dated January 1955, and the company is known to have advertised in *Games & Toys* magazine during 1958. Interestingly, the 1955 catalogue was acquired in Eastern Europe, and inside it was a reference to export pricing. Furthermore, the Hercules logo included 'Made in England', so it is possible that Critchley Bros concentrated on export sales for its products.

Model

No.905 Tractor Set

This was described in the catalogue as 'Comprising a realistic tractor, drawing a timber wagon loaded with natural wood log. Sturdily built, painted in bright colours. Each set boxed singly'. Although we do not know the exact dimensions, we assume it was similar in size to the locomotive pictured on the front of the catalogue and therefore quite large, as it would have been aimed at a similar age range of children. The tractor had a metal eyelet at the front to which string could be attached in order to pull the toy along.

1. Front cover of the 1955 Critchley Bros catalogue.

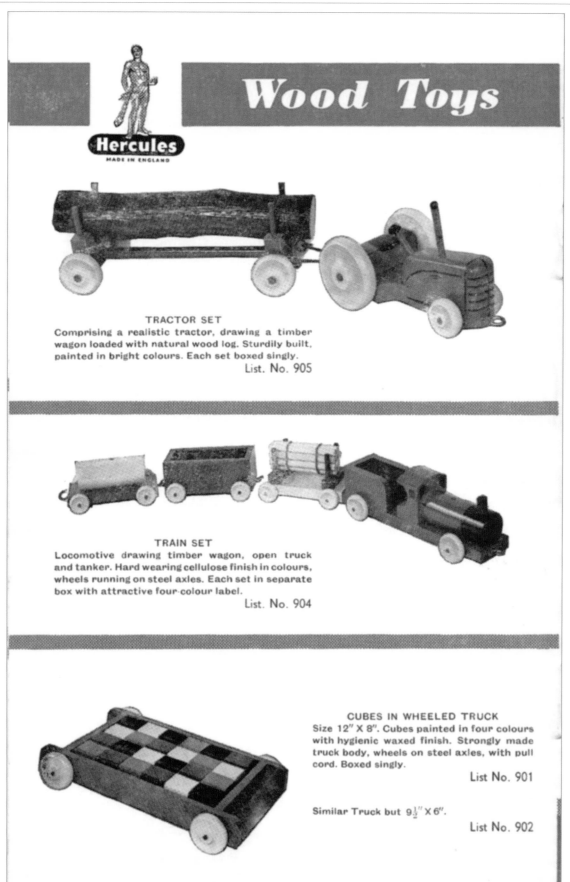

Hercules
MADE IN ENGLAND

Wood Toys

TRACTOR SET
Comprising a realistic tractor, drawing a timber wagon loaded with natural wood log. Sturdily built, painted in bright colours. Each set boxed singly.

List. No. 905

TRAIN SET
Locomotive drawing timber wagon, open truck and tanker. Hard wearing cellulose finish in colours, wheels running on steel axles. Each set in separate box with attractive four-colour label.

List. No. 904

CUBES IN WHEELED TRUCK
Size 12" X 8". Cubes painted in four colours with hygienic waxed finish. Strongly made truck body, wheels on steel axles, with pull cord. Boxed singly.

List No. 901

Similar Truck but 9½" X 6".

List No. 902

1st. Jan. 1955   657 HE (18000)

2. Back cover of the catalogue, showing the Tractor Set.

Hitchin Components

Company History

Hitchin Components Ltd was recorded in telephone directories at Astwick Road, Stotfold, Bedfordshire in 1968 and 1969 and at Fen Road in Stotfold in 1972. Local enquiries have not produced any more information about the company, and it does not show up in the records of Companies House. At this stage we cannot be sure whether Hitchin Components made complete tractor models or just the wheels and tyres for them.

Models

Tractor with large red metal seat

We have seen one large white wooden tractor with an enormous red metal seat, red plastic hubs, black rubber tyres and a steering wheel made from thick twisted bare metal wire (**1** and **2**). The tractor was steerable through the same wire which formed a steering column down to a thinner external wire connected to the front axle. *HITCHIN COMPONENTS LTD HERTS* was moulded on the rear tyres. Blue lines were printed on the engine block to give a minimal impression of an engine, but the grille was left blank. The seat may have been intended for a particular toy figure which was over-large for the tractor.

Tractor with cab

Wooden toy tractors with red plastic hubs and rubber tyres on the rear wheels marked *HITCHIN COMPONENTS HERTS,* but with the *LTD* deleted, were being made as recently as about 2000 (**3** to **5**), but we can find no evidence that the company was in business that long, so these tyres could have been from old stock.

1. Hitchin Components white wooden tractor with red metal seat, red plastic hubs, rubber tyres and bare metal steering wheel (230 mm).

2. Hitchin Components tractor, rear view of (1) showing the drawbar with hook and yellow spherical plastic spacers between tractor body and wheels.

3. *Hitchin Components wooden tractor painted in green, with a black grille and cream cab interior. The rear tyres marked* HITCHIN COMPONENTS HERTS *(198 mm).*

4. *Hitchin Components close-up of a rear tyre of the tractor in (**3**). Slight traces showing where* LTD *has been deleted from the mould can still be seen.*

5. *Hitchin Components wooden tractor and trailer painted in dark blue, with white cab and trailer interior. Tractor as (**3**) but with long black wooden exhaust. The rear tractor tyres as (**3**) and (**4**). The trailer tyres without writing.*

Hobbies

Company History

From the 1890s to the 1960s Hobbies Ltd was the leading supplier of fretworking equipment, designs and materials. Their head office and factory was at East Dereham in central Norfolk. The company had shops in London, including at 78a New Oxford Street and 87 Old Broad Street, and at various times in 11 cities and towns around the U.K. The company also had agents and stockists in many parts of the world. Their highly successful marketing was based on two publications, *Hobbies Weekly* and *Hobbies Handbook*, which became the *Hobbies Annual* after the Second World War. These were widely distributed. By 1938 the *Handbook* had grown to 284 pages full of advertisements for tools and gadgets the handyman could buy. The company was able to claim in their 1958 *Annual* that their products were 'sold the world over'.

The making of fretwork toys, household ornaments and small pieces of furniture became a way of life for boys and men alike, all of whom waited for the regular delivery of their *Hobbies Weekly*.

A history of the company can be found in *The Hobbies Story* by Terry Davy (1998) which covers the lives of the people who worked at the Dereham factory as well as the company's products. Sadly, the factory site is now occupied by a large supermarket and a car park, and the old station opposite, much used by the company in its heyday, is now a heritage steam railway. How times have changed!

The company was started by John Skinner who formed J.H. Skinner & Co. in the 1880s to sell fretworking tools and machines which he imported from the U.S.A. and Germany. John Skinner bought timber from Stebbings sawmill in Dereham, and in 1884 he started the first ever experiments on the production of plywood, which came to play such an important part in the fretwork industry. In 1887 J.H. Skinner & Co. published a 32-page catalogue of fretwork designs and tools, and in 1895 the Hobbies Publishing Co. launched *Hobbies Weekly* to promote a wide range of pastimes, including John Skinner's products. This greatly increased fretworking sales, and in 1897 the two companies were amalgamated as Hobbies Ltd. Also in 1897 John Skinner started to manufacture his own fretsaws, with adverts in *Hobbies Weekly*, and the new company quickly became the largest distributer of fretwork tools and supplies worldwide. By the early 1950s the circulation of *Hobbies Weekly* had reached 75,000 and *Hobbies Annual* sold 175,000 copies a year. However, by the late 1950s and 1960s sales declined as people became more affluent and able to afford luxury goods and did not feel the need to make them. In 1964 the company was taken over by Industrial and Commercial Holdings Ltd and the decline then accelerated. *Hobbies Weekly* ceased distribution in 1965, and in 1968

Hobbies Ltd went into voluntary liquidation after which all the company's records were unfortunately destroyed, except for a large quantity of photographs which were rescued. These are used liberally throughout Terry Davy's *The Hobbies Story*.

From 1922 the phone number for Hobbies Ltd was Dereham 15! The 1971 telephone directories showed a change of address within Dereham from Norwich Road to Elvin Road. In 1978 the copyright to the Hobbies trademark and designs was bought by an ex-employee, and from 1979 the company was known as Hobbies (Dereham) Ltd. It later moved to Raveningham in Norfolk.

Models

All the Hobbies fretwork designs came printed on thin sheets of paper so that the cutting lines could be pressed through onto sheets of plywood. The customer usually bought the designs and the plywood sheets as a set together.

Model farm vehicles were mostly in two series of fretwork sets advertised in *Hobbies Annual* from at least 1958 to 1962 (**1** and **2**), and there was also a separate no.3420 Horse and Cart in the 1962 *Hobbies Annual* which 'will delight the heart of any small child' (**3**). So far, we have found an assembled example of the wagon from Series 8 (**4**) but none of the others.

Model List

Model
Simple Model Patterns Series 1
Horse and cart
Tractor
Model Farm Figures Series 8
Tractor
Tumbrel
Wagon
Horse roller
Timber wagon
Plough
Horse rake
No.3420 Horse and cart

Further Reading

Anon, undated, *The Art of Fretwork*. Hobbies Ltd. (A booklet distributed with some fretwork outfits).

Anon, 2010, *Hobbies Small Wood Projects: Splendid projects for industrious chaps!* Ammonite Press. (Nostalgic compilation of articles originally published in Hobbies magazines between 1920 and 1964).

Davy, Terry, 1998, *The Hobbies Story*. Nostalgia Publications.

Hobbies Handbook pre-war and *Hobbies Annual* post-war (full range of publication dates not shown in Davey (1998)).

Hobbies Weekly 1895-1965.

Simple Model Patterns

These grand little toys and models can be made from a few panels of wood—details of which are given on page 146. The models vary from 4 ins. to 8 ins. long and are easy of construction. Patterns printed full size for making a Horse and Cart, Steam Roller, Noah's Ark and Animals, Lorry, Aeroplane, Liner and Tractor.

PRICE **2/6** PACKET

SERIES 1

DESIGNS NOT SUPPLIED SEPARATELY

1. Hobbies advertisement on p.135 of the 1962 Hobbies Annual for the Series 1 Simple Model Patterns, which included a horse and cart and a tractor.

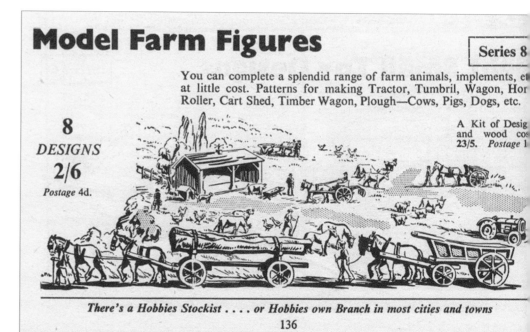

Model Farm Figures

Series 8

You can complete a splendid range of farm animals, implements, et at little cost. Patterns for making Tractor, Tumbril, Wagon, Hor Roller, Cart Shed, Timber Wagon, Plough—Cows, Pigs, Dogs, etc.

8 DESIGNS 2/6 Postage 4d.

A Kit of Desig and wood cos 23/5. Postage 1

There's a Hobbies Stockist . . . . or Hobbies own Branch in most cities and towns

136

2. Hobbies advertisement on p.136 of the 1962 Hobbies Annual for the Series 8 Model Farm Figures illustrating a tractor, tumbrel, wagon, horse roller, timber wagon and horse rake. A plough is listed but not illustrated.

No. 3420
Horse and Cart

Will delight the heart of any small child. Just 14 ins. long and easily made from kit of design with instructions, planed wood, stripwood, round rod, etc. 9/11. Postage 1/6. Note:- The wheels are cut with the fretsaw from wood supplied in kit. Design and Instructions separately, 1/-. Post 2d.

There's a Hobbies Stockist . . . . or Hob

130

3. Hobbies advertisement for the no.3420 Horse and Cart on p.130 of the 1962 Hobbies Annual

4. Hobbies assembled fretwork farm wagon as illustrated in (**2**) (200 mm).

Husky and Corgi Juniors

Company History

In 1964 The Mettoy Company Ltd (see the Mettoy chapter) introduced a range of small diecast model vehicles as competition to Lesney's Matchbox Series. Known as Husky Models, they were sold exclusively through F.W.Woolworth stores in the U.K. and U.S.A. They were also available through other retailers in markets where Woolworth did not operate. The range was at first manufactured at the Mettoy factory in Swansea, then production was gradually transferred to a new site at Stanley Road, Northampton in 1967 and 1968.

Husky catalogues were published approximately annually. For 1964, 1965 and 1966 these were in the form of fold-out leaflets, showing models up to no.25, no.29 and no.32 respectively. Booklet catalogues were produced for 1966–67 (with a copyright date of 1966) and 1968–69 (with a copyright date of 1968).

The Woolworth's contract came to an end in 1969 and the series was then rebranded by Mettoy as Corgi Juniors to join their Corgi Toys family (see the Corgi chapter in Volume 1). The models were modified to remove the *HUSKY* name from the dies, often leaving a blank panel as evidence of where the name had been cast. Some of these unbranded models appeared in the last of the Husky blister packs, and then the models were issued in Corgi Juniors blister packs with a *CORGI JUNIORS* paper label on the base. Soon the dies were modified again to add the *CORGI JUNIORS* name to the castings, and within the space of a year or two there were further modifications to add low-friction Whizzwheels to many of the models, although most of the farm items were not affected. These changes, combined with the fact that most models changed colour for the Corgi Juniors range, make the period 1968 to 1972 very interesting for collectors, and there are some rare variations to be found.

Corgi Juniors were given an E prefix to their model numbers on retailers' order forms and on Mettoy's internal documentation, to distinguish them from the main Corgi range which had a C prefix. Sometimes the E prefix was also used on packaging and in catalogues.

Packaging

The models were always sold in hanging blister packs which allowed them to be clearly visible in shop displays and was fresh and innovative compared with Matchbox toys in their old-fashioned boxes. Corgi Juniors did appear in conventional end-opening boxes for the Japanese market during the 1970s, but we do not know exactly which years they were available. From 1972 some Corgi Juniors models were put in pairs into Twin Packs, another new idea which Matchbox subsequently followed. There were also two farm gift sets, both of which are rare. The blister packs can be categorised as follows.

Type 1. The logo on the backing card was the head of a Husky dog, which featured prominently in white on a red background in the top left-hand corner above *HUSKY models.* For the American market, some cards had a pre-printed price of *39c*. On the rear of the card was a checklist of all the models available, and there were numerous variations of the checklist as new models were added to the series. We have seen checklists which went up to no.24, no.25, no.26, no.27, no.29, no.32 or no.36, the last of which had several sub-variations as new models replaced existing numbers.

Type 2. Around 1968 the backing card was changed to yellow with the word *HUSKY* in white on a red panel. The reverse of the card showed a line drawing of some of the models in the series. For the American market, the backing cards had a pre-printed price of *49c*, and for stock control purposes these packs had model numbers which were increased by 50 from the normal number, thus the Farm Trailer became no.83 rather than the usual no.33. Also, the American packs had no printing on the reverse of the card.

Type 3. In 1970, when the models were branded as Corgi Juniors, the backing cards became taller, allowing room for a framed picture of the model which could be cut out as a Collector's Card and stuck in a Collector's Album purchased separately. *CORGI* was in large black lettering followed by a smaller *JUNIORS*. American versions continued to have a pre-printed price on the card. Some of these cards had *WHIZZWHEELS* branding. A later variation of this type had the card reduced in height again with *JUNIORS* in black outline lettering, and this style was also used for the first Twin Packs.

Type 4. In 1974 the backing cards changed to orange, blue or red with a corgi dog in outline beside the words *CORGI JUNIORS* and with a flowing multi-coloured racing stripe ending behind the model. Twin Packs also appeared with this style of card. Sometimes the model numbers had an E prefix.

Type 5. A new *CORGI junior* (singular) bulls-eye theme for the backing card appeared in 1977. The Junior part of the name disappeared in 1979.

Models

Husky No.33 Farm Trailer with Calves (1967–69)

The first farm items in the series were a Volvo Tractor (no.34) and livestock trailer (no.33). The trailer was supplied with four tan-coloured plastic calves moulded in one piece with a plastic platform, representing straw on the floor of the trailer. The trailer and drawbar were a single casting with a separate let-down tailgate which formed a sloping ramp. The hubs were yellow plastic with black plastic tyres, and cast under the trailer was *HUSKY FARM TRAILER MADE IN GT BRITAIN*. The model was at first painted olive-green, changed to turquoise around 1968 (**1** and **2**).

Corgi Juniors No.53 Farm Trailer with Calves (1971)

The livestock trailer was discontinued as a single model before the introduction of Corgi Juniors branding, and no.33 became a Jaguar E-Type, introduced in 1970. However there is a Japanese leaflet for 1971 which shows the trailer numbered 53. Presumably it was decided to re-introduce the model, but a different number had to be used since the Jaguar had taken the number 33. We have never seen the trailer packaged as no.53, so we do not know whether it was actually released, and if so whether it was just for the Japanese market or worldwide. However from 1972 to 1973 the trailer was included in Twin Pack no.2501 with the Zetor Tractor, and in Gift Set no.3019 (see below) so examples of the Corgi Juniors version without packaging are reasonably easy to find. It was painted orange, and cast underneath was *MADE IN GT BRITAIN CORGI JUNIORS FARM TRAILER*. The wheels were the same as the Husky version, having yellow plastic hubs and black tyres (**3**).

Husky No.34 B.M.Volvo 400 Tractor (1967–69)

B.M. stands for Bolinder-Munktell, a Swedish manufacturer purchased by Volvo in 1950. The model consisted of separate body and chassis castings which were riveted together before painting, and *BM VOLVO* was cast on each side of the bonnet. An early example, possibly a pre-production model, has been found with a separate black plastic air cleaner and a hole where presumably there was also a black plastic exhaust (**4**). These small components were quickly eliminated by including the air cleaner and exhaust with the body casting. At first they were given black spray-painted trim (**5**), which was later omitted (**6**). The tractor was painted red with silver grille and headlights, and had yellow plastic hubs with black tyres. The hubs were riveted directly to projections from the chassis casting without using separate steel axles. Cast underneath the gearbox on the base was *HUSKY BM VOLVO TRACTOR MADE IN GT BRITAIN*.

Corgi Juniors No.34 B.M.Volvo 400 Tractor (1970–72)

The tractor continued into the Corgi Juniors range in the same colour scheme as the Husky model. At first there was a blank panel on the base where *HUSKY* had been removed from the die, together with a white label with *CORGI JUNIORS* in black lettering (**7**). The chassis casting was later modified to add *CORGI JUNIORS* along the length of the chassis. *BM VOLVO TRACTOR MADE IN GT BRITAIN* continued to be cast under the gearbox (**8**).

Husky No.8 Tipping Farm Trailer (1968–69)

This trailer had a black plastic hydraulic cylinder with a metal ram which allowed it to stay in the tipped position by friction. The bare metal hubs were at first smooth (**9**), then had a five-spoke pattern (**10**), both with black tyres. Cast underneath was *HUSKY models MADE IN GT. BRITAIN*. Usually the trailer had a yellow chassis and red body, but with the change to Corgi Juniors branding, the colours were changed to a dark blue chassis with orange body, and some of the orange models still had *HUSKY models* cast underneath (**11**).

Corgi Juniors No.8 Tipping Farm Trailer (1970–71)

The trailer continued into the Corgi Juniors range in orange with a dark blue chassis and five-spoke bare metal hubs. *HUSKY models* had been removed from the die and replaced with a *CORGI JUNIORS* paper label (**12** and **13**). This version was only available for a short time, packed in a type 3 blister pack with Collector's Card, and it had been discontinued by the time the 1970 Collector's Album was published, so there was no space in the album for the model's Collector's Card! However it may have continued to be available in the Japanese market, as it was included in the 1971 Japanese leaflet mentioned above under no.53. The trailer was re-introduced in 1972 in Twin Pack no.2502 with the Massey Ferguson Tractor, and in Gift Set no.3019 (see below). This version reverted to the colours of red body and yellow chassis, but with black plastic Whizzwheels (of which there were several variations) on thin low-friction axles. Cast underneath was *CORGI JUNIORS MADE IN GT BRITAIN PAT APP 3396/69* and *WHIZZWHEELS* was cast on top of the transverse bar at the front of the chassis (**14**). This version continued in Twin Pack no.2516 with the Zetor tractor (introduced in 1975), and in 1978 the colours were changed to a yellow body and metallic dark green chassis to match the metallic dark green tractor (**15**).

Husky No.43 Massey Ferguson 3303 Tractor with blade (1969)

This tractor was a smaller scale than the Volvo tractor, and included an angled dozer blade with movable arms that could be held in the raised position by spring pressure from the plastic interior moulding. The Husky version always had a yellow body, yellow arms and blade, red plastic engine block, exhaust and cab interior, bare metal base and black plastic wheels. *MASSEY FERGUSON* was cast on the arms on each side, and at first *3303 TRACTOR HUSKY* models *MADE IN GT BRITAIN* was cast underneath (**16**), then *HUSKY models* was removed from the casting but the model was still in Husky packaging (**17** and **18**).

Corgi Juniors No.43 Massey Ferguson 3303 Tractor with blade (1970–80)

The first Corgi Juniors version was identical to the Husky model with no brand name cast on it, but with the addition of a *CORGI JUNIORS* label (**19**). The base was soon amended to move the position of *3303 TRACTOR* and add *CORGI JUNIORS* to the casting (**20**), and then the colour of the arms and blade were changed to red (**21** to **24**). This colour change also occurred very quickly as it is shown in the 1970 catalogue. Some examples had a silver painted base rather than unpainted, and others were a more lemon-yellow colour (**22** and **23**). In 1980 the colours changed to orange with black arms and blade, with either a red or yellow plastic interior moulding (**25** and **26**). The model was included in Twin Pack no.2502 (see below) and also in several Road Construction Gift Sets which are outside the scope of this book.

Corgi Juniors No.4 Zetor 5511 Tractor (1970–79)

The Zetor tractor was a new addition to the range in 1970. This usually had an orange body, red base and black plastic wheels, exhaust, seat and steering wheel (**27** to **31**). We have also seen one example with an unpainted base (**32**). Cast underneath the chassis was *CORGI JUNIORS ZETOR 5511 TRACTOR,* and *MADE IN GT. BRITAIN* was cast under the front axle support. The cab had no window glazing. In 1978 the colours changed to a metallic dark green body and white base (**33**).

Corgi Juniors Twin Pack No.2501 Zetor Tractor and Farm Trailer with Calves (1972–73)

This Twin Pack (not illustrated) included the no.4 Zetor tractor and no.53 livestock trailer, both in orange. See above for details of the individual models.

Corgi Juniors Twin Pack No.2502 Massey Ferguson Tractor with Angledozer and Tipping Trailer (1972–73)

This Twin Pack contained the no.43 Massey Ferguson tractor with a red blade and no.8 Tipping Trailer in red and yellow with Whizzwheels (**34**). See above for details of the individual models.

Corgi Juniors Twin Pack No.2516 Zetor Tractor and Tipping Trailer (1975–78)

On a 1975 leaflet for retailers, the contents of Twin Packs nos.2503 and 2516 were transposed, so that the Zetor Tractor and Tipping Trailer were shown as no.2503 and the Land Rover and Horsebox were no.2516. In fact the Land Rover and Horsebox Twin Pack had been introduced a few years earlier as no.2503, and this continued, while the Tractor and Trailer was a new Twin Pack issued as no.2516 Farm Set. At first the tractor was orange and the trailer red and yellow (**35**), then from 1978 the tractor was metallic dark green with a white chassis, and the trailer was yellow with a metallic dark green chassis. See above for details of the individual models.

Farm Gift Sets

There were also two farm gift sets, both of which are very rare.

Husky No.3010 Farm Buildings and Six Die Cast Scale Models (1968 or 1969?)

This set consisted of six standard Husky models together with two dark brown plastic open-fronted farm buildings, all contained in a blister pack with a yellow backing card similar to the type 2 packaging for individual models. The models in the set were no.5 Willys Jeep, no.8 Tipping Trailer, no.33 Farm Trailer with Calves (in turquoise), no.34 Volvo Tractor, no.38 Horse Box and no.39 Jaguar XJ6. The plastic buildings had no identification moulded on them. Since the models were all in standard colours, the rarity of the set lies entirely in the packaging. It did not appear in any catalogues and was probably produced for one year's Christmas sales only (**36**).

Corgi Juniors No.3019 Agricultural Gift Set (1972–73)

This set was contained in a conventional Gift Set window box with a plastic tray. On the rear of the box was an illustration of a farm scene including all the contents of the set. Two plastic farm buildings were included, identical to those in the earlier set no.3010, and there were also six Corgi Juniors models in standard colours as follows: no.4 Zetor Tractor, no.8 Tipping Trailer (with Whizzwheels), no.16 Land Rover (with Whizzwheels), no.38 Horse Box (with Whizzwheels), no.43 Massey Ferguson tractor (with red blade) and no.53 Farm Trailer with Calves (in orange) (**37** and **38**).

1. Husky no.33 Farm Trailer and Calves, olive-green, in a type 1 blister pack (69 mm).

Model List

Serial No.	Model	Dates
Husky Models		
8	Tipping Farm Trailer	1968–69
33	Farm Trailer with Calves	1967–69
34	B.M.Volvo 400 Tractor	1967–69
43	Massey Ferguson 3303 Tractor with blade	1969
3010	Farm Buildings and Six Die Cast Scale Models	1968 or 1969?
Corgi Juniors		
4	Zetor 5511 Tractor	1970–79
8	Tipping Farm Trailer	1970–71
34	B.M.Volvo 400 Tractor	1970–72
43	Massey Ferguson 3303 Tractor with blade	1970–80
53*	Farm Trailer with Calves	1971
2501	Twin Pack containing Zetor Tractor and Farm Trailer with Calves	1972–73
2502	Twin Pack containing Massey Ferguson Tractor with blade and Tipping Trailer	1972–73
2516	Twin Pack containing Zetor Tractor and Tipping Trailer	1975–78
3019	Agricultural Gift Set	1972–73

*Catalogued as no.53, but it is uncertain whether this was issued as a single model as well as in Twin Pack no.2501.

Further Reading

Force, Dr.Edward, 1984, revised 1997, *Corgi Toys*. Schiffer Publishing Ltd.

Manzke, Bill, 2004, *Corgi Juniors & Husky Models*. Schiffer Publishing Ltd.

Van Cleemput, Marcel R., 1989, *The Great Book of Corgi 1956-1983*. New Cavendish Books.

2. Husky Farm Trailer with Calves, as (**1**) but turquoise, in a type 2 blister pack for the U.S.A. market numbered 83 (increased by 50 from the normal number 33).

3. Corgi Juniors Farm Trailer with Calves, orange. Catalogued as no.53 in 1971 but we are uncertain as to whether it was issued as a single model, so this example may be from Twin Pack no.2501 or Gift Set no.3019 (69 mm).

4. *Husky no.34 B.M.Volvo 400 Tractor, red, possibly a pre-production model with a separate black plastic air cleaner and a hole for (presumably) a black plastic exhaust which is missing (54 mm).*

5. *Husky no.34 B.M.Volvo 400 Tractor, as (4) but with black painted air cleaner and exhaust cast with the body of the model, in a type 1 blister pack.*

6. *Husky no.34 B.M.Volvo 400 Tractor, as (5) but red exhaust and air cleaner, in a type 2 blister pack.*

7. *Corgi Juniors no.34 B.M.Volvo 400 Tractor, red, CORGI JUNIORS label underneath (54 mm).*

8. *Corgi Juniors no.34 B.M.Volvo 400 Tractor, as (7) but CORGI JUNIORS cast underneath, in a type 3 blister pack. Note the prices for the U.S.A. and Canada printed on the card.*

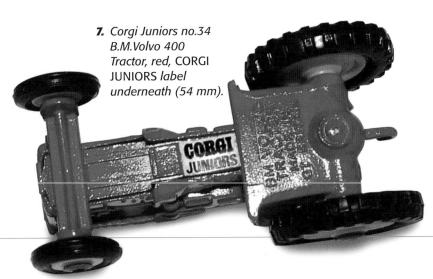

9. *Husky no.8 Tipping Farm Trailer, red body, yellow chassis, smooth bare metal hubs, in a type 1 blister pack (70 mm).*

10. *Husky no.8 Tipping Farm Trailer, as (9) but with five-spoke bare metal hubs, in a type 2 blister pack.*

11. *Husky no.8 Tipping Farm Trailer, as (10) but orange body and dark blue chassis.*

12. *Corgi Juniors no.8 Tipping Farm Trailer, orange body and dark blue chassis, CORGI JUNIORS label underneath (70 mm).*

13. *Comparison of the Husky (left) and Corgi Juniors (right) Tipping Trailers, showing that the HUSKY models lettering was removed from the die and replaced by a CORGI JUNIORS label.*

14. *Corgi Juniors Tipping Farm Trailer, red body, yellow chassis, fitted with Whizzwheels, CORGI JUNIORS cast underneath, sold in Twin Packs nos.2502 or 2516 or Gift Set no.3019 (70 mm).*

15. *Corgi Juniors Tipping Farm Trailer, as (14) but yellow body and metallic dark green chassis, sold in Twin Pack no.2516.*

16. *Husky no.43 Massey Ferguson 3303 Tractor with blade, yellow body, arms and blade, red plastic interior, HUSKY models cast underneath (73 mm).*

17. *Husky no.43 Massey Ferguson 3303 Tractor with blade, as (16) but HUSKY models removed from the die, in a type 2 blister pack.*

18. *Underneath view of the blister packed model in (17), showing the absence of any brand name on the model.*

19. *Corgi Juniors no.43 Massey Ferguson 3303 Tractor with blade, as (18) but CORGI JUNIORS label underneath.*

21. *Corgi Juniors no.43 Massey Ferguson 3303 Tractor Shovel, as (20) but red arms and blade, in a type 3 blister pack.*

20. *Corgi Juniors no.43 Massey Ferguson 3303 Tractor with blade, as (19) but CORGI JUNIORS cast underneath.*

22. *Corgi Juniors no.43 Massey Ferguson 3303 Tractor with Angledozer, as (21) but lemon-yellow body, in a type 4 blister pack.*

23. *Corgi Juniors no.43 Massey Ferguson 3303 Tractor with Angledozer, as (22) but in a type 5 blister pack with CORGI junior on the card.*

24. *Corgi Juniors no.43 Massey Ferguson 3303 Tractor with Angledozer, as (21) but in a type 5 blister pack without junior on the card.*

25. *Corgi Juniors no.43 Massey Ferguson 3303 Tractor with Angledozer, as (24) but orange body with black arms and blade.*

26. *Corgi Juniors no.43 Massey Ferguson 3303 Tractor with Angledozer, as (25) but yellow plastic interior.*

27. Corgi Juniors no.4 Zetor 5511 Tractor, orange with red base, in a type 3 blister pack (56 mm).

28. Corgi Juniors no.4 Zetor 5511 Tractor, as (**27**) but in a type 4 blister pack with an orange background.

29. Corgi Juniors no.4 Zetor 5511 Tractor, as (**28**) but with a blue background to the blister card.

30. Corgi Juniors no.4 Zetor 5511 Tractor, as (**29**) but with a red background to the blister card.

32. Corgi Juniors no.4 Zetor 5511 Tractor, as (**31**) but unpainted base.

31. Corgi Juniors no.4 Zetor 5511 Tractor, as (**30**) but in an end-opening box for the Japanese market. Note that the side of the box states Die-cast scale model with WHIZZWHEELS, but the tractor had conventional wheels and axles.

33. Corgi Juniors no.4 Zetor 5511 Tractor, as (**32**) but metallic dark green with white base.

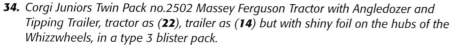

34. Corgi Juniors Twin Pack no.2502 Massey Ferguson Tractor with Angledozer and Tipping Trailer, tractor as (**22**), trailer as (**14**) but with shiny foil on the hubs of the Whizzwheels, in a type 3 blister pack.

35. Corgi Juniors Twin Pack no.2516 Zetor Tractor and Tipping Trailer, tractor as (**27**), trailer as (**14**), in a type 4 blister pack.

36. Husky Gift Set no.3010 Farm Buildings and Six Die Cast Scale Models, rare set containing two plastic farm buildings and six standard Husky models in a type 2 blister pack. The plastic bubble has separated from the backing card, so the set has been wrapped in non-original plastic by its owner to keep the items in place.

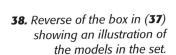

37. *Corgi Juniors no.3019 Agricultural Gift Set containing two plastic farm buildings and six Corgi Juniors models in a window box.*

38. *Reverse of the box in (37) showing an illustration of the models in the set.*

Johillco

Company History

To John Hill & Co. goes the honour of making the first British diecast toy tractor models, in 1932. The origins of the firm are somewhat vague, because the earliest street directory reference for them was in 1920 when they were at 11, 12, 16 and 17 Parkfield Street, Islington, London N1. The directory entry mentioned 'metal soldiers, also metal castings for all branches of the toy trade'. However an article in *Games & Toys* magazine (February 1932) stated that the firm was established in 1898 by the proprietor Mr F.H. Wood, and in the same magazine in January 1954 they advertised that they had been manufacturers of hollow-cast metal toys for over 50 years. According to Norman Joplin (writing in Dean, 2012), John Hill & Co. registered designs for toy soldiers from 1914 onwards. In 1928 the firm moved to Britannia Works, Tyssen Street, Dalston, London E8, but in June 1929, *Games & Toys* reported that they had acquired a new toy soldier factory of 27,000 sq. ft at 64 Essex Road, Islington, London N1. In August 1929, John Hill & Co. advertised in *Toy Trader* that they were moving to 2–22 Britannia Row, which was the same premises as 64 Essex Road, Britannia Row being a side turning off Essex Road. This factory was also known as Britannia Works. John Hill & Co. had used the brand name 'Britannia Series' since 1922 or earlier, and it was a happy coincidence that the new factory had an appropriate street address. The 'Johillco' brand name was used from 1932 or possibly earlier. The 1932 article stated that the firm had over 400 staff and a four-floor factory, which had by then been extended to 50,000 sq. ft.

The tractor models came about because of Johillco's willingness to undertake sub-contract castings for other manufacturers. In 1929 the toy wholesale firms of Bedington, Liddiatt & Co. Ltd and Eisenmann & Co. Ltd jointly marketed models of the Golden Arrow record car in several sizes. The smallest of these was a lead miniature to retail at 6d., which the wholesalers arranged to be manufactured by Johillco. It must have been very successful and sold in large numbers, because the toy is easy to find today. This venture was repeated with models of two more land speed record contenders, Kaye Don's Silver Bullet (1930) and Malcolm Campbell's Bluebird II (1931), again offered as lead models made by Johillco and as larger tinplate models by other manufacturers. At this time Bedington, Liddiatt & Co. Ltd was the U.K. importer for the American Tootsietoy range of diecast vehicles, with which they enjoyed a huge success in 1930 and 1931, despite the deepening world recession. There was nothing similar in the way of metal miniature cars and trucks produced in the U.K., and so confident were Bedington Liddiatt of strong sales that they took a four-page colour insert advertising Tootsietoys in the May 1931 issue of *Games & Toys*. The success was short lived, however, because faced with a currency crisis the new National Government suspended the gold standard from 21 September 1931, so that the value of the pound fell from a fixed rate of $4.86, going as low as $3.23 before stabilising around $3.40, and after a general election in October 1931, the Import Duties Act 1932 increased the tariff on imported toys from 10% to 25%. The result of these two factors was to increase the price of American toys in the U.K. market by about two-thirds, and just as important a threat to American imports was a campaign by the press and politicians to encourage the public to 'Buy British'.

Enterprisingly, Bedington Liddiatt embraced this campaign and decided to manufacture Tootsietoys in the U.K., but without asking permission of the American firm! They arranged for much of the 1931 Tootsietoy range to be copied by Johillco, and the previous business between the two firms was no doubt invaluable in getting the project off the ground. Bedington Liddiatt advertised 'British made Tootsietoys' in *Games & Toys* from February through to August 1932, and the models were blatantly sold in boxed sets marked *TOOTSIETOY* but also *MADE IN ENGLAND*. From September 1932 the *Games & Toys* advert suddenly changed to 'Miniature Transport Toys', and new boxes marked *MINIATURE TRANSPORT TOYS* were produced for the sets. No doubt Tootsietoy's lawyers had threatened legal action over Bedington Liddiatt's abuse of the Tootsietoy trade mark, but although it was possible to prevent the use of the Tootsietoy name, the law is rather weaker when it comes to the copying of designs. Johillco continued to produce their Tootsietoy copies right through the 1930s with impunity. Initial production may have been exclusive to Bedington Liddiatt, but by 1934 Johillco were offering the toys alongside their other lead vehicles and figures. They were also boxed in Johillco boxes and included in the 'pink catalogue' of 1938 or 1939 (see below).

Four of the Tootsietoy copies are relevant to us – a tractor, usually said to be a model of a Huber, a crawler tractor, a box-like trailer and a road-making implement with interchangeable scraper and rake. The trailer and road scraper/rake could be drawn by either tractor. The road scraper/rake is perhaps not strictly a farm toy, but we decided to include it because it was drawn by the farm tractors and is a rare and interesting toy. Tootsietoy produced two versions of the Huber tractor, as a farm tractor and as a U.S. Army gun tractor. The latter had the driver's position moved forwards and a large box was cast behind him to represent an ammunition caisson, and this was the version copied by Johillco. The single Johillco model did service both as a gun tractor (painted khaki) and as a farm tractor (painted red or green). In the listing and photographs we have made it clear how to distinguish the Johillco models from the original Tootsietoys.

As a manufacturer of hollow-cast lead soldiers, John Hill & Co. was a direct competitor of Britains Ltd, and followed Britains by introducing civilian farm and zoo figures during the 1920s. There were two all-lead horse-drawn farm items, a four-wheeled cattle float and a two-wheeled horse (or cattle) float. Both of these were included in the well-known pre-war Johillco catalogue, of which a reproduction was available some years ago, often called the 'pink catalogue' by collectors, from the colour of the cover. A letter from John Hill & Co. dated October 1939 (reproduced in Dean, 2012, p.391) mentioned a catalogue 'issued to you last August', which could be taken to mean either August 1938 or August 1939. We are quite sure that this referred to the pink catalogue. Also, the pink catalogue could not be earlier than 1938, confirmed by John Hill & Co.'s phone number. Research in the telephone directory archives revealed that their number changed to CANonbury 3218 in 1938, and this was the number shown on a covering letter accompanying the 'new illustrated Catalogue'. Another catalogue showing Johillco farm items, but with '"Britannia Series" MODEL FARM YARD ACCESSORIES' on the cover, was sold at auction in 2012. This catalogue can be dated to around 1934, again because of a change of phone number. The catalogue contained typewritten lists of 'Latest Additions' on the headed paper of distributor Lawrence, Haigh & Co. Ltd of Manchester, and the printed phone number on the headed paper had been altered by a typewritten amendment, changing it from CITy 5468 to CENtral 3468. The telephone directories showed that this change in phone number occurred during 1934, thus dating the catalogue fairly precisely. The 'Latest Additions' included the horse-drawn floats, so these must also date from around 1934.

There was also a range of horse-drawn farm implements made of tinplate, wood and wire, but using hollow-cast lead Johillco horses and figures. It is not known whether these were sold by Johillco or if they were the products of another manufacturer who bought in the Johillco horses and figures to complete their own models. They did not appear in the catalogues. Some boxed examples of these models were sold at auction in 2011, but it has been proved that the box labels were not original. The tinplate, wood and wire models were probably sold loose, since no genuine boxed models have been found, and the dates of production are unknown. We assume they are from the 1930s.

After the Second World War, a new company John Hill & Co. (Metal Toys) Ltd was incorporated to acquire the Johillco moulds and trade mark. The company re-started production in Burnley, Lancashire, however none of the pre-war farm vehicles was re-introduced.

Models

Tootsietoy Copies

No.617 Farm Tractor

This model was painted red, green or khaki, with a black chassis and red wheels (**1** to **3**). *MADE IN ENGLAND* was cast above the front axle (**4**). The Tootsietoy tractor had *TOOTSIE TOY* cast up inside the bonnet and the military version also had lettering cast on the ammunition caisson, which was not copied on the Johillco version. The Johillco tractors, trailer and road scraper/rake did not appear in the Johillco catalogues, but we know the number for the tractor from an original box, and the model was also pictured in *Games & Toys* February 1934 (**5**). The individual box was a pinkish-red end-opening box. Unfortunately we could not locate one to photograph, so instead have shown the similar box for the Anti-Aircraft Gun no.606. Johillco's Tootsietoy copies were sold by Bedington, Liddiatt & Co.Ltd in various sets, whose contents exactly copied Tootsietoy sets, and the Johillco tractor (in khaki) was included in a Field Battery Set with several guns, in *MINIATURE TRANSPORT TOYS* packaging. This almost certainly dates the Johillco tractor to 1932. Tootsietoy produced a Farm Tractor Set containing their civilian tractor, box trailer, road scraper/rake and Ford Model T truck (**6** and **7**), and although we have not seen a British-made version of the set, it would not be surprising to find one, because Johillco did produce British versions of the contents (**8**). Also, the pink catalogue called the Ford truck 'Farm Lorry', suggesting it originated in the farm set, however the model is not within the scope of this book.

Johillco also supplied models to be packed in sets by other companies, including Bell (Toys & Games) Ltd and J.Randall (Toys & Games) Ltd (see the Merit chapter), although we have not seen any sets from these two firms which included farm vehicles. The tractor and trailer did appear in a box labelled *KIDDYCRAFT TOYS*, which was probably the brand name of another wholesaler (**9**). The tractor and road scraper/rake were still available in 1938 when they were mentioned as part of a 'Small Joytime Set' in the Winter 1938 mail order catalogue from Hobday Bros.Ltd (**10**).

Farm Trailer

Probably first issued in 1932, this model was painted in distinctive lemon-yellow (**8**) or orange-yellow colours (**9**), both with red wheels, and had no identification cast on the model. The Tootsietoy trailer always had *TOOTSIE TOY* cast above the axle – this can be hard to see (**11**).

Road-Mending Machine

This was probably also first issued in 1932, and was still available in 1938, as mentioned above. It was painted red with grey or lemon-yellow wheels, and khaki scraper and rake pieces (**8** and **12**). The Johillco model had *BRITISH MADE* cast under the driver's seat position, whereas the Tootsietoy version had *TOOTSIE TOY* cast in the same place. The Johillco and Tootsietoy scraper and rake pieces were almost identical in colour and design, and it is quite hard to tell them apart. The overall height of the Johillco pieces was slightly greater – the Johillco scraper was 31 mm high and rake 30 mm high, whereas the Tootsietoy scraper was 30 mm high and rake 29 mm high.

Crawler Tractor

Again, this model was not included in the Johillco catalogues so we do not know the catalogue number. It probably dates from 1932. It was painted red or light green with unpainted or black wheels (larger on the rear axle) and had white rubber tracks (**13** and **14**). It is rather rare to find an example with original tracks, as the rubber has usually perished. *MADE IN ENGLAND* was cast underneath. The Tootsietoy version was always marked with the maker's name. There is another copy of the Tootsietoy by Charbens, similar to the Johillco but slightly larger (see the Charbens chapter in Volume 1) (**15**).

Cattle and Horse Floats

The cattle and horse floats were listed in the pre-war pink catalogue as follows (each individually boxed):

No.388 Cattle Float and Horse

No.389 Cattle Float, Horse and Show Cow

No.404 Horse Float and Horse

No.405 Horse Float, Horse and Show Horse.

The retail prices were 2s. each for the models with the show cow or horse, and 1s. 9d. for the empty floats. Number 389 was the only item illustrated (**16**), and was the four-wheeled float, hence it might be reasonable to assume that the horse floats numbered 404 and 405 were the two-wheeled model, otherwise numbers 388 and 404 would be identical. Further evidence for this is that both floats had their dies modified to add additional lettering on the sides, suggesting that both were available concurrently, rather than one succeeding the other. However a box for the two-wheeled float as a Hereford Bull Cart has been recorded (Joplin and Dean, 2005, p.127), and confusingly it has the number 388 rubber-stamped on the box (**17**). More usually, the floats came in lift-off lid boxes covered with shiny red paper, but without any label to identify the contents.

Our photographs show the floats loaded with either the Johillco bull or the cow. The bull can be found in brown and white or in black, both with additional hand-painted detailing on top of the basic colours. At first he had *J HILL & CO* and *COPYRIGHT* cast underneath, and later *ENGLAND* was added to the casting. The bull was listed separately in the Johillco catalogues as no.500F. The cow was painted brown and white or black and white, with additional paint trim, and had *COPYRIGHT* cast underneath, later with *J.HILL & CO* and *ENGLAND* added. She was numbered 307 when sold separately. The large horse shown in (**18**) is possibly the show horse referred to above. This was a heavily-built shire horse with feathered legs, wearing a belly band, and was numbered 500A in the catalogues. It was painted brown with additional paint trim and had *COPYRIGHT J HILL & CO ENGLAND* cast underneath.

The cart horse went through various changes of cast lettering, and either *COPYRIGHT* or *COPYRIGHT ENGLAND* or *JOHILLCO ENGLAND* can be found cast underneath. It was painted white, medium brown or dark brown with considerable additional paint detail, and was available separately as no.310. According to the catalogues, the drover figures were not included in the sets; the walking drover waving a red handkerchief was numbered 416, while the seated figure was sold as no.375 Aged Villager.

Four-wheeled Cattle Float

This model can be found in two paint styles, either (i) a dark grey wagon with light grey interior and red wheels with yellow trim (**19**), or (ii) a medium grey wagon with red wheels (**20**). Either *COPYRIGHT* was cast on the side rails on each side, or *J.HILL & Co. COPYRIGHT* was cast on the right side rail and *COPYRIGHT MADE IN ENGLAND* on the left side rail.

Two-wheeled Horse or Cattle Float

The cart was dark grey with light grey (**21**) or medium grey interior (**22**), and had red wheels with yellow trim. Either *COPYRIGHT* was cast on the side rails on each side or *J.HILL & Co. COPYRIGHT* was cast on the right side rail and *COPYRIGHT MADE IN ENGLAND* on the left side rail.

Tinplate, Wood and Wire Models

The Johillco horse was common to these models, painted white, medium brown or dark brown with additional hand-painted trim. Cast underneath the horse was either *COPYRIGHT* or *COPYRIGHT ENGLAND* or *JOHILLCO ENGLAND.*

Horse and Plough

The plough was made of thick wire, carrying tinplate mouldboards, and had vastly over-sized lead wheels. At the rear the wire formed two handles, and the hollow-cast lead ploughman was listed separately by Johillco as no.291 Farm Labourer, intended to push a wheelbarrow. At the front, the wire formed a hook to which the thinner wire of the traces was attached. The Johillco horse was in cart harness, which would not have been seen in reality when pulling a plough. The plough was painted red or gold, or both, or unpainted, and had fourteen-spoke wheels (of which there were two sizes) painted red, yellow or dark green (**23** to **29**).

Horse and Roller (narrow)

Either the wooden roller and wire shafts were unpainted, or the roller was dark blue with red wire (**30** and **31**).

Horse and Roller (wide) with seat

The wooden roller was painted red, and the tinplate seat was red or gold, with the seat at the level of the shafts (**32**). Another version had the wire shafts shaped to raise the seat, and the roller and seat were dark green (**33**). The wire shafts were unpainted.

Horse and Roller (wide) with seat and metal flap

The two-part wooden roller was painted red, the tinplate seat was yellow, and the flap and wire shafts were unpainted (**34**).

Horse and Roller with Rake

This had a narrow wooden roller, wire shafts and rake, all left unpainted (**35**).

Horse and Hay Rake with seat

This had a red, dark green, unpainted or gold tinplate seat, and unpainted wire shafts and rake. It was fitted with fourteen-spoke lead wheels painted pale yellow, dark green or red; or larger fourteen-spoke wheels painted red; or twelve-spoke wheels painted red or yellow (**36** to **41**).

Horse and Reaper

The sails rotated as the model was pulled along, driven by the wheel on the left side acting on the wooden roller. There were two main variations:

(i) With tinplate sails - the tinplate, wood and wire reaper was painted red, with red or dark green lead wheels (**42** and **43**);

(ii) With pierced steel strips as sails – we have found seven different colour combinations for this version, described in the photo captions (**44** to **49**).

Horse and Water Cart

This had a black painted wooden water tank with red tinplate top, gold seat, unpainted shafts and perforated metal water sprayer. The lead wheels were painted red, light yellow or dark green (**50**).

Horse and Timber Wagon

There were wire shafts and loops for retaining the log, a wooden chassis and bolsters, all painted grey-green. The lead wheels were painted red. The wire shafts were more complicated than on the other models, allowing the possibility of a second tandem horse (**51**).

Horse and Perforated Cylinder on wheels

We are not at all clear what this implement was meant to be – perhaps a seed drill? It consisted of a cylinder of perforated metal with wooden ends, a tinplate seat above, and the usual wire shafts and lead wheels. Colours were red or blue for the cylinder with a gold or unpainted seat and red, light yellow or dark green wheels (**52** and **53**).

Horse and Rotary Plough

This was a blue and unpainted tinplate, wood and wire plough with red lead fourteen-spoke or twelve-spoke wheels (**54** and **55**).

Rotary Plough (Tractor-drawn)

This was similar to the previous model but with the wire fashioned into a drawbar so that it could hook into the towing eye of a tractor. It is the only tractor-drawn example of these tinplate, wood and wire models that we have seen, but it is conceivable that other tractor-drawn implements were made. The rotary plough was made of red and unpainted tinplate, wood and wire, with dark green lead wheels (**56**).

Model List

Serial No.	Model	Dates
Tootsietoy Copies		
617	Farm Tractor	1932–38
	Farm Trailer	1932?
	Road-Mending Machine [Scraper/Rake]	1932?–38
	Crawler Tractor	1932?
Cattle and Horse Floats		
388	[Four-wheeled] Cattle Float and Horse	1934–40
389	[Four-wheeled] Cattle Float, Horse and Show Cow [or bull]	1934–40
404	[Two-wheeled] Horse Float and Horse	1934–40
405	[Two-wheeled] Horse Float, Horse and Show Horse [or bull or cow]	1934–40
Tinplate, Wood and Wire Models		
	Horse and Plough	1930s?
	Horse and Roller (narrow)	1930s?
	Horse and Roller (wide) with seat	1930s?
	Horse and Roller (wide) with seat and metal flap	1930s?
	Horse and Roller with Rake	1930s?
	Horse and Hay Rake with seat	1930s?
	Horse and Reaper	1930s?
	Horse and Water Cart	1930s?
	Horse and Timber Wagon	1930s?
	Horse and Perforated Cylinder on wheels	1930s?
	Horse and Rotary Plough	1930s?
	Rotary Plough (Tractor-drawn)	1930s?

Further Reading

Dean, Philip, 2012, *Johillco – Second to None*. Published by the author.

Joplin, Norman, 1993, *The Great Book of Hollow-Cast Figures*. New Cavendish Books.

Joplin, Norman and Dean, Philip, 2005, *Hollow-Cast Civilian Toy Figures*. Schiffer Publishing Ltd.

Johillco pre-war 'pink catalogue', 1938 or 1939, reproduction (publisher unknown).

Seeley, Clint and Newson, Robert, *A History of Pre-War Automotive Tootsietoys*. Published online at http://www.tootsietoys.info/.

1. *Johillco no.617 Farm Tractor with the box for no.606 AA Gun (the tractor box was similar), red tractor with black chassis and red wheels (80 mm).*

2. *Johillco Farm Tractor, as (1) but green (driver's head broken).*

3. *Johillco Farm Tractor, as (2) but khaki, sold in a military set as a gun tractor.*

4. *Johillco Farm Tractor, as (1), underneath view showing MADE IN ENGLAND above the front axle.*

The "Johillco" Metal Toys.

The manufacturers of these well-known British cast metal toys are Messrs. John Hill and Co., whose factory is situated at 2-22 Britannia Row, Essex Road, Islington, London, N.1. They produce for the wholesale and shipping trade only, and their range is extremely wide and varied.

First and foremost we should mention that they have many lines in toy soldiers, sailors, marines and various corps. They also have a wide range of special lines, such as Roman warriors, Roman chariots, cowboys, Indians, Zulus and race-horses. In the domestic field they produce tea sets, kitchen sets, laundry sets and toilet sets, while attention should also be called to their railway sets, farm sets and zoological subjects.

The illustrations we reproduce, however, will show the variety of their productions. These include tanks, farm tractors, omnibuses and tram-

cars. It is interesting to record that Messrs. John Hill and Co. make a special feature of the fact that all "Johillco" toy motors, lorries and buses are now fitted with rubber tyres.

Good sales were reported last year with the "582 Roman Gladiators' Triumphal Procession," a special line of the firm, while other specialities include "speed cops," despatch riders and motor-cars with moving wheels.

Most of the lines mentioned, if not all, are supplied in attractive sets, and can be obtained in bulk for retailing singly. It is also interesting to point out that customers of the firm can have castings made to their own designs.

* * *

5. An article on Johillco from Games & Toys *magazine, February 1934, illustrating the Johillco Tractor.*

Cousin Tom's Farm . . . . .

Cousin Tom has a great big farm
With fields and orchards so pretty,
He grows lots of wheat and good things to eat
For the boys and girls in the city.

His Tootsietoy Tractor is great
For plowing the rich black fields
And his truck and box cart, all do their part
To furnish our daily meals.

Farm Tractor

A dandy farm outfit. Tractor, box trailer, Ford truck and road scraper with an additional rake arm.

Price **50**c

6. Tootsietoy Farm Tractor Set advertised in the consumer booklet 'Let's Make Believe with Tootsietoys 1932'. This sort of consumer advertising was highly innovative in the 1930s and helps to explain Tootsietoy's huge success.

7. *Tootsietoy Farm Tractor Set (made in U.S.A. 1928–32). The set illustrated has been together from new, having been bought from the grandson of the original owner.*

8. *Johillco issues of the same models as (**7**), likely contents of a British-made Farm Tractor Set. Tractor as (**1**), Trailer in lemon-yellow with red wheels, Farm Lorry [Ford Model T] in metallic dark blue, Road-Mending Machine red with lemon-yellow wheels, khaki scraper and rake pieces.*

9. *Johillco Farm Tractor and Trailer, Tractor as (1), Trailer as (8) but orange-yellow, lift-off lid box with* KIDDYCRAFT TOYS *on the label. Kiddycraft Toys was probably a wholesaler's brand name and the label illustrates other Johillco vehicles and aircraft (Tractor 80 mm, Trailer 67 mm).*

Small Joytime Set. Consists of express delivery van, mail van, road-mending machine, road-mending truck, general utility truck, large delivery van, anti-aircraft gun on lorry, searchlight on lorry, farm tractor, safety coach, monoplane.

No.
S3608/401 ... ... per set 6/-

10. *Detail from the Winter 1938 mail-order catalogue of Hobday Bros.Ltd describing a set of Johillco vehicles, including the Farm Tractor and Road-Mending Machine.*

11. *Comparison of the Tootsietoy Farm Trailer (left) with* TOOTSIE TOY *above the axle and the Johillco copy (right) without lettering.*

12. *Johillco Road-Mending Machine with interchangeable scraper and rake, red with grey wheels, khaki scraper and rake (glue repair to front wheel) (91 mm).*

13. *Johillco Crawler Tractor, red with black wheels and white rubber tracks (69 mm).*

14. *Johillco Crawler Tractor, as (13) but light green, unpainted wheels, missing the rubber tracks, driver's head broken off.*

15. *Comparison of the Johillco Crawler Tractor (right) and the larger version by Charbens (left).*

389, measures approx. 8½in. long.

16. *No.389 Cattle Float, Horse and Show Cow pictured in the Johillco 'pink catalogue'. The catalogue dates from 1938 or 1939 and is available in reproduction.*

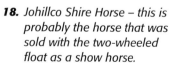

17. *Johillco Two-wheeled Horse or Cattle Float, boxed as no.388 Hereford Bull Cart, two-tone grey with red-and-yellow wheels (210 mm with ramp down)* (photo by Philip Dean).

18. *Johillco Shire Horse – this is probably the horse that was sold with the two-wheeled float as a show horse.*

19. *Johillco Four-wheeled Cattle Float with Bull, dark grey with light grey interior, red-and-yellow wheels, COPYRIGHT cast on each side (210 mm with ramp down).*

20. *Johillco Four-wheeled Cattle Float with Bull, medium grey with red wheels, J.HILL & Co. COPYRIGHT cast on the right side rail and COPYRIGHT MADE IN ENGLAND on the left side rail (210 mm with ramp down).*

21. *Johillco Two-wheeled Horse or Cattle Float with Cow, dark grey with light grey interior, red-and-yellow wheels, COPYRIGHT cast on each side (210 mm with ramp down).*

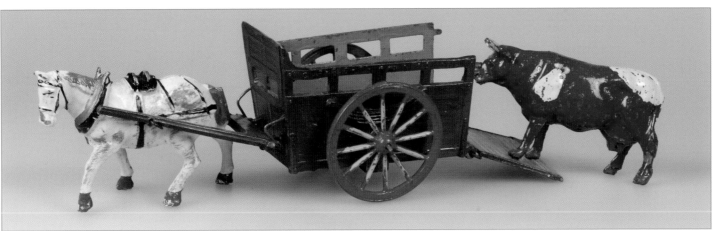

22. *Johillco Two-wheeled Horse or Cattle Float with Cow, dark grey with medium grey interior, red-and-yellow wheels, J.HILL & Co. COPYRIGHT cast on the right side rail and COPYRIGHT MADE IN ENGLAND on the left side rail (210 mm with ramp down).*

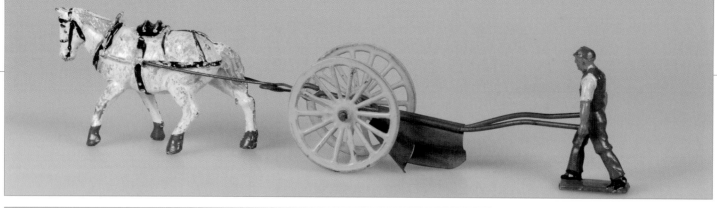

23. *Johillco Horse and Plough, unpainted with yellow wheels (220 mm).*

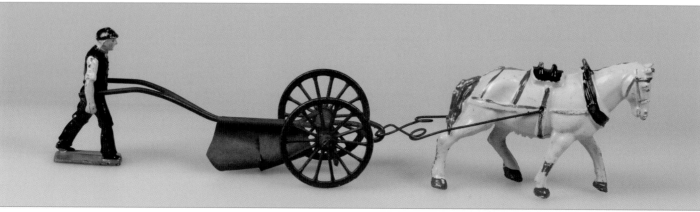

24. *Johillco Horse and Plough, as (23) but dark green wheels.*

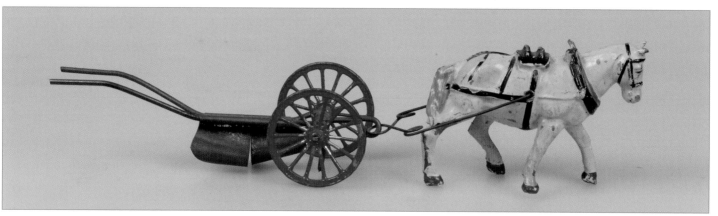

25. *Johillco Horse and Plough, as (24) but red wheels.*

26. *Johillco Horse and Plough, red with red wheels (220 mm).*

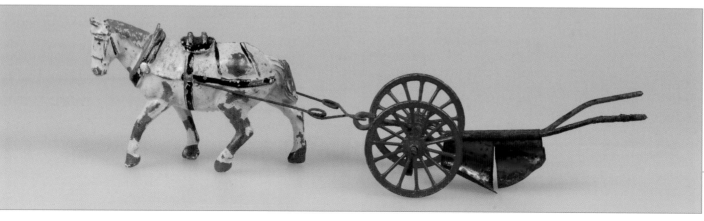

27. *Johillco Horse and Plough, red and gold with red wheels (220 mm).*

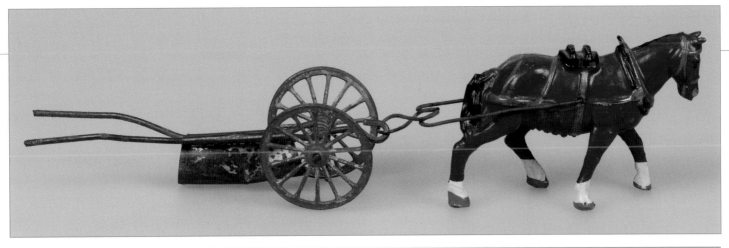

28. *Johillco Horse and Plough, gold with red wheels (220 mm).*

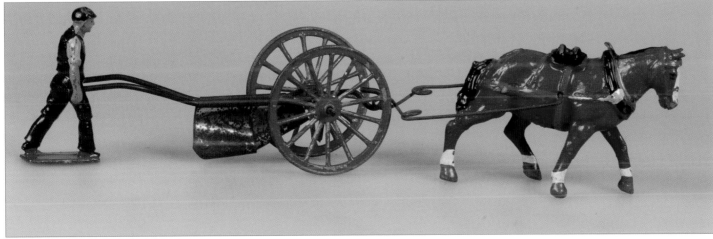

29. *Johillco Horse and Plough, as (28) but larger wheels.*

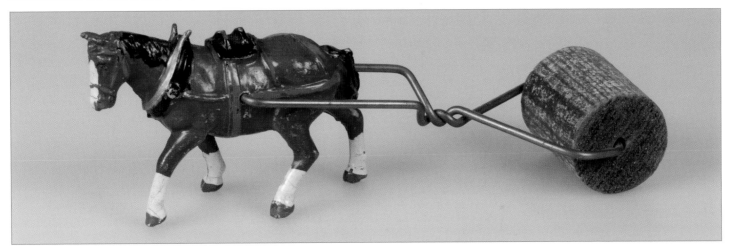

30. *Johillco Horse and Roller (narrow), unpainted (160 mm).*

31. *Johillco Horse and Roller (narrow), as (30) but red and dark blue.*

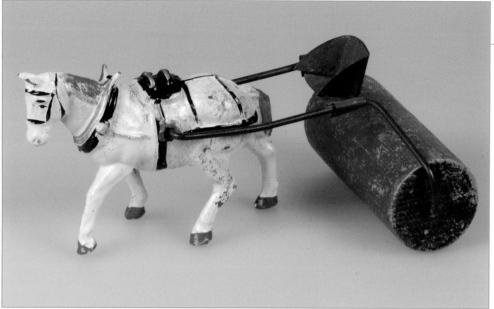

32. *Johillco Horse and Roller (wide) with seat, red (125 mm).*

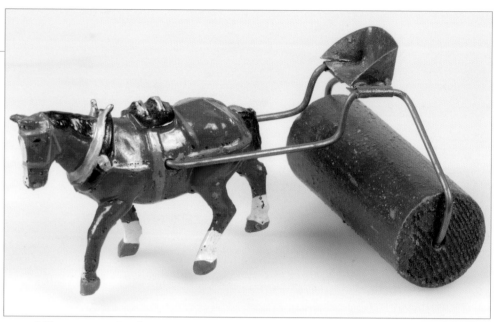

33. *Johillco Horse and Roller (wide), as (32) but higher level seat, dark green.*

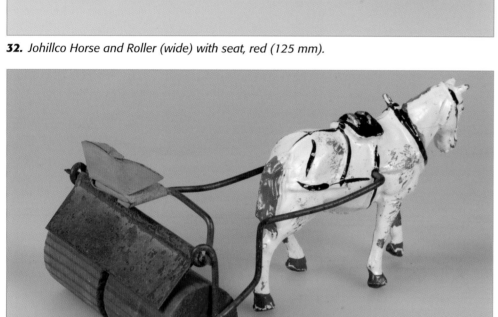

34. *Johillco Horse and Roller (wide) with seat and metal flap, red with yellow seat (135 mm).*

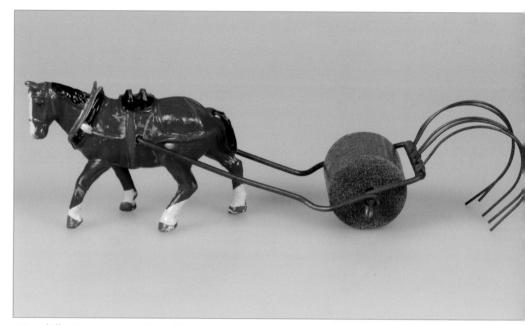

35. *Johillco Horse and Roller with Rake, unpainted (185 mm).*

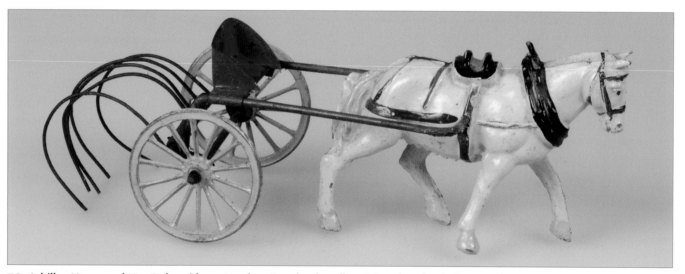

36. *Johillco Horse and Hay Rake with seat, red seat and pale yellow 14-spoke wheels (155 mm).*

37. *Johillco Horse and Hay Rake with seat, as (36) but dark green seat and 14-spoke wheels.*

38. *Johillco Horse and Hay Rake with seat, as (37) but unpainted seat and red 14-spoke wheels.*

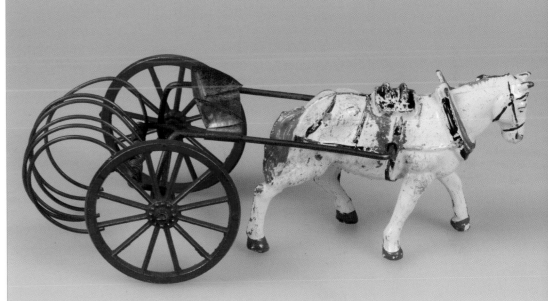

39. *Johillco Horse and Hay Rake with seat, as (38) but gold seat and larger 14-spoke wheels.*

40. *Johillco Horse and Hay Rake with seat, as (39) but red 12-spoke wheels.*

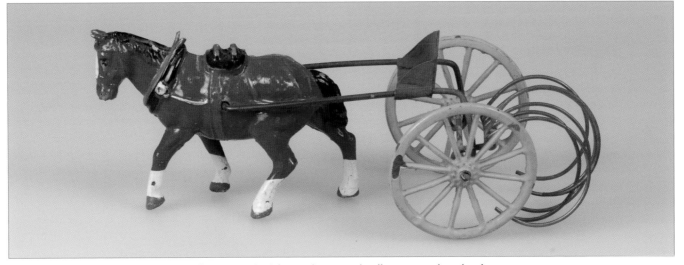

41. *Johillco Horse and Hay Rake with seat, as (40) but red seat and yellow 12-spoke wheels.*

42. *Johillco Horse and Reaper, tinplate sails, red with red wheels (150 mm).*

43. *Johillco Horse and Reaper, as (42) but red with dark green wheels.*

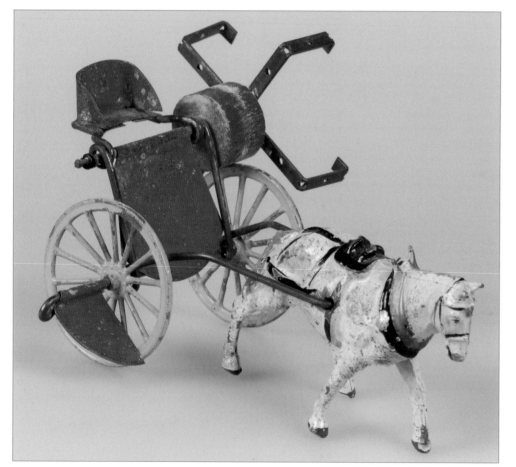

44. *Johillco Horse and Reaper, pierced steel sails, red with red seat and light yellow wheels (150 mm).*

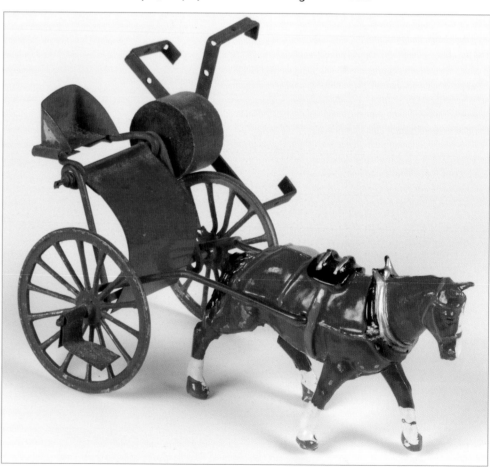

45. *Johillco Horse and Reaper, as (44) but with gold seat and red wheels. This version also exists with green wheels (not illustrated).*

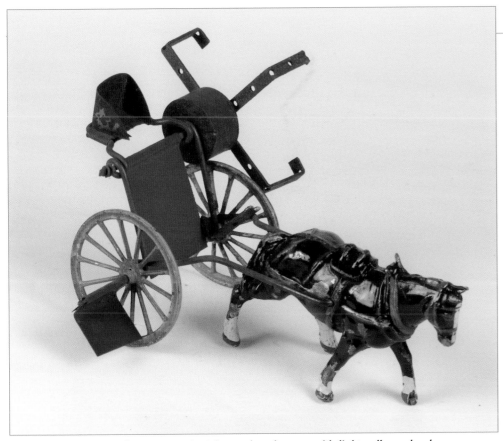

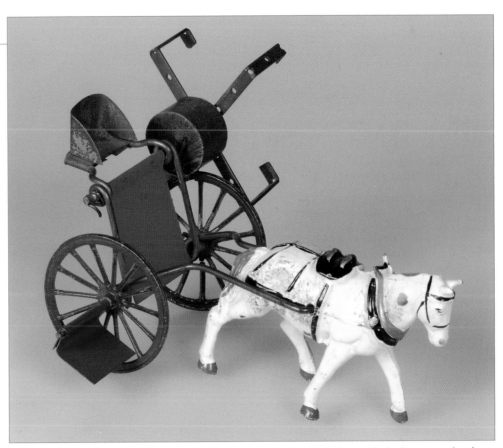

46. Johillco Horse and Reaper, as (**45**) but red and green with light yellow wheels.

47. 47. Johillco Horse and Reaper, as (**46**) but red, green and blue with dark green wheels.

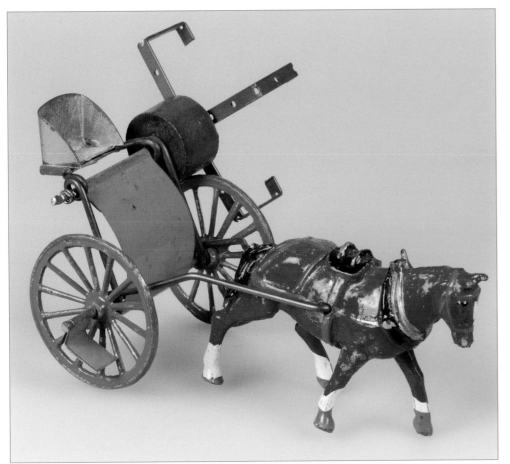

48. Johillco Horse and Reaper, as (**47**) but orange and blue with green wheels.

49. Johillco Horse and Reaper, as (**48**) but light blue and red with yellow wheels.

50. *Johillco Horse and Water Cart, black wooden tank with red tinplate top and red wheels. The seated driver is the Johillco Aged Villager figure (140 mm).*

51. *Johillco Horse and Timber Wagon, grey-green with red wheels* (photo by Philip Dean).

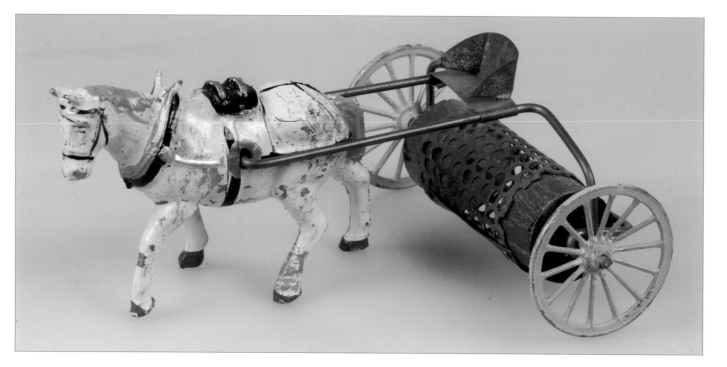

52. *Johillco Horse and Perforated Cylinder on wheels, red with light yellow wheels and unpainted seat (128 mm).*

53. *Johillco Horse and Perforated Cylinder on wheels, blue with dark green wheels and gold seat (128 mm).*

54. *Johillco Horse and Rotary Plough, blue and unpainted with red 14-spoke wheels (150 mm).*

55. *Johillco Horse and Rotary Plough, as (**54**) but 12-spoke wheels (photo by Philip Dean).*

56. *Johillco Rotary Plough (tractor drawn), red and unpainted with dark green wheels (100 mm).*

Kayron Playthings, A.V.H. Farm Toys & Olson Farminit Toys

Company History

The Kayron brand name was used by a company called Metal Tube Products, and they, A.V.H. and Olson seem to have been closely inter-related firms from the Woking area of Surrey. Metal Tube Products was based at 129A Courtenay Road, Woking during 1948 and 1949. Olson (Farminit) Ltd, was registered as a company in 1948, and appeared in telephone directories, listed as Diecasters, at 127A Courtenay Road during 1950 and 1951. So, Metal Tube Products and Olson were based in adjacent premises, possibly from 1948 onwards, but certainly by 1950.

A.V.H. Productions Ltd, Wholesalers, was in street directories at Send Road, Send (a part of Woking), Surrey from 1948 to1949, and in *Kelly's Trade Directory* at Duke Street, Woking between 1948 and 1950. We have been unable to find out what the letters A.V.H. stand for, or indeed why the brand name of Kayron was chosen by Metal Tube Products.

The products of each company, when they can be positively identified, are almost identical, though the products of Olson Farminit are generally larger scale than similar products from Kayron and A.V.H. Because Kayron and A.V.H. used the same model numbers and box end labels they are unlikely to have been selling the same models at the same time.

We have seen two different advertisements by Moko in *Games & Toys* magazine, for Olson Farminit products. The first advert was dated February 1948 and although it showed an Olson horse-drawn rake, there was no accompanying description and no mention of Olson or Farminit.

The second Moko advert, dated May 1948, showed one of 'THE "FARMINIT" SERIES of FARM IMPLEMENTS' as one of their 'many exclusive

"Moko" lines', but again there was no mention of Olson.

We have also found a boxed Olson Farminit item with the label stating *SOLE DISTRIBUTORS for S., S.E. & S.W. A.V.H. PRODUCTIONS LTD. 1, DUKE STREET, WOKING, SURREY (27)*. We deduce this to mean that A.V.H. was the sole distributor of Olson products for the South, South-East and South-West of England. While we know from directory research that A.V.H. occupied the Duke Street address from 1948, it is unlikely that A.V.H. were sole distributor at the same time that Moko were promoting the range as one of their exclusive lines. Consequently we deduce that A.V.H. started distributing the Olson Farminit range only at some point after May 1948.

Advertisement by Moko for the Hay Rake in Games & Toys *February 1948*

Advertisement by Moko for the Olson Farminit series wire wagon as in (1) in Games & Toys *May 1948*

A simple explanation for this apparently complicated situation seems to be that Metal Tube Products and their neighbours Olson created a range of metal farm toys in the mid to late 1940s, probably taking advantage of the increased availability of metal after the Second World War. They used the Kayron brand for the range they sold themselves, and the Farminit brand for items distributed by Moko. Possibly the use of the Farminit brand proved more successful than the Kayron brand, but for reasons currently unknown, Metal Tube Products disappeared from phone directories in 1950, and Olson appeared. A.V.H. took on the two ranges from mid-1949, re-launching the Kayron brand under their own A.V.H. label and with new box artwork, while taking over the distribution of the Olson products. Whatever the correct explanation, the whole enterprise does not seem to have been successful as no trace of any of the companies can be found after 1951, other than the possibly coincidental appearance of a metal-working company, the Courtenay Road Welding Works, at 127A Courtenay Road, the previous address of Olson (Farminit) Ltd, in the 1957 Woking Street Directory.

Models

Note: In the photo captions all length measurements exclude the horse, since there is uncertainty about how some of these implements were harnessed to horses.

The products of these three companies are a large and little recognised group of farm toys, which forms a greater range of toy farm implements than any other manufacturer at the time, even Britains. Doubtless it is the robust nature of the toys that has led to so many surviving complete, despite the companies seeming to have only been in business for approximately four years. Olson (Farminit) Ltd was described as diecasters in the 1950–51 telephone directory, so it is probable that they were the source of the diecast wheels. However, because none of the products is marked, when they are found unboxed it is difficult to be certain which of the three companies a specific item should be attributed to. As the range is so broad it is possible that some of the items shown were not made by any of the three companies, but there are many similarities across the ranges of their products that we have illustrated, such as the materials used, very similar tinplate or diecast wheels, similar methods of manufacture, and the use of the same model horses. Consequently we are confident in attributing the products to one of the three companies, though to try and give an explanation why these small companies were responsible for such a broad range would involve much speculation, which course we have decided against.

Additionally, there may have been a further brand for models of a clockwork tractor made from aluminium sheet (**101** to **107**), a wire trailer (**102** and **103**) and a reaper (**101**), manufactured with the wire bent around at the joints, rather than soldered. Unfortunately the box label in (**102**) is missing, so this is inconclusive.

It is likely that there are some of these companies' products still to be found.

Horse-drawn Implements

The models described in this chapter were mostly horse-drawn implements made of heavy-gauge wire soldered at the joints, and sold under the brand names 'Kayron Playthings', 'A.V.H. Farm Toys' or 'Olson Farminit Toys', in addition to the Farminit items sold by Moko.

Kayron Playthings

The products of Kayron and A.V.H. were in many respects identical except for the box labels, and the companies used the same model numbers. The Kayron boxes had an un-illustrated label with *"KAYRON"PLAYTHINGS MINIATURE FARMING IMPLEMENTS RANGE INCLUDES: HAY RAKES,*

KAYRON PLAYTHINGS MINIATURE FARMING IMPLEMENTS *box label.*

ROLLERS, SEED DRILLS, PLOUGHS, CULTIVATORS, BARNS ETC., and along the bottom of the label was *METAL TUBE PRODUCTS, WOKING 2650.*

End labels on the lids had the model number and name. Under that was *MADE IN ENGLAND.* All the implements listed for sale on the lids were horse-drawn.

A.V.H. Farm Toys

The A.V.H. labels simply had *A.V.H FARM TOYS* above a farmyard scene showing a tractor in an open-ended shed to the right, a hay rake standing outside an empty cart shed to the left, a plough next to it and in the middle a wire cart being pulled by a horse led by a farm worker.

The tractor depicted in the picture was probably the tinplate model tractor in (**92**).

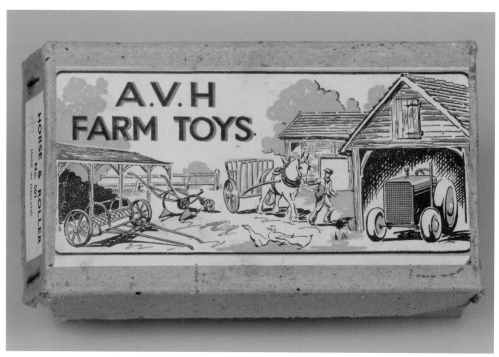

A.V.H. FARM TOYS *box label.*

Olson Farminit Toys

The Olson label had a ploughing scene with a pair of horses and *OLSON FARMINIT TOYS* in the bottom right-hand corner.

OLSON FARMINIT TOYS *box label.*

Olson Farminit models were often larger than Kayron and A.V.H. products and were usually blue and yellow and had seats on some of their implements made from circular washers (**29**). None of the Olson labels we have seen describes the contents of the boxes.

Horses

All three companies' horse-drawn implements used one of two model horses. The smaller horse-drawn vehicles had horses measuring 85 mm long, which were either hollow-cast lead or solid aluminium. The larger vehicles were drawn by hollow-cast lead or solid aluminium horses which were 115 mm long. While the large horse had a walking pose, the smaller horse was standing, and both horses were cast with a bridle, collar and cart saddle. The smaller horse was similar to Charbens Farm Horse, catalogue no.307 (see the Guide to Harnessed Horses). When found made of hollow-cast lead, the smaller horse is sometimes marked *COPYRIGHT* around the horse's belly band and sometimes it has no markings. Charbens are known to have marked *COPYRIGHT* on a number of their farm items, so it is likely that the marked version is by Charbens, and the unmarked by Kayron *et al.*

The larger hollow lead horses were either heavy castings or much lighter castings that used less material. The lighter version had slightly more detail in the casting than the heavy version, including a mane, and was almost definitely another Charbens product, catalogue no.312 Large Horse, which was listed in the 1953 Charbens catalogue at four times the price of item no.307, the Farm Horse (see pp.194-5 in the Charbens chapter).

It seems unlikely that the aluminium horses were produced by Charbens, as they never used aluminium in any of their other toys. Also the heavy casting of the large horse would have been very wasteful of precious lead, and would not have been produced by an experienced hollow-casting manufacturer such as Charbens. We conclude that Kayron *et al* either produced the horses themselves, or bought them in from another manufacturer, and then Charbens bought the moulds when those companies ceased production. This also explains why Charbens did not themselves use either horse with any of their own implements.

Although the Kayron, A.V.H. and Olson ranges of implements and vehicles were extensive, they were made of diecast metal, heavy metal wire or tinplate, and for such small enterprises it would have been very ambitious to also cast their own aluminium or lead horses, so perhaps these are clues as to why the venture was not a success.

Horse Harness

All three brands used the same unique system for harnessing the smaller horses to their ranges of implements, which consisted of strips of soft plastic, with small paper fasteners at the joins, forming a loop around the horse's neck and a loop around the horse's tail to fit over a wire hook on the implement. When the implement had shafts, the harness also had small loops each side of the horse to hold the shafts in place (**3**). On implements without shafts the link between horse and implement was limited to the rear loop (**19**).

It is interesting that the two 1948 Moko advertisements for Olson Farminit implements in *Games & Toys* showed horses and their implements but without harness. The shafts on most of the Olson implements were set at an angle, and this held the horse without requiring a harness (**29** and **75**). However, we have found two Olson large horses with what are probably original harnesses (**46** and **82**). These harnesses were made of thin brown leather in a similar design to the plastic version used for the smaller horse but without the strap round the rear of the horse.

Wheels

Across the ranges a variety of wheels were used. There were two types of tinplate wheel. The most common were hollow wheels made from two pieces of pressed tinplate and found in a number of sizes, as in (**7**), (**92**) and (**103**), and a less common version of the large wheel was made from two pieces of tinplate riveted together (**104** and **105**). The diecast wheels had either eight spokes or were smaller six-spoked versions (**1**), and were either smooth-rimmed, as in (**1**), or had a central ridge on the rim with a series of projections between the ridge and the edges of the rim (**15** and **33**). For lack of any very suitable word, we describe these latter wheels as 'lugged'.

The Range of Implements

There were many different horse-drawn implements sold by these three companies, and a high proportion of them have yet to be found in boxes. It therefore seems best to describe them according to implement type, and to attribute to a brand only those implements for which we have boxes.

Wagon

The all-wire wagon, advertised by Moko in *Games & Toys* in May 1948 (see the Moko advert), is one of the hardest items to find. The advert described it as a part of 'THE "FARMINIT" SERIES of FARM IMPLEMENTS', and it was designed for the large Olson horse. It had red smooth-rimmed spoked wheels (**1**), with smaller wheels on the front on a turntable that swivelled, and the wagon was painted in the usual Olson combination of blue and yellow.

A quite different wagon had a tinplate floor and sides. It was yellow with a blue swivelling turntable, red side rails and small green smooth-rimmed, spoked wheels of the same size (**2**).

The most common wagon was the hay wagon with tinplate floor, fixed high wire raves at both ends, side rails, a swivelling turntable and small tinplate wheels. The wagon was green (**3**), light blue (**4** and **5**), dark blue (**6**) or red (not illustrated) and the tinplate wheels were red (**3**, **5** and **6**) or green (**4**).

Tip Cart

The tip carts had tinplate floors, sides and backs and wire fronts and side rails. They were in blue with yellow tinplate wheels (**7**), or dark green with light blue tinplate wheels (**8**).

Tinplate and wire Cart

Making carts with wire sides saved on the use of metal, but they must have been hopeless playthings, because the contents could so easily fall out. They came in green (**9** and **12**), dark blue (**10**) or light blue (**11**), with tinplate wheels which were dark blue, red or bare metal (**9** to **12**).

Cart with wire floor

The carts that were made entirely of wire were usually light blue, and the smooth-rimmed or lugged spoked wheels were red (**13**, **15** and **16**) or green (**14**). The cart in (**16**) was a larger version for use with the big Olson horse. The wheels on the small carts had six spokes and those on the larger carts had eight, as (**16**). Although the A.V.H. box label did show a wire cart, it differed from these models

Small two-furrow Plough
(Kayron and A.V.H. No.504 Horse & Plough)

By modelling a rather unrealistic form of two-furrow plough, the manufacturers created an implement which was self-standing and ideal for making grooves in the sand pit. The small solid diecast front wheels were always bare metal and the frames came in a wide range of colours. The versions we have seen were red (**17** and **22**), green with yellow shares (**18**), green (**19**), green with red shares (not illustrated), yellow with red shares (**20**) or blue (**21**).

Single-furrow Plough

There was a rare, more realistic, single-furrow plough with coulter and mouldboard, which clearly did not stand up unsupported. This had a red frame, red coulter, blue mouldboard and green front wheel, hook and handles (**23**). The coulter used on this was the plough share on the previous two-furrow examples.

Large two-furrow Plough

There was also a heavy, realistic, two-furrow plough which was possibly tractor-drawn. It had mouldboards and a lever on the top to control plough depth. We have found examples in blue and yellow (**24**) or in red (**25**) with two green lugged spoked wheels. The example in (**25**) was missing its rear wheel.

Two-furrow Plough or Root Lifter

A larger version of the small two-furrow plough was sold by Olson. While superficially it just looked like a bigger version of the two-furrow plough, the shares were actually turned inwards so much that it was more like a root-crop lifter. We have found it in light blue (**26**), dark blue (**27**) and red (**28**) with green shares and smooth-rimmed or lugged spoked wheels.

Disc Cultivator

This yellow-and-blue Olson implement had three discs in front and two behind. They were diecast and were mounted vertically under a diecast metal, rectangular frame. The cultivator had a circular seat made from a washer and was set on a pair of lugged spoked wheels. These implements can hardly be called disc harrows, so disc cultivator appears to be a better description. The wheels were red (**29**) or green (**30**).

Disc Harrow

This was a realistic pair of disc harrows, each with six green discs set within a red frame (**31**). Set above the frame on a pole was a seat, and there was a depth lever to the right of the seat. The hook, which could have been for towing by tractor or to attach to a horse harness, was kept clear of the ground by a bare metal wheel similar to that found on the small two-furrow ploughs.

Spike-toothed Cultivator

Three rows of curving spikes were set under a triangular frame with wheels either side controlled by a height adjustment lever. The two examples illustrated were of different sizes: (**32**) was 130 mm long with two spikes in the middle row, and (**33**) was 150 mm long with four spikes in the middle row. In (**32**) the frame was red and the lugged wheels with spokes were green. In (**33**) the model had been repainted dark blue with red spikes, but probably was originally light blue. On both models the hook for the harness at the front was identical to that on the disc harrow in (**31**).

Roller (Kayron and A.V.H. No.501 Horse & Roller)

Rollers were popular items for Kayron and A.V.H. A solid metal cylinder was held under a rectangular wire frame. The cylinders were usually green (**34**, **35**, **38**, **41** and **43**), light blue (**37** and **40**) or less frequently, red (**36**, **39** and **42**). Olson made bigger versions of the roller, to be pulled by the larger horse, but with wooden cylinders. We have found one with the usual light blue and yellow frame (**44**) and another with a red and yellow frame (**45**). Olson also made a smaller roller with a red cylinder and a blue and yellow frame with circular seat, with the cylinder cast in two halves (**46**). There was also a roller with wooden cylinder that had a seat and long thin shafts with bulbous terminals (**47**), as on the incomplete hay rake in (**65**).

Seed Sower with seed holes and four coulters (A.V.H. No.502 Horse & Seed Sower)

A.V.H. sold a seed drill they called a 'Seed Sower' with small holes in the trough which fed into a row of four coulters below. In front of the coulters were small discs which first made grooves in the ground into which the coulters dropped the seed, and then behind the coulters were blades set at an angle to fill in the grooves. It was a clever device which was free-standing, supported by the four front discs (also used as the front wheels on the small two-furrow ploughs) and by the blades behind (also used as shares on the small two-furrow ploughs). The holes for the seed were too fine for sand but would have taken grains of salt.

These drills were usually on tinplate wheels, but a few were on spoked wheels. The drills were green, dark blue or light blue (**48** to **55**). The lugged spoked wheels were red (**48**) or green (**49**), and the tinplate wheels were red, green or dark blue (**50** to **55**).

Seed Sower without seed holes and with eight coulters

Olson sold a smaller drill without holes in the trough and with a row of eight coulters underneath. As the coulters were blind, there was no need to make grooves or to have a device to fill in the grooves! The drills were usually light blue (**56** to **59**), with red tips to the coulters and with red (**56** to **58**) or green (**59**) smooth-rimmed or lugged spoked wheels. In an unusual variation the seed trough was painted red inside (**59**). We have not seen any of these drills with tinplate wheels.

Hoe for root crops

The hoe for root crops had five v-shaped horizontal hoe blades set on a frame (**60**) and they could be lifted up into a transport position (**61**). There were also two pairs of vertical discs which were probably for depth control. This was not a common toy and was probably fairly expensive. The examples we have seen were green with red smooth-rimmed spoked wheels, yellow handles and shafts, and one version had the discs painted blue (**62**), which was possibly not original.

Mower

Mowers were also a rare item in this series, and we have seen just one (**63**). It had a high seat already seen on the disc harrows in (**31**) and was rather crude with a fixed height control lever. It was red and blue with green spoked wheels.

Hay Rake with seat

Hay rakes were the commonest item found in the series, and they usually did not have seats. We have seen one by Olson in blue and yellow which had a circular seat made from a washer, small red lugged wheels and a lever to raise the tines (**64**).

An incomplete example of a hay rake (**65**) had a seat similar to those on the disc harrows and mower (**31** and **63**), with spoked wheels, nineteen tines and long slender shafts with unusual bulbous terminals, all in green except for the red spoked wheels (broken in this example).

Hay Rake without seat (Kayron and A.V.H. No.503 Horse & Hay Rake)

In 1948 Moko promoted the hay rake in *Games & Toys* without a mention of Kayron, A.V.H. or Olson (see the Moko adverts), so it is possible some were sold in Moko packaging, although none has been found so far. The hay rake came with smooth-rimmed, spoked wheels (**66** to **75**) or with tinplate wheels (**76** to **81**). The usual (smaller) version (**66** to **73** and **76** to **81**) is perhaps best illustrated by (**69**). On this example, the set of light green tines was hinged on a light green frame with two hoops supporting a horizontal bar on which an orange lever pivoted. The lever was hooked onto one of the tines, and as pressure was applied to the lever, the tines were all raised.

On the wider version of the hay rake (**74** and **75**), which was the one actually illustrated in the 1948 Moko advertisement, the lever was finished with a loop which was connected to the tines by string (missing in these examples). One variant had wheels with fewer and thicker spokes than on the smaller hay rakes (**75**). Olson made another version which allowed the tines to be raised by a lever attached directly to the bar on which the tines were fixed (**82** and **83**).

In (**66**) and (**67**) the box carried a label with *HORSE & HAY RAKE*, but (**68**) was just *HAY RAKE* and *No.507* had been changed by hand to *503*.

Hay Tedder

This was the most weird and wonderful toy we have seen (**84** and **85**). It had two pairs of three-pronged reciprocating forks which rose and fell in pairs, powered by a fine wire spring belt, driven off a wheel on the axle. It ran on large spoked, smooth-rimmed wheels as on hay rakes (**66** to **73**) and on gigs (**86** to **90**). The tedder had green forks, yellow wheels and shafts, and a red frame and mechanism, which included a depth lever to the right of a foot rest.

Gig

Gigs were made of tinplate set on a wire chassis and coloured red (**86**), light blue (**87**), dark blue (**88**) dark green (**89**) or mid green (**90**), with blue (**86**), yellow (**87**), or red large smooth-rimmed spoked wheels (**88** to **90**). No human figures are known with this series, and no doubt there were plenty of other makes of drivers of roughly the right scale available to use on the gig, such as the Britains seated figures (see the Britains chapter).

Tractors and Tractor-drawn Implements

We illustrate two types of tractor and a small number of unboxed tractor-drawn implements which were related to the series in various ways, including the method of manufacture and the materials used. The tractor in (**91**) to (**93**) was surely the one illustrated on the A.V.H. box lid. The clockwork tractors in (**101**) to (**107**) were possibly from another manufacturer, as they were made from aluminium sheet, although their wheels are similar to those used by Kayron *et al*. The boxed set in (**102**) included a wire four-wheeled trailer which was similar to the Kayron, A.V.H., Olson series, but the wires used on this trailer were bent around each other at the joints, rather than soldered. The implements illustrated in (**91**) and (**94**) to (**103**) all had downward-pointing hooks which fitted into holes at the rear ends of both tractors.

Small tinplate Tractor without mudguards

This small tinplate tractor without mudguards (**91** to **93**) was found with a hay rake (**91** and **94**), suggesting it was part of the Kayron/A.V.H. series. It was also similar to the tractor illustrated on the A.V.H. box. The body of the tractor was light brown with red tinplate wheels, seat, steering wheel, air cleaner and exhaust, and had a pierced tinplate grille.

Hay Rake

The hay rake in (**91**) and (**94**) was based on the horse-drawn version, but two rather thin shafts were drawn together with a third thicker central shaft to form a drawbar. The lever to raise the tines was extended down the drawbar to provide some control from the tractor. The rake was green with light brown tines and with yellow tinplate wheels.

Simpler tractor conversions of the horse rake can be seen in (**95** to **97**) where the horizontal bar to take the hook for the horse's harness was dispensed with and the two shafts were joined at the ends to form a drawbar with a downward hook to attach to the tractor. The two-part tinplate wheels with flat rims were not used on horse rakes, but they were the wheels used on the tractors in (**101**), (**102**), (**104**) and (**105**). The colours on these rakes were either uniformly light blue (**95**) or green (**96** and **97**), and not a mixture of colours as we have seen in horse-drawn versions.

Wire Cart

The Olson wire cart for tractors (**98**) was a straightforward conversion from the horse-drawn equivalent in (**13**) with green lugged wheels with spokes.

Disc Harrow

This set of harrows (**99**) used discs which were different from those used on the horse-drawn items, and the discs were attached to a simple wire frame with a tractor hook.

Roller

The horse roller converted easily from shafts to a tractor hook. The roller on this example was metal and painted an unusual red-brown (**100**).

Reaper

We have found only two examples of this rare model, both of which were tractor drawn versions (**101**), consequently we do not know if the model was ever made as a horse-drawn item. The wire pieces used to make the implement were partly soldered together and partly bent into position, as on the wire trailers in (**102**) and (**103**). The wheels were identical in shape and colour to those used on the tractor-drawn hay rakes in (**95**) and (**96**), but both tractor and implement in (**101**) have the addition of elastic bands around the outer rims. It is not known if the elastic bands are original. The wire sails were possibly made to turn as the implement moved, by means of an elastic band between one of the ground wheels and the smaller wheel located on the arm on which the sails rotated.

Large clockwork tinplate Tractor with mudguards

It is fortunate that one example of this clockwork tractor has been found in a boxed set with a four-wheeled wire trailer (**102**), although the box has no label. The rear tinplate wheels with their flat rims were identical in style to the wheels on the hay rakes in (**95**) to (**97**), and the reaper in (**101**), making it possible to link this tractor to the series of wire implements. The scale is correct for those implements to be used with this tractor, while they are rather large for the tractor in (**91**) to (**93**). The motor was wound on the left hand side, just in front of the rear wheel. The body of the tractor was either light or dark blue (**101** to **104**, **106** and **107**) or red (**105**). The mudguards and seat were either green (**102** and **105** to **107**) or red (**101**, **103** and **104**), and the seats had slots for flat brown tinplate drivers (**102** and **106**). The tinplate wheels were either the usual hollow two-piece rounded shapes front and rear (**103**, **106** and **107**) or the rear wheels were constructed in two parts riveted together (**101**, **102**, **104** and **105**). The wheels were either red (**101**, **102** and **104**), dark blue (**105**) or black (**103**, **106** and **107**). The steering wheels were made from bare metal brass washers.

Four-wheeled wire Trailer

On this wire trailer (**102** and **103**) the horizontal wires were bent around the vertical corner wires, and there is no sign that solder was used. Nevertheless, the construction was quite firm. We have seen the trailer in light green with blue wheels (**102**), and light blue with red wheels (**103**).

Model List

Items found in labelled boxes

Serial No.	Model
Kayron	
501	Horse & Roller
503	Horse & Hay Rake
504	Horse & Plough
A.V.H.	
501	Horse & Roller
502	Horse & Seed Sower (with holes and four coulters)
503	Horse & Hay Rake
504	Horse & Plough
Olson	
	Wagon
	Wire Cart (two sizes)
	Wire Cart for Tractor
	Two-furrow Plough or Root Lifter
	Disc Cultivator
	Seed Sower (without holes and eight coulters)
	Horse & Hay Rake with seat
	Horse & Hay Rake (large) without seat

Full list of models

	Model
Soldered wire horse-drawn implements	
	Wagon (three designs)
	Tip Cart
	Cart with tinplate floor
	Cart with wire floor (two sizes)
	Small two-furrow Plough
	Single-furrow Plough
	Large two-furrow Plough – possibly tractor-drawn
	Two-furrow Plough or Root Lifter
	Disc Cultivator
	Disc Harrow
	Spike-toothed Cultivator
	Roller (four designs)
	Seed Sower (with seed holes and four coulters)
	Seed Sower (without seed holes and with eight coulters)
	Hoe
	Hay Mower
	Hay Rake with seat (two designs)
	Hay Rake without seat (two sizes)
	Hay Tedder
	Gig

	Model
Tractors and Tractor Implements	
	Small tinplate Tractor without mudguards
	Hay Rake (two designs)
	Wire Cart
	Disc Harrow
	Roller
	Reaper
	Large clockwork tinplate Tractor with mudguards
	Four-wheeled wire Trailer with the wire bent round at the joints

Further Reading

Kelly's Directory, for Woking, editions for 1948 and 1949-1950.

Woking Street Directory, 1957.

1. *Olson four-wheeled wire Wagon in light blue with swivelling turntable, red smooth-rimmed spoked wheels and yellow shafts (318 mm).*

2. *Tinplate Wagon in yellow with light blue swivelling turntable, green smooth-rimmed spoked wheels and yellow shafts (203 mm).*

3. *Tinplate and wire Hay Wagon in green with red tinplate wheels, dark blue swivelling turntable and shafts (220 mm).*

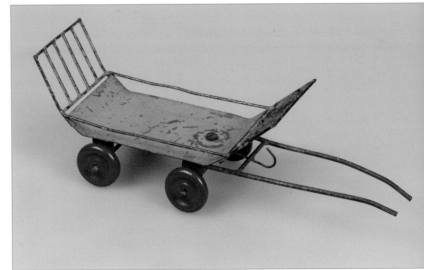

4. *Tinplate and wire Hay Wagon as (**3**) but light blue with green tinplate wheels and red swivelling turntable and shafts.*

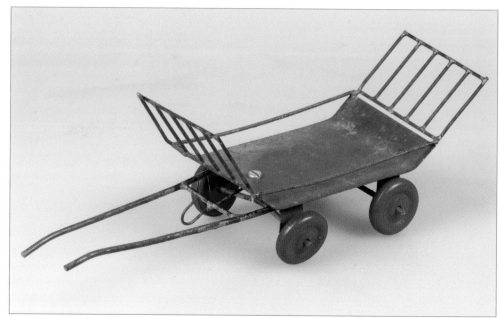

5. *Tinplate and wire Hay Wagon as (**4**) but with red tinplate wheels, green swivelling turntable and yellow shafts.*

6. *Tinplate and wire Hay Wagon as (**3**) but dark blue with green swivelling turntable and shafts.*

8. *Tip Cart as (**7**) but dark green with light blue tinplate wheels.*

7. *Tip Cart in blue with yellow tinplate wheels and yellow shafts (159 mm).*

10. *Tinplate and wire Cart as (9) but dark blue with bare metal tinplate wheels.*

9. *Tinplate and wire Cart in green with dark blue tinplate wheels and yellow shafts (150 mm).*

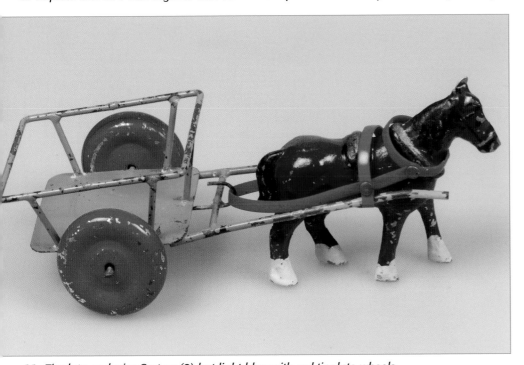

11. *Tinplate and wire Cart as (9) but light blue with red tinplate wheels.*

12. *Tinplate and wire Cart as (9) but with red tinplate wheels.*

13. Olson boxed wire Cart in light blue with red smooth-rimmed spoked wheels and yellow shafts (160 mm).

14. Wire Cart as (**13**) but with green spoked wheels with lugs.

15. Wire Cart as (**13**) but with red spoked wheels with lugs and red shafts.

16. Olson boxed large wire Cart in light blue with large red smooth-rimmed spoked wheels and yellow shafts (210 mm).

17. Kayron boxed no.504 Horse & Plough small two-furrow plough in red with red shares and yellow handles (110 mm).

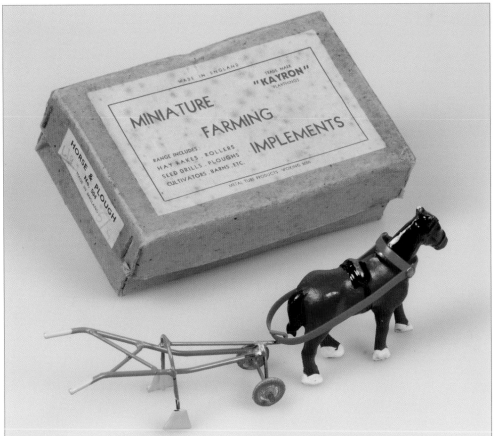

18. Kayron boxed no.504 Horse & Plough as (**17**) but green with yellow shares.

19. Kayron boxed no.504 Horse & Plough as (**17**) but green with green shares.

20. *A.V.H. boxed no.504 Horse & Plough as (17) but yellow with red handles.*

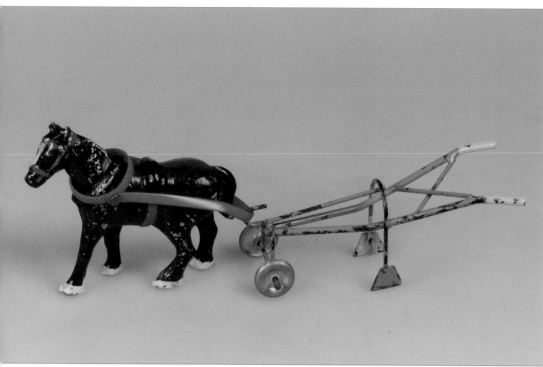

21. *Small two-furrow Plough as (17) but blue with blue shares.*

22. *Small two-furrow Plough as (17) but with orange handles.*

23. *Single-furrow Plough in red with green front wheel, red coulter, blue mouldboard and green handles (148 mm).*

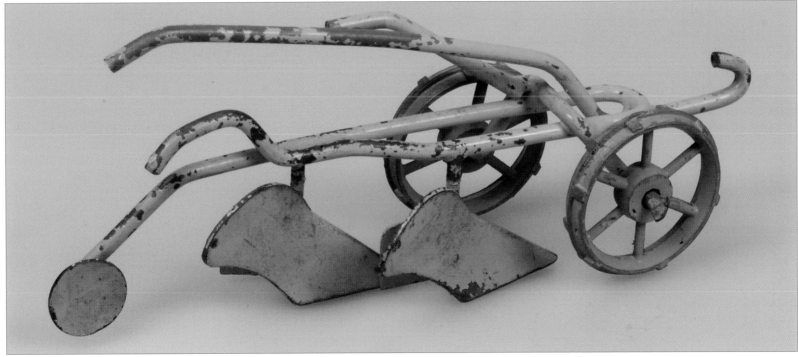

24. *Large two-furrow Plough in light blue and yellow with two green spoked wheels with lugs (148 mm).*

25. *Large two-furrow Plough as (24) but red with dark green wheels (one wheel missing).*

26. *Olson boxed two-furrow Plough or Root Lifter in light blue with dark green spoked wheels with lugs and green handles (145 mm).*

27. *Olson boxed two-furrow Plough or Root Lifter as (**26**) but darker blue with dark green smooth-rimmed wheels .*

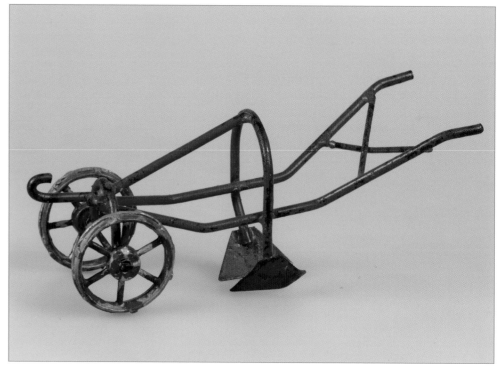

28. *Olson two-furrow Plough or Root Lifter as (**26**) but red.*

29. *Olson boxed Disc Cultivator with seat, in light blue with red spoked wheels with lugs and yellow shafts (150 mm).*

30. *Olson boxed Disc Cultivator as (29) but with dark green wheels.*

31. *Disc Harrow with green discs, red frame, bare metal front wheel, black seat and green front hook (132 mm).*

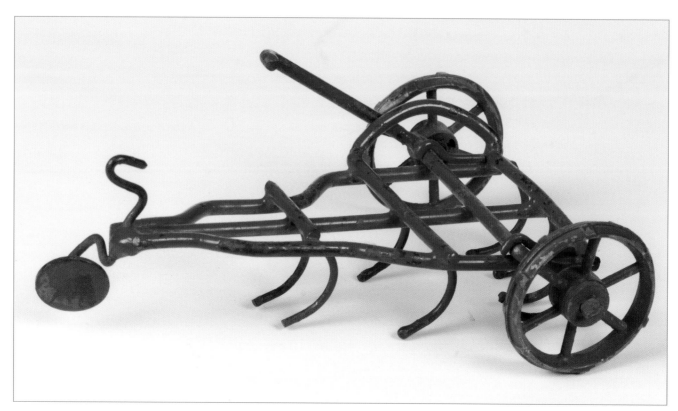

32. *Spike-toothed Cultivator with red frame, green spoked wheels with lugs, green front wheel and front hook (130 mm).*

33. *Spike-toothed Cultivator, repainted, with four spikes in the middle row. The bars holding the spikes were over the frame in (32), and under it in this model (150 mm).*

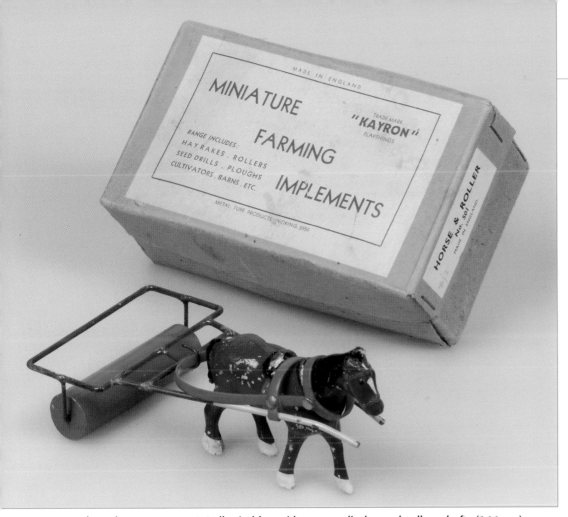

34. *Kayron boxed no.501 Horse & Roller in blue with green cylinder and yellow shafts (144mm).*

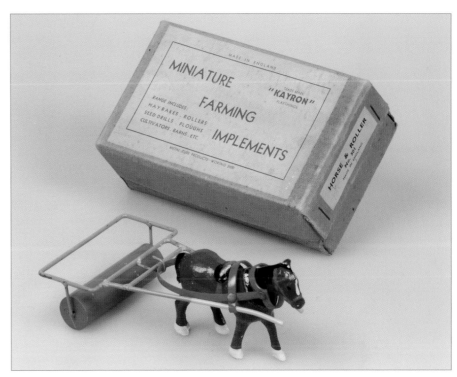

35. *Kayron boxed no.501 Horse & Roller as (**34**) but in light blue.*

36. *A.V.H. boxed no.501 Horse & Roller as (**34**) but green with a red cylinder.*

37. *A.V.H. boxed no.501 (with 505 crossed out) Horse & Roller as (**34**) but green with light blue cylinder.*

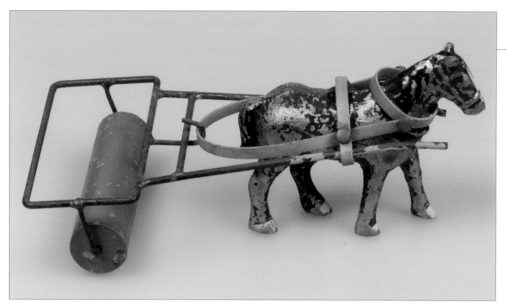

38. *Roller as (34) but dark blue.*

39. *Roller as (34) but yellow with red cylinder and red shafts.*

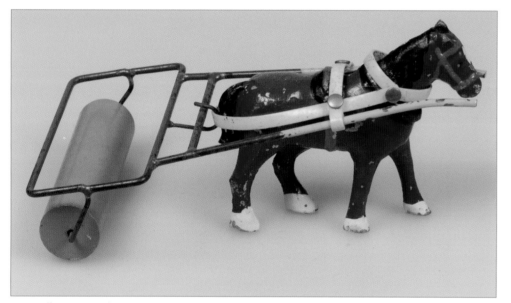

40. *Roller as (34) but green with light blue cylinder.*

41. *Roller as (34) but yellow with light blue shafts.*

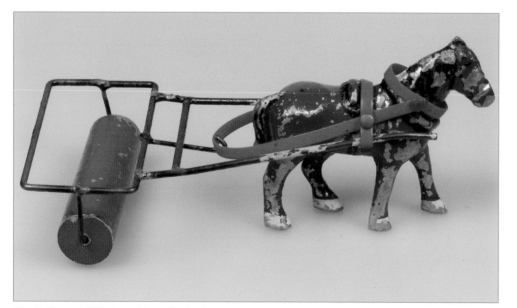

42. *Roller as (34) but with red cylinder.*

43. *Roller as (34) but red with white shafts.*

44. *Large Roller in light blue with a bare wooden cylinder and yellow shafts (170 mm).*

45. *Large Roller as (44) but red.*

46. *Probably Olson Roller with seat in light blue with red cylinder and (bent) yellow shafts. Note that the horse has a thin brown leather harness (130 mm).*

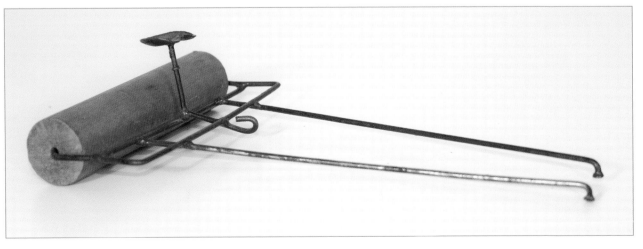

47. *Roller with long shafts with bulbous terminals. Wooden cylinder with red frame and seat (190 mm).*

48. *A.V.H. boxed no.502 Horse & Seed Sower with holes. Dark blue seed trough, with red spoked wheels with lugs and red shafts (140 mm).*

49. *Seed Sower as (48) but green with green spoked wheels with lugs and yellow shafts (with most of the paint on the shafts missing).*

50. *Seed Sower as (48) but green with red tinplate wheels and yellow shafts.*

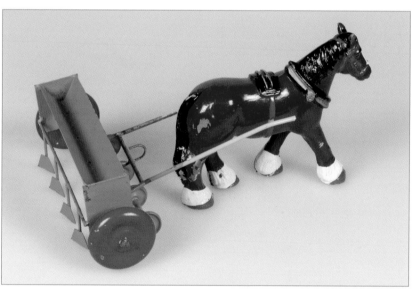

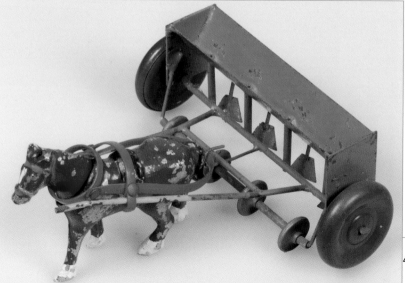

51. *Seed Sower as (48) but light blue with green tinplate wheels.*

52. *Seed Sower as (48) but green with blue tinplate wheels and yellow shafts.*

53. *Seed Sower as (48) but with green tinplate wheels and yellow shafts.*

54. *Seed Sower as (**48**) but with red tinplate wheels and yellow shafts.*

55. *Seed Sower as (**48**) but light blue with red tinplate wheels and orange shafts*

56. *Olson boxed Seed Sower without holes in the blue seed trough, with red spoked wheels with lugs, red tips to the coulters and yellow shafts (130 mm).*

57. *Seed Sower as (**56**) but with blue shafts with red tips.*

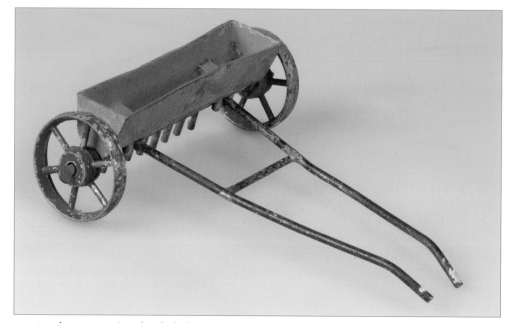

58. *Seed Sower as (**56**) but light blue with red smooth-rimmed spoked wheels*

59. *Seed Sower as (**56**) but with red interior to seed trough and dark green smooth-rimmed spoked wheels.*

60. Hoe for root crops in green with black discs and hoe blades, red smooth-rimmed spoked wheels and yellow shafts and handles (265 mm).

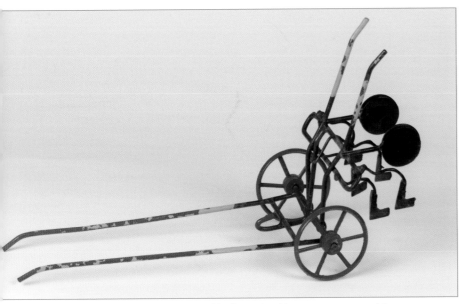

61. Hoe for root crops as (**60**) with blades raised into transport position.

63. Mower in red and blue with green smooth-rimmed spoked wheels (180 mm).

62. Hoe for root crops as (**60**) but with dark blue blades and discs.

64. Olson boxed Hay Rake with seat, with yellow tines, blue frame and seat, red spoked wheels with lugs and yellow shafts (126 mm).

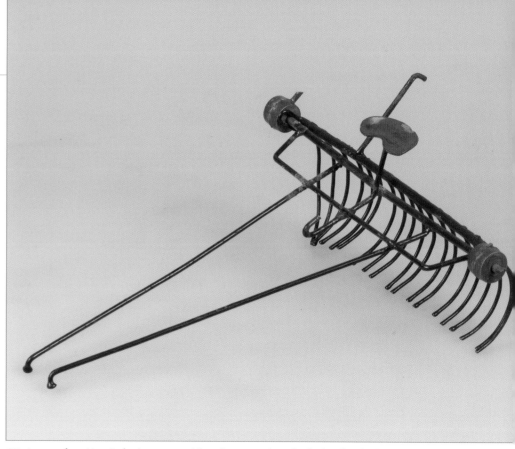

65. Incomplete Hay Rake in green with red seat and spoked wheels which are missing except for the hubs. The shafts were long and very thin with bulbous terminals, as (**47**), and the rake had nineteen tines, unlike anything else in the Kayron/A.V.H./Olson tradition (195 mm).

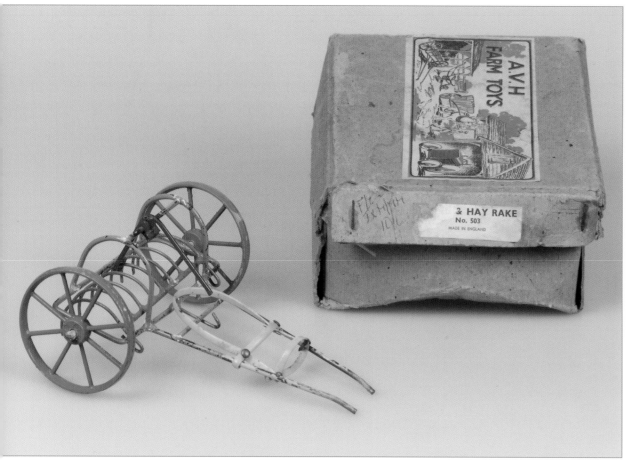

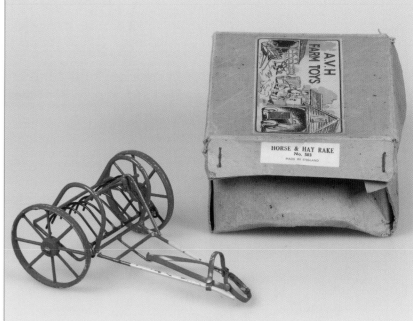

67. A.V.H. boxed no.503 Horse & Hay Rake as (**66**) but with dark blue tines, red frame and wheels and brown lever.

66. A.V.H. boxed no.503 Horse & Hay Rake with light green tines, light blue frame, red lever, large light green smooth-rimmed spoked wheels and yellow shafts (165 mm).

68. A.V.H. boxed no.503 Hay Rake (with 507 crossed out and replaced in pen with 503), as (**66**) but with dark yellow lever and wheels.

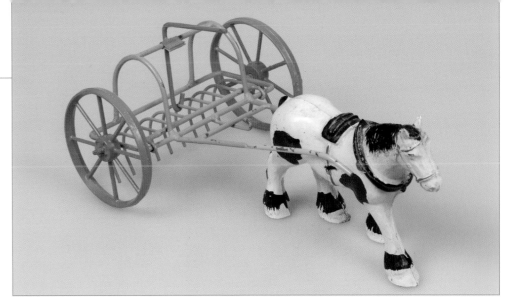

69. Hay Rake as (**66**) but with light green frame, orange lever and wheels and large horse.

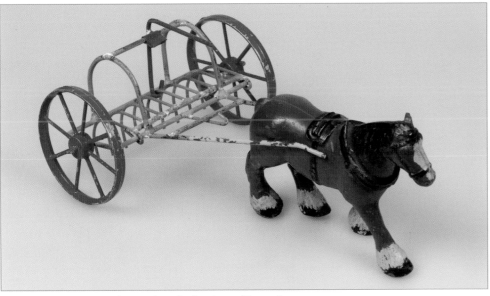

70. Hay Rake as (**66**) but with red wheels and large horse.

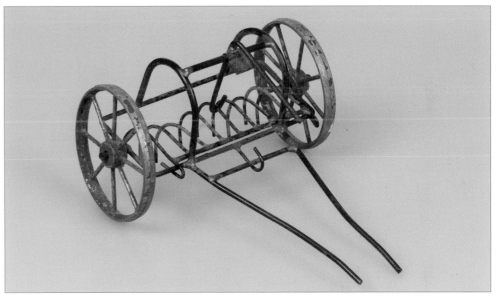

71. Hay Rake as (**66**) but with orange tines, green frame, bare metal lever and red wheels.

72. Hay Rake as (**66**) but with dark blue frame and red wheels.

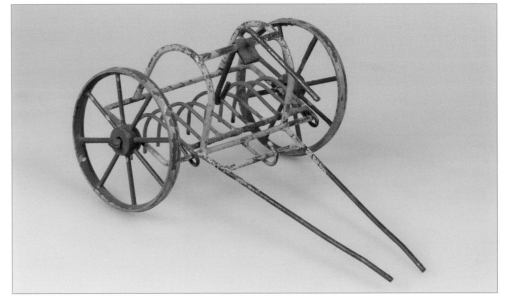

73. Hay Rake as (**66**) but with bare metal tines, yellow frame, bare metal lever and green wheels.

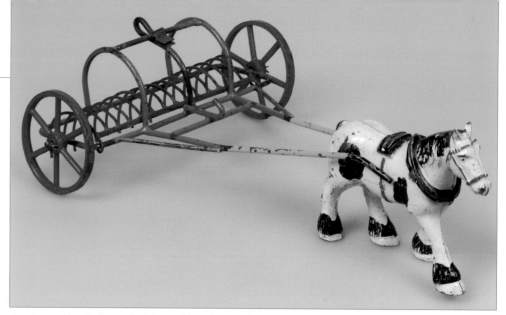

75. *Large Hay Rake as (**74**) but with red tines, light brown frame and red lever.*

74. *Olson boxed large Hay Rake, 165 mm wide, with yellow tines, light blue frame, dark green lever, red spoked wheels and yellow shafts with large horse (235 mm).*

77. *Kayron boxed no.503 Horse & Hay Rake as (**76**) but with red tines, dark green frame, light blue lever and orange shafts.*

78. *Hay Rake as (**76**) but with yellow tines and light blue frame.*

76. *Kayron boxed no.503 Horse & Hay Rake, as (**66**) but with red tinplate wheels and dark blue frame (150 mm).*

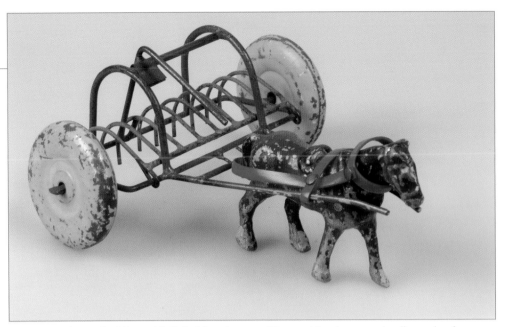

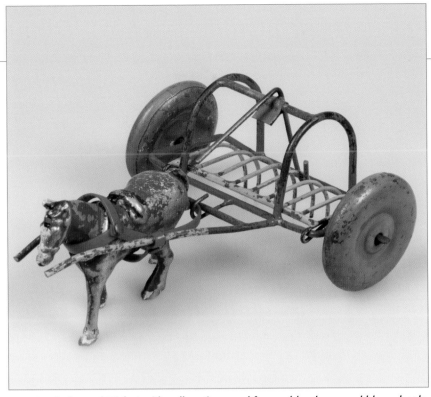

79. *Hay Rake as (**76**) but with yellow tines, red frame, blue lever and blue wheels.*

80. *Hay Rake as (**76**) but with light blue tines, red frame, blue lever and yellow wheels.*

81. *Hay Rake as (**76**) but with red tines, light green frame and lever.*

82. *Olson large Hay Rake as (**74**) but with different lifting mechanism and red tines and shafts. The harness made of thin brown leather as on (**46**).*

83. *Olson large Hay Rake as (**82**) showing elevation lever.*

84. *Hay Tedder with reciprocating green forks, red frame and mechanism, yellow wheels and shafts, and black seat and foot rest (280 mm).*

85. *Hay Tedder rear view of (**84**).*

86. *Gig with red tinplate body, large blue wheels and yellow shafts (162 mm).*

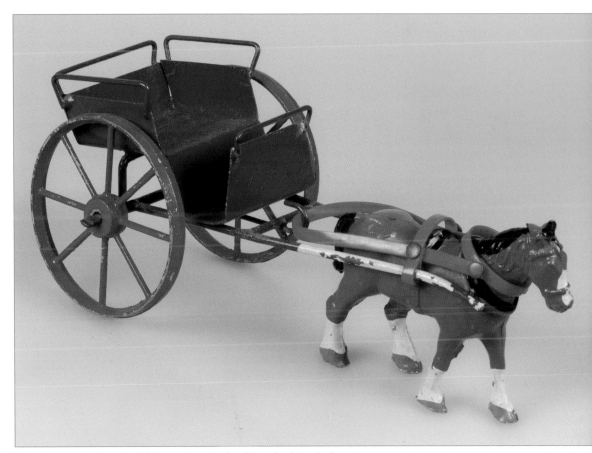

88. *Gig as (86) but dark blue with red wheels and white shafts.*

87. *Gig as (86) but light blue with yellow wheels and red shafts.*

89. *Gig as (86) but dark green with red wheels and bare metal shafts.*

90. *Gig as (89) but mid green with yellow shafts.*

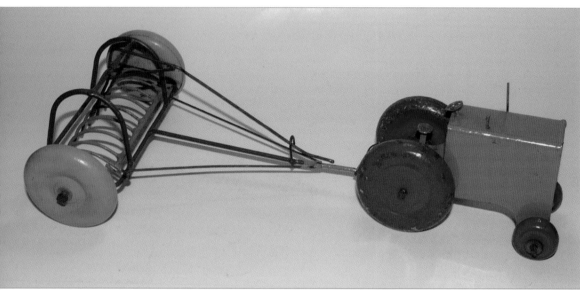

91. *Small tinplate Tractor described in (**92**) found with this hay rake described in (**94**).*

92. *Small light brown tinplate Tractor with red wheels, seat, steering wheel, exhaust and air cleaner very similar to the tractor illustrated on A.V.H. box lids. The radiator grille was a separate piece of tinplate and the seat was a large brass paper fastener, both soldered in place (112 mm).*

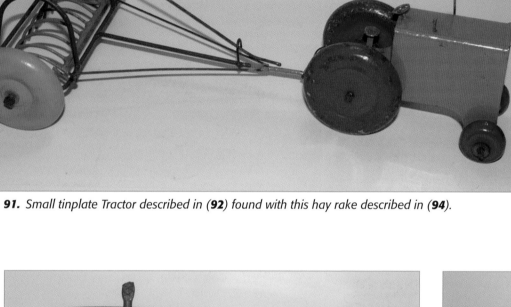

93. *The underside of (**92**).*

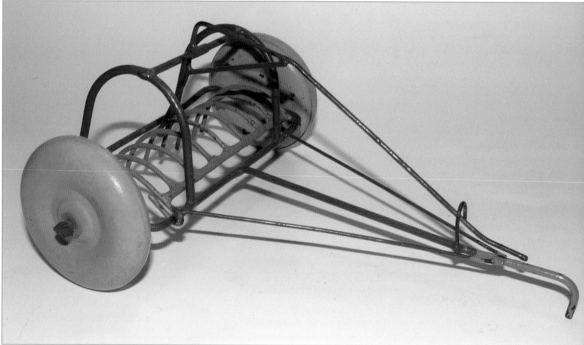

94. *Hay Rake for a tractor with light brown tines, green frame, drawbar and lever and yellow tinplate wheels (168 mm).*

95. *Hay Rake for a tractor in light blue with red tinplate wheels with flat rims (156 mm).*

96. *Hay Rake as (95) but light green.*

97. *Hay Rake for a tractor as (96) but with blue wheels and shafts joined in a Y to form the tractor hook (188 mm).*

98. *Olson boxed wire Cart for a tractor, in light blue with green lugged wheels with spokes and yellow tow bar with red hook (140 mm).*

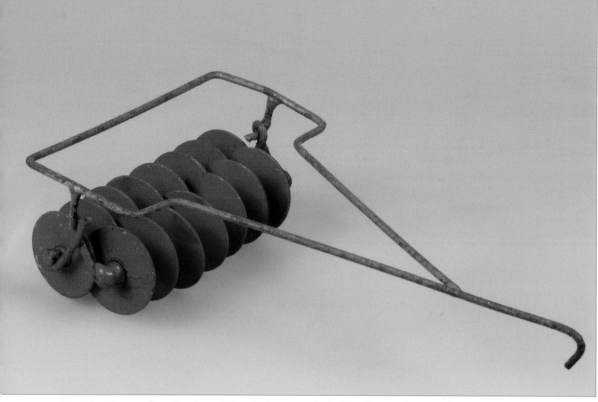

99. *Disc Harrow for a tractor, with red discs on a light blue frame (182 mm).*

100. *Roller for a tractor, with a red-brown cylinder on a blue frame (115mm).*

101. *Reaper with blue frame and sails, and red wheels made in two sections with flat rims, pulled by an aluminium clockwork tractor. Light blue tractor with red mudguards and seat, brass washer steering wheel, large flat-rimmed wheels at rear and smaller, hollow tinplate wheels at front. Note that both sets of flat-rimmed wheels have the addition of elastic bands around the outer rims (tractor 130 mm; reaper 230 mm).*

102. *Boxed set of an aluminium clockwork Tractor pulling a wire four-wheeled trailer. Tractor as (**101**) but with green mudguards and seat, with a light brown push-in driver. Wire four-wheeled trailer in light green with light blue hollow tinplate wheels (tractor 130 mm; trailer 163 mm).*

103. *An unboxed set of an aluminium clockwork Tractor pulling a wire four-wheeled trailer, as (102), but tractor with red mudguards and seat, and black hollow tinplate wheels, but missing steering wheel and driver, and trailer in light blue with red wheels.*

104. *Aluminium clockwork Tractor as (101).*

105. *Aluminium clockwork Tractor as (101) but red with green mudguard and seat, and blue wheels.*

106. *Aluminium clockwork Tractor as (103) but with green mudguards and seat, and light brown driver.*

107. *Aluminium clockwork Tractor as (106) but light blue, missing driver and steering wheel.*

Company History

The family business of O. & M. Kleemann Ltd commenced production of plastic toy products in 1938, using six injection moulding machines purchased from the American manufacturer, Reed Prentice. The company was actually started by Oscar Kleemann in the eastern district of the City of London in 1907, and continued to be based there until the outbreak of the Second World War. The firm changed its name to Kleemann Bros in 1909 and subsequently to O. & M. Kleemann in 1919, the brother being Max Kleemann. It became a limited company during 1935. Between the two world wars O. & M. Kleemann acted as importers as well as manufacturers of household products such as combs, ashtrays and tea sets, using cellulose acetate and one of the earliest forms of synthetic plastic, Bakelite. These early products can still be found, as the U.K.-produced Bakelite items were often marked with the Kleemann brand name of Kabroloid.

In 1940, O. & M. Kleemann moved its production out of London to a factory in Welwyn Garden City, Hertfordshire, while retaining a head office and sales department in London. It was in the new factory that the recently acquired injection moulding machines were installed, along with compression moulding machines which were required for the Bakelite products. The volume production of toys now commenced.

During the Second World War the company produced combs for the British armed forces, and radio parts which took advantage of the excellent insulating properties of Bakelite. The period immediately after the Second World War was one of great expansion for O. & M. Kleemann, and in 1946 the company moved most of its production to an old munitions factory in Aycliffe, Co. Durham, as it outgrew the factory in Welwyn Garden City. Harry Kleemann, the son of Max, joined the family firm in 1949 after his discharge from the army, and in the early 1950s visited a Dutch company called Luxor Plastics to discuss the continuing exchange of moulds and ideas that had developed between the companies. Harry was impressed by the latest injection moulding machines that Luxor used, because 'The secret of these fast machines was that they used a single cavity that was water-cooled. In addition, the products were made of wafer-thin plastic. I can't remember the size of the slices, but it was approximately 1/22nd of an inch… their production method was a sensation' (Young, S. Mark et al., 2001).

The machines were made by an American company called Pyro Plastics, and Harry was so impressed that his cousin Derrick, who was Kleeware's managing director at that time, visited Pyro and bought six of the machines, which were rapidly installed at the Aycliffe factory. The visit to Pyro Plastics also led to Kleeware borrowing some moulds, which were run in the U.K. for a few weeks until a year's product had been made. The moulds were then returned to Pyro and Kleeware paid a royalty for the quantity produced. Kleeware also borrowed moulds from the Ideal Plastics Corporation, and some toys marketed by Kleeware have been found marked Ideal. Even some of the Kleeware packaging has obvious origins in the original American toy maker's products.

Similar mould-sharing arrangements were common between a number of U.K. and U.S. plastic toy manufacturers (see the T.N. Thomas chapter). Additionally, Harry Kleeman recalled visiting F.W. Woolworth stores in the U.S. in the late 1950s and buying suitcase-loads of toys which he brought back to the U.K. for Kleeware to copy. As the American plastic toy makers Ideal, Marx, Renwal and Thomas Manufacturing were all major suppliers to F.W. Woolworth in the U.S., and plagiarism was common among the U.K. manufacturers, this was another reason why so many U.K.-produced toys of the 1950s and early 1960s were similar to North American products.

During the decade or so after the Second World War, the firm of O. & M. Kleemann became one of the U.K.'s major plastic products manufacturers, and at its busiest point the factory at Aycliffe had 50 injection moulding machines and 900 staff, which included six or seven in the design and development department. The company exported its products around the world, and in an ironical twist, supplied both toy and house-ware products to the U.K. arm of F.W. Woolworth. In addition to the vast toy range which included construction kits, military and civilian vehicles and buildings, space ships, medieval castles, dolls and doll's house accessories, the company produced a range of household products, from sunglasses to tea sets and cutlery, from coat hangers to cigarette boxes, all sold under the Kleeware brand.

Alongside the manufacture of plastic products, the company had developed a raw materials business, initially from sales of Kabroloid, but later of its polystyrene-based product, Kleestron. In 1957, O. & M. Kleemann bought Erinoid, the major U.K. manufacturer of casein, a protein-based plastic made using milk curds. The company's focus now moved to its raw materials production, and in 1959 the Kleeware business was sold to its major competitor, Rosedale Associated Manufacturers (see the Tudor Rose chapter). The name of the Aycliffe factory operation was changed from Kleemann to Kleeware Plastics Ltd, and production moved to Glamorgan in the mid-1960s.

Models

The Kleeware range contained a number of farm-related vehicles and implements, all of which seem to have been based on American originals made by the Ideal Plastics Corporation. All the vehicles shown were made of a hard polystyrene-based plastic, and no items that can be positively attributed to Kleeware have been found made of polythene. This was possibly because Kleeware became part of Rosedale Associated Manufacturers in 1959, and it was only in the late 1950s that polythene was used extensively for toys, but probably also because Kleeware was so successful with its polystyrene-based products. Probably for the same reasons, Kleeware never made use of blow-moulding technology.

Kleeware toys are quite often found without any marking, other than *Made in England* in a roundel. We believe this to be because, while the original maker's brand mark had been removed by Kleeware, they did not bother replacing it with their own brand mark. While we are aware of the Kleeware version of the farm toy set shown, all the vehicles in (**1**) are actually Ideal originals. The set comprised a crawler tractor, an American-type corn drill, a grass cutter, a hay rake, an industrial-type bottom-unloading dump-wagon and a bulldozer blade, the arms of which slotted into holes on either side of the tractor's skirts. The crawler tractor was in blue with yellow driver and red tracks, which were actually only for show as the tractor had four rollers underneath. The corn drill had a blue chassis, two red hoppers, a green seat and yellow mechanism, all on four plastic wheels on a single metal axle. The grass cutter had a green chassis, red knife and mechanism, with a blue bar, all mounted on black rubber wheels on a metal axle. The red dump-wagon had a green mechanism, set on two yellow wheels on a metal axle; the hay rake was all in red with black plastic wheels on a metal axle and the bulldozer blade was red.

The colours of the Kleeware versions were generally different from the Ideal originals. The Kleeware tractor was completely orange, including the tracks and driver; the hay rake was yellow rather than red although it also had black wheels, and the dump-wagon was all-green including the wheels. Both the Ideal original implements and their Kleeware counterparts had working mechanisms, with the exception of the bulldozer blade.

Further Reading

Plastic Warrior, 2003, *The Plastic Warrior Guide to UK Makers of Plastic Toy Figures.* Plastic Warrior Publications.

Plastiquarian, the magazine of the Plastics Historical Society, No. 9, Winter 1991 edition.

Young, S. Mark et al., 2001, *BLAST OFF! Rockets, Robots, Ray Guns and Rarities from the Golden Age of Space Toys!* Dark Horse Comics, Inc.

1. *Ideal Crawler Tractor in blue with red tracks. Implement set consisting of corn drill, grass cutter, dump-wagon, hay rake and bulldozer blade (tractor 123 mm). All items are Ideal Plastic Corporation originals of models subsequently made by Kleeware.*

Kondor

Company History

In July 1948, an editorial piece (reproduced here in full) appeared in the trade magazine *Games & Toys* describing a new type of clockwork mechanism for toys, developed by G. Oxonford & Co:

'New Clockwork Mechanisms

The "Kondor" spring wire clockwork mechanisms are up-to-date developments in the field of clockwork motors. The "Kondor" utilizes piano or any other gauge spring wire. The wire, which takes the place of the flat spring, is coiled on an expanding driver which acts concertina-wise, opening out lengthwise as the coil of wire is wound up and closing as the spring unwinds. This means that as the coil of wire is relieved of all end tension or pull, it cannot deform and has almost indefinitely long life. Breakage is substantially impossible as a winding detent prevents any risk of overwinding or overstressing the spring. The power output from the spring itself is practically 95 per cent of the available energy put in, owing to the elimination of all losses due to the patented expanding drum on which the wire is coiled. The coil may have any reasonable form and wire of any gauge may be used, making it suitable for all types of spring motors, whether for a cheap mechanism for low priced toys or heavy duty types for such purposes as model locomotives to haul heavy loads. The "Kondor" motor for which world patents are pending is manufactured by G.Oxonford & Co., Fulwood House, Fulwood Place, High Holborn, W.C.1 who will be pleased to supply details of this new and interesting clockwork motor.'

It seems clear that Oxonford decided that they needed to produce a toy to show the capabilities of their new patented design of clockwork motor, and the Kondor tractor was the result. This was a heavy zinc diecasting, which no doubt the heavy-duty Kondor motor was quite capable of powering, but it did not make the most of the strength of zinc castings which could be much thinner and lighter. Also, while the Kondor spring may have been unbreakable, surviving models are often let down by the gearing failing to operate correctly. As far as we know, there were no more Kondor toys after this one, although some other manufacturers did use the Kondor motor, for example there was a plastic roadster made by Rovex Plastics Ltd marked *"KONDOR" SPRING* underneath.

Model

The Kondor Tractor

The diecastings for the tractor were made by sub-contractors (since Oxonford was not a diecasting firm), and it is known that the wheels were made by Kemlows Die Casting Products (see the Master Models chapter), because Kemlows' accounting ledgers for the period have survived. In July 1948 the accounts record that Kemlows charged G. Oxonford & Co. £60 for a die, and paid the same amount to Shindler Engineering Co. Ltd of Hatton Garden, London EC1 who were the actual makers of the die that would be run on Kemlows' diecasting machines. This was probably a single die with two or four cavities for the different front and rear wheels on the tractor. Six sample sets of wheels were supplied to Oxonford in September 1948, but the accounts record nothing further for over a year until October 1949, when Oxonford paid five guineas for some alterations to the die. Deliveries of the finished castings commenced in January 1950 and continued till June of that year. This gives us a precise date of issue for the tractor of 1950. A total of 276,800 wheel castings were invoiced at £3 per thousand. Assuming that Kemlows made both the front and rear wheels of the tractor, and therefore four of their castings were required per model, this means that Oxonford could have made a maximum of 69,200 tractors. This quantity seems entirely reasonable and is consistent with other toys where details of production numbers are known.

KONDOR and *PAT No.8194/48* were cast underneath the tractor. It was painted yellow, orange, red or light green with gold trim on the grille and engine and black on the seat. The wheels were black with rear wheel centres matching the body colour, and the steering wheel was a sheet brass pressing. No driver was provided. The toy was packed in an end-opening box with red and black printing (**1** to **6**).

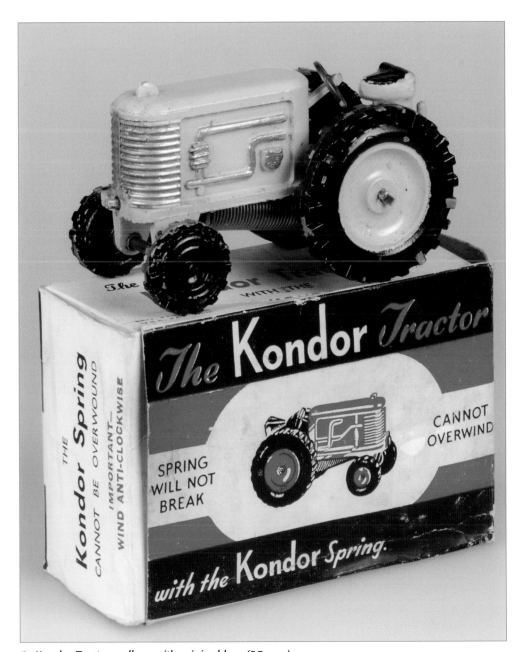

1. Kondor Tractor, yellow, with original box (95 mm).

2. Kondor Tractor, as (*1*) but orange.

3. Kondor Tractor, as (*2*) but red, with original box.

4. *Kondor Tractor, as (**3**) but light green.*

5. *Kondor Tractor, rear view of the tractor in (**4**).*

6. *Kondor Tractor, underneath view showing the coiled spring clockwork mechanism.*

Kraftoyz

Company History

Kraftoyz was recorded in telephone directories from 1946 to 1950 at "Merriechest", Park Lane, Saffron Walden, Essex, and our knowledge of the output of this manufacturer is so far limited to two wooden tractors with tinplate wheels.

Models

The two wooden tractors we have found (**1** and **2**) were plain with little decoration save for transfers on both sides of the engine block reading *KRAFTOYZ SAFFRON WALDEN*. The background to the lettering was gold surrounded by a red frame, and the word *KRAFTOYZ* was in red. *SAFFRON WALDEN* underneath was in a smaller black type. The radiator grille was a transfer with black lines on a silver background placed either horizontally (**1**) or vertically (**2**). The steering wheel was a painted disc of wood. The exhausts were both wooden dowels set into the bonnet and painted the same colour. The radiator cap was a short dowel in one and a nail in the other. The seat was also of wood, the same colour as the bodywork.

The tinplate wheels were fixed on with nails. The front axle could be rotated on its one fixing nail. The back wheels appear to be of the same size and manufacture as the front tinplate wheels on the Dragon Toys wooden tractor (see the Dragon Toys chapter). There was no tow hook. The price of *2/5d* was written underneath (**2**) in pencil, quite expensive for such a basic item.

1. *Kraftoyz wooden tractor painted orange with a yellow face to the steering wheel, yellow tops to the exhaust and to the radiator cap, which was made with a nail, and blue tinplate wheels. The transfer for the radiator grille had horizontal lines (103 mm).*

2. *Kraftoyz wooden tractor painted light blue with a red face to the steering wheel and yellow tinplate wheels. The transfer for the radiator grille had vertical lines. The radiator cap was made from a short dowel. The nail attaching one of the rear wheels has been replaced by a screw, and the exhaust is broken.*

Company History

Leeway was a trade mark of a company named Patterson Edwards. Although no details are known of the early years of the company, it is thought to have been established in 1892 and was based in premises on Lee High Road, Lewisham, London SE13. By 1919, Patterson Edwards were considered to be one of the major U.K. wheeled-toy makers. It was held in the same esteem as G. & J. Lines – a contemporary London-based company that made a similar range of products (see the Tri-ang chapter) – and was one of the founding members of the Toy and Fancy Goods Federation in April 1923, sitting on its representatives council.

Although the company could not match the expansion of Lines Bros (see the Tri-ang chapter) which had 1,000 employees in 1931, by 1925 Patterson Edwards had 250 employees, and 300 by 1931, although these numbers fluctuated depending on the season.

A government report into the perambulator trade of 1924-25 found that in 1923 Patterson Edwards made 458 doll's prams for every employee, compared with 181 each for G. & J. Lines and 625 each for Lines Bros. Interestingly, the report found that Patterson Edwards' products, including its pedal cars, were at the cheaper end of the toy market, and its rocking horses were also inferior in quality to the products of the two firms owned by the Lines family.

Originally the company's products were all wooden, but gradually the use of metal increased, mirroring the evolution of products at Lines Bros. At the British Industries Fair in 1929, Patterson Edwards was listed as a manufacturer of 'Perambulators, Steel Folders, Invalid Chairs, Toy Prams, Toy Motor Cars, Toy Bicycles and Tricycles, Horses, Wheelbarrows, and Strong Toys generally' (we presume toy motor cars to have been pedal cars, and horses to have been rocking horses).

Around 1925, Patterson Edwards moved into Hurst Lodge in Lee High Road and built a factory in the grounds, named the Manor Works. Patterson Edwards survived the Second World War – although it is not known what the company produced during the conflict – and exhibited at the British Industries Fair in 1947. The Leeway trade mark is thought to have been registered in 1955, but for the 1947 British Industries Fair, Patterson Edwards advertised in the Toys and Games Section as 'Manufacturers of "Leeway" Baby Carriages, Folding Baby Cars, Push Chairs and Invalid Chairs. Wheeled and Mechanical. Metal and Wood Toys. "Supreme in Quality, Foremost in Design".'

The 'Lee', in Leeway probably referred to the company's premises on Lee High Road, so it is quite possible that the Leeway trade mark was registered prior to the Second World War, but until further information comes to light

it is impossible to be precise. Note that the 'L' was actually a £ sign.

During the 1950s, the company started production of large, generic, pressed-steel toy vehicles such as panel trucks and flat-bed lorries, and additionally pedal cars and floor toys such as trains. Patterson Edwards' catalogues from the 1960s showed a broad range of its traditional wooden rocking horses, in addition to products based on steel tube, such as children's desks and chairs, toy scooters, trikes and garden toys, as well as the large pressed-steel toy ranges of pedal cars, baby carriages, and toy vehicles.

The company moved to new premises in Orpington in 1970, but it is not known if this reflected the expansion or contraction of the business. However, by the mid-1970s Patterson Edwards had closed. Intriguingly, we have found a

Patterson Edwards 1949 British Industries Fair advertisement.

toy in packaging marked *LEEWAY-SELCOL PLASTICS DIVISION*. The toy was a plastic guitar very similar to other musical instruments sold by Selcol (see the Fairchild chapter), so perhaps the Leeway brand was bought by Selcol after the demise of Patterson Edwards.

Model

Wooden Horse and Cart

Patterson Edwards' Leeway wooden Horse and Cart (**1** and **2**), at 18 inches (460 mm) long, was probably intended to hold a large doll. The cart was made of varnished pine with a plywood base, and had heavy side pieces to which the side panels were nailed through the plywood base. The side pieces also supported a cross member – to which the metal axle was held by heavy-gauge staples – and the shafts. It had a top-rail at the front and a separate tailboard held by wooden pegs on chains that slotted into metal brackets on each side (**2**). Across the middle of the tailboard was the Patterson Edwards logo of *£eeway REGD. MADE IN ENGLAND*. The scrolling tails of the letters *L* and *Y* enclosed the company's initials of *P* and *E*. The wheels were spoked spider wheels with rubber tyres, similar to the wheels that Patterson Edwards used on their larger baby carriages.

The horse was wooden, with a hollow body made from sheet wood and solid wooden legs, neck and head. It stood on a wooden platform, painted green, which had four small metal disc wheels with rubber tyres. The horse was painted a dapple grey with black hooves. The bridle and collar were detailed in red, and it had a mane and tail made of sheepskin. There was a leather pad tacked to its back, and leather straps both over and under the horse. The straps were permanently fixed to the shaft on one side but could be unclipped from the other shaft, and this allowed the cart to be tipped. Small chains were attached between the front of the horse's body and the shafts each side to allow the horse to pull the cart along.

The horse and cart were very similar to pine horses and carts made by Lines Bros over a similar period from the 1920s to the early 1960s. Because both cart and horse were on metal wheels with rubber tyres we assume that they were one of Patterson Edwards' more expensive models, as the cheaper models were likely to have had solid wooden wheels. We know that Lines Bros also made a range of products that used metal wheels with rubber tyres, or solid wooden wheels, depending on the price of the horse and cart. But anecdotal evidence suggests that Leeway products were always considered inferior to the Lines Bros equivalents. It is difficult to date the horse and cart, because, although it was probably manufactured in the 1950s, the design had not changed significantly from the 1920s. It is likely that the wooden models were phased out by the company during the late 1950s, and Patterson Edwards' catalogues from the 1960s show their range of metal toys, pedal cars and baby carriages, in addition to wooden rocking horses.

Further Reading

Brown, Kenneth D., 1996, *The British Toy Business, A History Since 1700.* The Hambledon Press.
Fawdry, Marguerite, 1990, *British Tin Toys.* New Cavendish Books
Grace's Guide to British Industrial History website: www.gracesguide.co.uk.

1. *Leeway Horse and Cart. Wooden cart on spoked metal wheels with rubber tyres. Pulled by a hollow-bodied horse on a wooden platform, set on metal wheels with rubber tyres (460 mm).*

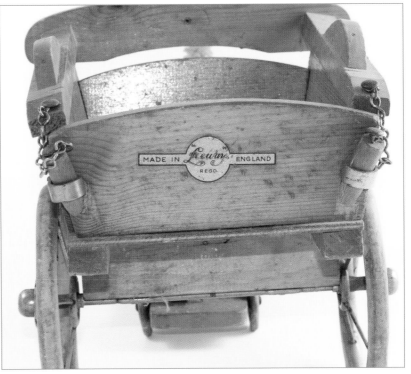

2. *Rear view of (**1**) showing the tailboard and the Patterson Edwards logo.*

Company History

The story of how Lesney Products started has been re-told so many times that it has become legendary. On returning to civilian life after the Second World War, Rodney Smith worked for Die Casting Machine Tools Ltd in Palmers Green, north London. D.C.M.T. manufactured diecasting machines but were also getting involved in toy production (see the D.C.M.T. chapter in Volume 1). Rodney and Leslie Smith (not related) used their service gratuities to buy a D.C.M.T. machine and set up on their own as diecasters in 1947, choosing the name Lesney Products by combining their first names. Their premises were a disused pub called The Rifleman, at 2 Union Row, Upper Edmonton, London N18. This street no longer exists but it was a turning on the east side of Fore Street, on the boundary between Upper Edmonton (N18) and Tottenham (N17).

Jack Odell was another D.C.M.T. employee who wanted to start business on his own account, and initially he rented space at The Rifleman from the Smiths, but it was not long before he joined them as a partner in Lesney Products. The first toys were made in 1948 as a sideline to the main business of diecasting industrial components.

From the first, Lesney produced high-quality diecastings, and their toys very rarely suffer from metal fatigue, unlike those of so many of their competitors from this era. The toys found a ready market with wholesalers and stores such as F.W. Woolworth, at a time when Dinky Toys were only available from appointed retailers.

In 1948 or early 1949 the firm moved to 1A Shacklewell Lane, Dalston, London E8, starting their long association with the Hackney area which lasted until the 1980s, and Lesney Products & Co.Ltd was incorporated as a limited company on 9 March 1949. Rodney Smith left the company around 1951, but in 1954 he set up a diecasting operation for Morris & Stone which in later years produced Budgie Toys (see the Budgie chapter in Volume 1). Lesney's big break occurred when they produced a miniature Coronation Coach in 1953, of which over a million were sold, paving the way for a new range of miniature models in 1954 – the Matchbox Series – which enjoyed phenomenal success.

The Coronation Coach, the Matchbox Series and the large Massey-Harris tractor were distributed in the U.K. market exclusively by J. Kohnstam Ltd, whose trade mark was Moko, hence the inclusion of the Moko name on some Lesney models and boxes. Kohnstam also put their Moko brand on various toys by other manufacturers, including some farm toys (see the Moko chapter).

Early Lesney Toys is the term given to Lesney's toy production from

before the Matchbox era, but also covers some items which were produced concurrently with the early Matchbox range. The Matchbox chapter deals with the company's later history.

Models

Early Lesney farm subjects consisted of just two models.

Caterpillar Tractor

This model was introduced in 1948, based on the real Caterpillar RD8/D8 tractor which had been introduced in 1935. The Lesney model was also available as a bulldozer version, but bulldozers are outside the scope of this book. A close copy was produced by Benbros (see the Benbros chapter in Volume 1), but the Lesney model can be distinguished by the open slats in the radiator grille (closed on the Benbros). Also the Benbros version was clearly marked with the maker's name underneath, whereas the Lesney model had no identification. The Lesney tractor has been found in five colours: yellow, dark green, orange, light green and red (**1** to **5**). In addition there was a rare version of the bulldozer in metallic blue, so maybe the tractor also exists in this colour. The Benbros tractor and bulldozer are quite common in metallic blue, but it is certain that the Lesney bulldozer in this colour does exist and is genuine, not just a mis-reporting of a Benbros model. Black rubber tracks were fitted, and Jack Odell is quoted as saying that the bought-in tracks cost more than the model itself! The separate driver came in six distinct colour shades which are shown in photo (**6**). The Caterpillar tractors were not individually boxed, but were supplied in plain card boxes, probably containing six items in each. In 1950 the bulldozer version was modified for use in a set with a prime mover and trailer, and the tractor was discontinued.

Massey-Harris 745D Tractor

This impressive model was launched at the 1954 toy trade fairs, concurrently with the Matchbox Series, which included (as no.4) a miniature version of the same tractor (see the Matchbox chapter). Advertising to the trade included a double-page spread in *Games & Toys* April 1954 contrasting the 'Dignity' of the large tractor with the 'Impudence' of the Matchbox models (**7**). The large tractor had *A LESNEY PRODUCT* and *MADE IN ENGLAND* cast inside the rear mudguards. There were no major variations of the model, although the red colour can vary in shade to a dark orange. Details were picked out in white and gold trim. The wheel hubs, front axle and steering linkage were painted cream, and the tyres were black rubber, with *MADE IN ENGLAND BY LESNEY* moulded on the rear tyres (**8** and **9**).

There were three versions of the box for this model – the first two differed only in the illustration of the tractor (**10** and **11**). The third box described it as *LESNEY "MAJOR-SCALE" SERIES No.1.* and dates from 1955 (**12**). It was intended to produce a series of large vehicles, of which the tractor would be the first, and some prototypes of other models were made, but the success of the Matchbox Series and the relative profitability of selling small toys in huge volume meant that the Major Scale Series was

abandoned and did not go beyond no.1. The tractor was still available in 1958, although it was never included in Matchbox catalogues. All three types of box had a similar rear face with white lettering on a red background (**13**), and a minor variation is that some boxes can be found with a sticker on the front reading *MOKO SOLE DISTRIBUTORS*, indicating distribution by J. Kohnstam Ltd.

Further Reading

McGimpsey, Kevin and Orr, Stewart, 1989, *Collecting Matchbox Diecast Toys The First Forty Years*. Major Productions Ltd.

1. Lesney Caterpillar Tractor, yellow with black rubber tracks (96 mm).

2. Lesney Caterpillar Tractor, as (1) but dark green.

3. *Lesney Caterpillar Tractor, as (2) but orange.*

4. *Lesney Caterpillar Tractor, as (3) but light green.*

5. *Lesney Caterpillar Tractor, as (4) but red.*

6. *Colour variations on the Caterpillar Tractor drivers – (left to right) dark brown, light brown, medium brown, pale brown, grey-brown and tan.*

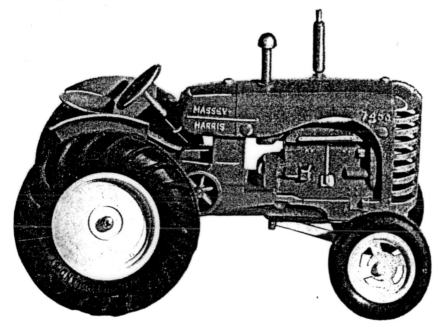

7. *Advertisement from* Games & Toys, *April 1954.*

8. *Lesney Massey-Harris Tractor, red with cream hubs, front axle and steering linkage (200 mm).*

9. *Lesney Massey-Harris Tractor, as (8).*

10. *Lesney Massey-Harris Tractor, first type of box.*

11. *Lesney Massey-Harris Tractor, second type of box with a different illustration from (10).*

12. *Lesney Massey-Harris Tractor, third type of box, mentioning the "MAJOR-SCALE" SERIES.*

13. *Reverse of the box in (10).*

Lipkin

Company History

With no previous knowledge of the toy trade, in 1946 Leon Raphael 'Bobby' Lipkin set up the wholesale company Pritchard Lambert Ltd to sell humming tops, but quickly diversified into selling other toys. Having established himself in this role, he was approached by S. Guiterman & Co. Ltd to act for them as a buyer. Guiterman had been major importers of American toys prior to the war, and needed to restore this business and to develop new business in Europe. Lipkin was quickly promoted to buying director and explored suppliers and markets in both America and Europe. Guiterman soon became the biggest importers of German toys.

In 1954 Bobby Lipkin set up the company Raphael Lipkin Ltd at 52 South Lambeth Road, London SW8, and began to market small plastic toys. The range of toys increased and Lipkin built on his experience to establish contacts with various trade outlets including a number of leading retail chains. Later he became involved with the U.K. manufacturer of a Scandinavian construction toy called 'Playsticks'. Lipkin quickly saw the potential of this toy and agreed to pay royalties so that he could move into producing it himself. It was very successful and in 1955 he introduced a similar toy called 'Build-up'.

He recognised that there was a market for well-designed and packaged quality toys, and having developed a good knowledge of plastics technology, quickly diversified and expanded his business. Early boxes had a logo with *AN LRL PRODUCT* incorporated into a figure of a running clown as in (**13**) (L.R.L. presumably standing for L.R. Lipkin). Later boxes had the *Pippin toys* logo of two apples with faces shown clearly in (**18**). Lipkin also had models made in Hong Kong.

From 1957 to 1969 Raphael Lipkin Ltd was listed in phone directories as Toy Manufacturers at 48-52 South Lambeth Road, London SW8. But in 1969 they were also listed as being at Morden Rd., Merton , London SW19. This address was also the address of Lines Bros.Ltd the makers of Tri-ang toys.

We believe that Raphael Lipkin Ltd was taken over by Lines Bros around 1969. This is supported by the fact that many of Tri-ang's Spacex series of science fiction models had packaging marked *Raphael Lipkin Ltd* as well as *Tri-ang,* or *A Pippin PRODUCT* and *ROVEX Tri-ang.* Furthermore, early boxes for the Nuffield tractor and trailer were marked *Lipkin,* with the *Pippin toys* logo, whereas later *Pedigree Playtime* boxes for the same tractor and trailer were marked *ROVEX Tri-ang.*

Models

Since no catalogues are available, the dates of issue of most Lipkin models are unclear.

Massey-Harris 745 Tractor

This tractor was produced in 1956 but we do not know how long production continued. It was almost identical, apart from the wheels, to a model made in Germany by Kuroco, suggesting that this toy might have originated there.

The model was moulded in plastic of the correct red colour, with yellow hubs and belt pulley, and with black rubber tyres and exhaust (**1**). It had *MASSEY-HARRIS* moulded on each side of the bonnet just in front of the steering wheel and *745* on the bonnet close to the grille. It was steerable via a centre pin from the red plastic steering wheel. It had several interesting details including rear wheel weights, American-style mudguards with lights, a power take-off housing with safety guard (not seen on many toy tractors at the time) and a revolving belt pulley. A simple drawbar allowed implements to be hitched to the tractor. It was sold in a red-and-white illustrated box which carried the message *You can take it apart and build it up again* and a set of instructions was issued (**2**). A six-cylinder inline injection pump was moulded on the left side of the engine. This would denote a six-cylinder engine, and the only Massey-Harris with this engine was the 744, which also had front wheels the same design as those on the model. This suggests that Lipkin badged their model incorrectly. (Lesney appear to have badged their Massey-Harris 745 model incorrectly as well.)

Model Farm Tractor or Countryman Farm Tractor

To extend the use of an obsolete mould, a later version of this model was produced in blue plastic with red hubs and belt pulley (**3**), and the Massey-Harris name was deleted from the sides. It was sold in a colourfully illustrated box, and simply called *model farm tractor*, although when sold with the tipping trailer it was called *countryman farm tractor* (**6**). Both these boxes carried the *Pippin toys* logo.

Massey-Harris 3 ton Trailer

Made to complement the Massey-Harris tractor, this three-ton Ferguson-style tipping trailer was available either on its own (**4**) or in a set with the Massey-Harris tractor (**5**). The red plastic trailer had yellow hubs and black rubber tyres, a tipping lever and an opening and closing drop-down tailboard. A later version was made in red plastic with blue hubs to go with the blue Countryman tractor (**6**).

When sold alone, the tipping trailer came in a white, green and black illustrated box with the *LRL* clown logo (**4**). When sold with the Massey-Harris tractor it came in a white, blue and black illustrated box, again with the clown logo (**5**), and when sold with the Countryman tractor it came in a much more attractive, colourfully illustrated box (**6**), this time with the *Pippin toys* logo.

Lanz Bulldog Tractor

This model of the 1955 Lanz D2416, with a single cylinder diesel engine, originated with the German toy manufacturer Rex who may have licensed production to Lipkin. Alternatively, Lipkin may have acquired the Rex moulds. The model was sold under the name *'PULL-APART' FARM TRACTOR*. All the *PULL-APART* models we have found still have *REX* moulded under the left-hand footplate. The tractor was moulded in red polythene with yellow hard-plastic hubs fitted with black rubber tyres, and a black plastic steering wheel (**7**) or with the seat and lights on the mudguards in yellow (**8**), or moulded in yellow with red hubs and lights (**9**) or with a red flywheel as well (**10**). The moulding was finer than in the Massey-Harris and the model could easily be taken apart and built up again, as indicated on the box. The front axle swivelled, so that the model could be steered, but it was not linked to the steering wheel. It had gear levers which made for a more realistic driving position, and clevis drawbars front and rear. The boxes carried an illustration of the tractor in work, a drawing of the assembled model, and instructions on how to assemble the tractor on the box end. They also carried the *LRL* clown logo.

Four-wheeled Trailer

A four-wheeled trailer in red polythene with yellow hubs and black rubber tyres was made to go with the Lanz tractor, and was sold with it as a set (**11**). The set was in a white, orange and black illustrated box which still carried the *'PULL-APART'* name and the *LRL* clown logo.

Massey Ferguson 780 Combine

Like Corgi, Lipkin produced the tanker version of this combine with the large unloading spout. The model was moulded in plastic of the correct red-and-yellow colours, with *Massey Ferguson* moulded into both sides of the main body (**12**). It had a ground-engaging wheel on the right-hand side connected to the reel, which caused the latter to rotate when the combine was pushed forwards, and the whole cutting head could be raised and lowered. A driver was supplied. The toy could be easily dismantled into 34 parts according to the end-opening box, which carried a colourful illustration of the combine at work. The box end (**13**) carried the name of the model and the wording *Sturdy Polythene Reel revolves Cutting table raises* in black, and *Takes apart into 34 pieces!* in red script. It also carried the *LRL* clown logo, with the clown having a white head.

Nuffield Universal Four Tractor

Produced around 1960 in about 1:16 scale, this was moulded in plastic of the correct red (**14**) or orange (**15**) colours. The main moulding was screwed together. The tractor had black rubber rear tyres with *DUNLOP FIELDMASTER 11-36* moulded round them, realistically moulded black rubber front tyres with *DUNLOP TRACTOR 6.00-16* moulded round them, and a black rubber exhaust. It could be steered by the red plastic steering wheel. The toy featured gear levers, belt pulley, a single headlamp on the right-hand side (an item often missing from surviving models) and a swinging drawbar. Either *NUFFIELD* (**14** and **17**) or *NUFFIELD UNIVERSAL FOUR* (**15**) was printed on either side of the bonnet in black. *BMC DIESEL* was moulded on the grille and filled with gold paint. There were two box variations both with lift-off lids. The first had a colourful picture of the tractor in use in a field setting with *Authentic Scale model of the WORLD FAMOUS NUFFIELD UNIVERSAL FOUR* printed on it and on the box end (**16**) a list of the features of the model: *TOUGH PLASTIC FULLY STEERING MASSIVE DUNLOP 'FIELDMASTER' TYRES CAN BE TAKEN APART AND PUT TOGETHER.* It also carried the *LRL* clown logo, but this time without the white infilling to the head. The second box (**17**) had a picture of the tractor against a black background. The lettering on the box was the same as that on the first except that the tractor was simply called the *NUFFIELD,* and the box end (**18**) no longer suggested that the tractor could be taken apart and put together. It carried the *Pippin toys* logo.

Tipping Trailer (for Nuffield Tractor)

The tractor was sold alone or with a Ferguson-style tipping trailer moulded in yellow plastic with black rubber tyres, which was longer than that sold with the Massey-Harris. The brake lever operated the tipping mechanism on the trailer, which had a drop-down tailboard and a drawbar jack. The set was sold in an end-opening box with an illustration of the tractor and trailer on the side (**19**) and with *Raphael Lipkin Limited*, the *Pippin toys* logo and reference number *1129* on the end (**20**). The set was now advertised as being made of *long life polypropylene*. The tractor and trailer were later sold in a blue *Pedigree Playtime* box with a picture of a child playing with the model (**21**), and with *MADE IN GREAT BRITAIN BY ROVEX Tri-ang* on the end (**22**).

Model List

Model	Dates
Massey-Harris 745 Tractor	1956-?
Model Farm Tractor or Countryman Tractor	1960s
Massey-Harris 3 ton Trailer	1956-?
Lanz Bulldog 'Pull-Apart' Tractor	late 1950s
Four-wheeled Trailer for Lanz Tractor	late 1950s
Massey Ferguson 780 Combine	late 1950s
Nuffield Universal Four Tractor	1960?-69
Tipping Trailer for Nuffield Tractor	1960?-69

Further Reading

'All Sides of the Fence', *Games & Toys*, April 1962, p.42.

'Tri-ang, Rovex, Pippin, Raphael Lipkin, Spacex - Blimey!', *Project Sword Toys blog*, 28 July 2009, website: www.projectswordtoys.blogspot.co.uk/2009/07/tri-ang-rovex-pippin-and-raphael-lipkin.html.

2. Instruction sheet for assembling the Massey-Harris 745.

1. *Lipkin Massey-Harris 745 in a white-and-red illustrated box. Tractor red plastic with yellow plastic hubs and belt pulley and a black plastic exhaust (167 mm)* (photo by Jim Russell).

3. *Lipkin Model Farm Tractor in a colourfully illustrated box. Tractor as (1) but blue plastic with red hubs and pulley wheel and lacking any reference to Massey-Harris.*

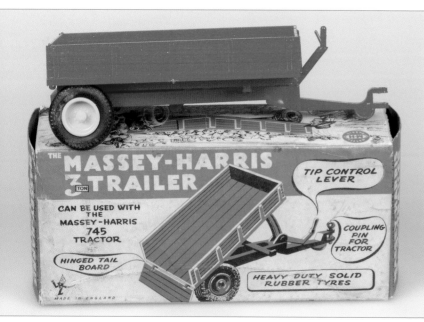

4. Lipkin Tipping Trailer in white-and-green illustrated box. Trailer in red plastic with yellow hubs and black rubber tyres (coupling pin missing) (250 mm).

5. Lipkin Massey-Harris 745 Tractor with 3 Ton Tipping Trailer in white-and-blue illustrated box. Tractor as (**1**) and trailer as (**4**). Tailgate missing from this example.

6. Lipkin Countryman Tractor and Tipping Trailer in colourfully illustrated box. Tractor as (**3**) and trailer as (**4**) but with blue hubs.

7. Lipkin Lanz Bulldog 'PULL-APART' Tractor in red-and-white illustrated box. Red polythene tractor with yellow hubs and Lanz emblem on grille, black plastic steering wheel and black rubber tyres (130 mm).

8. *Lipkin Lanz Bulldog* 'PULL-APART' *Tractor as* (**7**) *but with yellow seat and lights.*

9. *Lipkin Lanz Bulldog* 'PULL-APART' *Tractor as* (**7**) *but yellow with red hubs, Lanz emblem and lights.*

10. *Incomplete Lipkin Lanz Bulldog* 'PULL-APART' *Tractor as* (**9**) *but with red flywheel.*

11. *Lipkin Lanz Bulldog* 'PULL-APART' *Tractor and Four-wheeled Trailer in illustrated box. Tractor as* (**7**) *but missing various attachments, trailer in red polythene with yellow hubs and black rubber tyres (trailer 230 mm).*

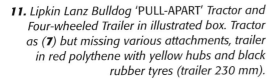

13. *End view of box in (**12**) listing characteristics of the model and showing the LRL clown logo with a white head.*

12. *Lipkin Massey Ferguson 780 Combine in colourfully illustrated box. Red polythene combine with yellow hubs and steering wheel and blue driver (187 mm).*

14. *Lipkin Nuffield Universal Four Tractor in colourfully illustrated box. Red polythene tractor with black rubber tyres (215 mm).*

15. *Lipkin Nuffield Universal Four Tractor as (**14**) but orange tractor.*

16. *End view of boxes in (14) and (15) listing characteristics of the model and showing the LRL clown logo with a black head.*

18. *End view of box in (17) listing characteristics of the model, no longer highlighting that it could be taken apart and put together again, and showing the Pippin toys logo.*

17. *Lipkin Nuffield Tractor as (14) but box with black background and tractor called Nuffield rather than Nuffield Universal Four.*

19. *Lipkin Nuffield Tractor with Tipping Trailer in colourfully illustrated box. Tractor as (14) with all-yellow plastic trailer with black rubber tyres (trailer 280 mm).*

20. *End view of box in (19) showing* made in England by Raphael Lipkin Limited *and with* Pippin toys *logo.*

22. *End view of box in (21) showing* ROVEX Tri-ang *trademark.*

21. *Lipkin Nuffield Tractor and Tipping Trailer as (19) but in* Pedigree Playtime *box.*

Luntoy

Company History

Luntoy was a brand name of the London Toy Company Ltd. Based on Hackney Road, Bethnal Green, London E2, between 1948 and 1965, the company was described as wholesalers and factors in the 1960-61 telephone directory.

While it is known that Luntoy sold wooden toys such as forts, they are usually associated with a range of hollow-cast lead toy figures of TV characters that were actually made for them by Barrett & Sons but sold individually boxed under the Luntoy brand. Luntoy purchased some of the Sacul moulds when that company closed and a range of lead knights was then sold under the Luntoy brand, also made for them by Barrett & Sons, using the ex-Sacul moulds.

Packaging from the 1960s stated that the *LONDON SERIES* of lead figures was *A "LUNTOY" PRODUCT*, and was *MADE IN ENGLAND BY THE MODEL DIVISION OF LONDON TOY CO. LTD. LONDON E2*. The company logo was actually the coat of arms of the City of London, with its motto *DOMINE DIRIGE NOS,* 'O Lord Direct Us' (**1**). The wooden horse and cart we have found is the only example known to us and unfortunately the manufacturer's label on the front of the cart adds no new information to our understanding of this company's product range.

Model

Wooden Horse and Cart

This is the only farm-related toy in the entire Luntoy range that we have found (**2**). From its naïve style it seems to have been aimed at a younger age group than Luntoy's lead toy products. It was a wooden cart pulled by a wooden horse set between shafts. Both horse and cart were on wheels to allow the toy to be pushed along the ground. The horse, shafts, cart and cart wheels were made from plywood, while the front wheels were turned pieces of wood. A small tin bell fixed below the front of the cart would ring as the cart moved. The cart was undecorated apart from a painted single red band; all four wheels were painted in the same red colour, the shafts were painted blue and the horse had a naïve paint scheme of grey body with black hooves, mane and tail. The horse's harness, belly band and saddle were all painted yellow. The shafts were permanently fixed in place to both horse and cart with small nails. On the front of the cart was a transfer stating *LONDON TOY COy. LTD A LUNTOY Product MADE IN ENGLAND* all set in a scroll which was a different logo from that on Luntoy lead toy packaging.

Further Reading

Joplin, Norman, 1993. *The Great Book of Hollow-Cast Figures.* New Cavendish Books.

1. *Luntoy lead soldier packaging, showing company logo and motto.*

2. *Luntoy Wooden Horse and Cart. Grey horse with black mane and tail and yellow harness pulling a brown cart on red wheels (330 mm).*

Company History

Prior to the First World War Louis Marx and his brother David both worked for Ferdinand Strauss of New York, manufacturers of mechanical tin toys. Louis served in the U.S. army during the war and then he and David set up Louis Marx and Co. in New York City in 1919. Initially they acted only as middlemen, marketing toys for other companies, and had no design or manufacturing facilities. However, they soon recognised the opportunity for identifying potential bestsellers, for developing ways of producing them more cheaply, and eventually for producing toys themselves. As early as 1922 they were in a position to buy patents and dies for Strauss animated toys, improve them and sell them successfully.

Another example of their success was the yo-yo, which they did not invent but which, having recognised its potential, they produced and marketed, eventually selling many millions. Louis had the drive and the design capability and David the operational expertise. Much of their success was due to their focus on quality and value.

Despite the recession they expanded rapidly during the 1930s, and in 1932 set up the U.K. subsidiary company Louis Marx & Co. Ltd. In that year they also incorporated Lumar Ltd, which was a trade name used by Marx. A factory was set up at the Waddam's Pool Works, Hall Street, Dudley, Worcestershire, in part of the former Bean Motor Works. The company had entries in telephone directories for Waddam's Pool Works from 1932 to 1947. The U.K. had become an important market for Marx products, and the Dudley factory allowed them to by-pass the 25 per cent duty imposed in 1932 on imported toys. Initially the U.K. company produced American-type trains, but a wider range of toys soon went into production including a clockwork climbing tractor.

During the Second World War, Marx companies supplied military equipment but after the war they returned to toy manufacture. As part of this, in 1948 the company moved U.K. production to a site, provided by the Government, on a new industrial estate established to help regenerate Swansea in South Wales. Telephone directories show Marx on Ystrad Road, Fforestfach Trading Estate from 1946 until 1981. They also had a London office at 16 Berkeley Street from 1949 until 1967.

Overall, Marx soon became the largest tinplate toy manufacturers in the world, expanding rapidly in the 1950s with factories in ten countries producing more than 5,000 different products. Very few of their farm-related toys were produced in Britain, and so most are outside the scope of this book. However, we do know that clockwork tinplate horses and carts and some tractors were produced in Swansea in the early 1950s,

but we have not been able to find any advertising or other published information to provide clear dates for their introduction.

In 1967, Marx sold the U.K. subsidiary to Dunbee-Combex for £1million (Brown 1996, p.179) which thereafter took the name Dunbee-Combex-Marx Ltd. In 1972 Marx sold the American company to Quaker Oats Co. After struggling to make Marx pay, Quaker Oats sold it, together with Fisher-Price, to Dunbee-Combex-Marx, who then continued to run Marx on both sides of the Atlantic until they were bankrupted in 1980. American Plastic Equipment of Florida acquired the Marx assets in 1982, resurrected the Marx name and re-issued many Marx toys, stimulated much by the demand from collectors. In 1995 a separate toy company, Marx Toy Corporation, was formed. So the Marx tradition has continued – but, unfortunately, not in Britain.

Louis Marx was a very successful innovator and manufacturer of toys over many years. His impact was recognised when, in 1985, he became the first person to be inducted into the American Toy Industry Hall of Fame, which recognises the contributions made by toy makers around the world, with a plaque which described him as 'The Henry Ford of the Toy Industry'.

The Marx Logo

The company trade mark was a circle with *MAR* superimposed on the centre of an *X* with *TOYS* underneath. The boxes for items made in Dudley had *MADE IN ENGLAND* below (**1**). But when the company moved to Swansea the reference to England was replaced with *MADE IN GT. BRITAIN* (or, less frequently, *MADE IN GREAT BRITAIN*) around the circumference (**2**). It is interesting that the company chose not to mention that their toys were made in Wales. The logo was printed on the boxes and on the tinplate models. On the horses and carts the logo was found on the underside of the horses. The large crawler tractors tended to have the logo at the rear above the tow hook, while the smaller versions and the wheeled tractors had the logo printed on the base. The exception was the Tricky Tommy tractor, on which *LOUIS MARX & CO. MADE IN GT. BRITAIN* was moulded into both black rubber front wheels and there was no logo.

On the final box in the sequence a little man was added incorporating another version of the logo as well (**26**).

Models

Climbing Tractor, Van and Stake Truck

The first British-made Marx tractor was a clockwork climbing tractor with white rubber tracks made in the 1930s at Dudley. One version in silver and red was sold in a set with a green four-wheeled van and a red stake truck, both of which had articulated front axles, tinplate wheels and white rubber tyres (**3** to **5**). The Marx logo was stamped above the tractor grille, and on the back was *LICENSED UNDER PATENT No. 1,334,539.* The van and stake truck had *LEP TRANSPORT* transfers on each side. The set was sold in a box (**3**) with cut-out cards to hold the various items, and it was illustrated in red and green on a white background. The wording on the box was *CLIMBING TRACTOR VAN AND STAKE TRUCK MADE IN ENGLAND BY LOUIS MARX & Co. LTD DUDLEY WORCS ONE OF THE MANY MARX TOYS – HAVE YOU ALL OF THEM?* This was the same wording later used on the Dual Horse & Cart.

Another version of this set, in the same box, had a bare metal tractor and the van and stake truck, which were without transfers, had black rubber wheels with *GIRARD U.S.A. BALLOON* moulded round them (**6**). Although the set was made in Dudley, the wheels were obviously imported.

Tinplate Horses and Carts

As far as we can ascertain, the tinplate clockwork horses and carts were all produced in Swansea in the 1950s. Early ones were small, tumbrel-type carts pulled by a single horse, with the clockwork motor under the cart. Two small wheels on the nearside foreleg of the horse allowed it to be pushed along. The horses were printed with a complete set of decorative cart harness, but the carts and wheels were very plain, and printed only in a simple, striated yellow-brown colour. The key was located in front of the right-hand wheel. Some were sold without a driver (**7**). Others were sold complete with seat and driver wearing a yellow smock and hat, holding string reins (**8**). These seated drivers had a very oriental appearance in the face and clothing. The models came in simple brown card boxes with *Mechanical Horse* printed on both sides in red and with an illustration of the horse and cart on the top. *LOUIS MARX & Co. Ltd. Swansea* was in larger letters on the side and the logo was on the end flap.

In later versions the cart was re-designed to a larger, more conventional rectangular shape with perforations stamped through the cart sides and back (**9** to **12**), the first probably being the cart with a single horse (**9**). Marx had made the same perforations in the 1930s on the four-wheeled stake truck towed by the early climbing tractor (**5** and **6**). The clockwork winder was now positioned behind the right-hand wheel (**9**). The horses retained decorative harness, and the wheels were given a chequerboard design, with *MARX SUPER* printed round the circumference. They all had seats and drivers.

The most adventurous of this series of toys was the so-called Dual Horse & Cart, which was the later cart, with driver, pulled by a pair of horses, and sold in a box with a colourful illustration (**10** and **11**). These carts have been found in blue and red (**10** to **12**). Printed on the box was *Dual HORSE & CART* in (**10**) and *DUAL HORSE & CART* in (**11**). It was numbered *423* on (**10**) but not on (**11**). The earlier box also carried *"One of the many MARX toys – have you all of them?"*

Later Tinplate Tractors

The clockwork climbing tractor of the 1930s was followed in the early 1950s by a smaller red-and-yellow printed tinplate tractor (**13**) made at Swansea. It had black-and-white tinplate rear wheels (also used on the Marx no.410 racing car) and one-piece rubber front drive-wheels with silver metal hubcaps also used on various cars. Moulded on the rubber tyres was *LOUIS MARX Co MADE IN GT. BRITAIN.* A tinplate clockwork crawler was also produced (**14**), basically the same as the previous tractor but with metal wheels and black, tapered-section, narrow, rubber tracks. This tractor was based on the original American Marx Midget Climbing Tractor except that the key was removable. Both toys were fitted with a push-in driver, a stop/go lever and a removable key fitted on the right-hand side. The crawler was also produced in a long-wheelbase version (**15**). The short-wheelbase version of the crawler was later produced with plastic wheels and flat-section, wide rubber tracks, either with the winding mechanism on the right-hand side (**16**), or on the left-hand side (**17**). This second version had a smaller key. The tow hooks on all these crawlers were simply hook-shaped sections of the rear panel stamped and bent out.

In the mid- to late 1950s Marx also produced a large green tinplate, clockwork crawler with a sparking action (called *Sparkling* by Marx) out of the red plastic exhaust pipe (**18**), based on a similar toy produced in America. This was printed in green, cream and red, with a push-in driver and a removable front-mounted bulldozer blade, described on the end flap of the box as a plough. This set was sold in a yellow end-opening box with illustrations of the tractor, and lettering in red and black. The box had *Sparkling CLIMBING TRACTOR STURDY CONSTRUCTION FOR PLAYTIME • HAULING • LOGGING • FARMING • BUILDING* (**18**).

A later rarer version was produced in yellow and red, with a revised printed tinplate pattern and with *MARX 5* printed on either side of the bonnet. These changes were not reflected on the box, which was green with grey-and-red illustrations, and red-and-green lettering (**19**).

A further version of the yellow crawler was powered by two batteries housed in a separate hand-held plastic control box (**20** and **21**). There were separate buttons on the control box for forward and reverse travel. The colourful and well-illustrated end-opening box described the toy as a *Remote Controlled Battery Operated ELECTRIC TRACTOR CLIMBS, STOPS AND REVERSES BY Remote Control.*

Tricky Tommy Tractor

In 1956 the press in America were extolling the 'Electronic Age of Toys'. Among the battery-operated toys listed (*Time* magazine, December 1956) was the Marx *ELECTRIC POWERED FARM TYPE TRICKY TOMMY, "The Big Brain" TRACTOR* with its bump-and-turn action. It travelled along until it met an obstruction, and then automatically reversed and turned in a new direction until it met another obstruction. This was described on the box as *"MYSTERIOUS • AUTOMATIC • STEERING ACTION!"* The Swansea-built version did not become available until 1958, and this had a red or green plastic body with yellow tinplate rear hubs and rubber front wheels. The motor was encased in thick tinplate. *REVERSIBLE DIESEL ELECTRIC* and *PAT. NO. 2587082* were moulded on both sides of the bonnet, and *TOOL BOX* over the battery holder on both sides of the tractor. It was sold in an end-opening box with a red illustration and blue, white and red lettering (**22** and **23**). More common later versions of this tractor were orange or blue and had plastic rear wheels. These were made in Hong Kong and so are outside the scope of this book.

Plastic and Tinplate Mechanically Operated Tractor

Two further Swansea-made tractors, possibly the last of the line, were clockwork but this time had the mechanism encased in a tinplate chassis with the tractor body and push-in driver made of plastic. The origins were American, but the U.K. version was again modified to take a detachable key. One version was a red crawler bulldozer with a detachable blade which is outside the scope of this book. The other was a blue, wheeled version (**24**) based on the crawler, but with Jeep-style front wheels and large black plastic rear wheels as used on the Marx Howitzer Cannon. It had rear mudguards glued to the footplates and a hole drilled in the rear of the air filter housing to take a steering wheel to replace the two levers on the footplate. It towed a Jeep-type trailer, in orange (**24**) or yellow (**25**) and had the same driver as the crawler. Both versions had a stop/go lever, and a printed tinplate panel showing engine detail, which was subtly different from that on the American version. The sparking mechanism fitted to the crawler was not incorporated into the wheeled version, but the detail was

of a spark ignition engine, although moulded into the plastic body on the left hand side was the word *Diesel*.

The yellow end-opening box featured both blue and red tractors described as *TRACTOR and TRAILER MECHANICALLY OPERATED.*

Model List

Serial No.	Model	Dates
	Climbing Tractor, Van and Stake Truck set	1930s
	Mechanical Horse clockwork horse-drawn cart with single horse, with or without a driver	1950s
	Clockwork large horse-drawn cart	1950s
423	Dual Horse & Cart clockwork large horse-drawn cart with two horses	1950s
	Clockwork short-wheelbase tractor	1950s
	Clockwork short-wheelbase crawler tractor	1950s
	Clockwork long-wheelbase crawler tractor	1950s
	Sparkling Climbing Tractor, large tinplate, clockwork crawler tractor with sparking action	Late 1950s
	Electric Tractor, large tinplate, battery-driven crawler tractor with remote control	Late 1950s
	Tricky Tommy battery-driven tractor	Late 1950s
	Mechanically Operated Tractor and Trailer, plastic/tinplate, clockwork tractor and plastic trailer	Late 1950s

Further Reading

Brown, Kenneth D., 1996, *The British Toy Business, A History since 1700.* The Hambledon Press.

Virtual Marx museum website: www.marxmuseum.com.

Louis Marx Swansea factory: www.knightsandvikings.co.uk/factory.htm.

*1. The early Marx logo on the boxes of the two Dudley-made climbing tractor sets in (**3**) to (**6**).*

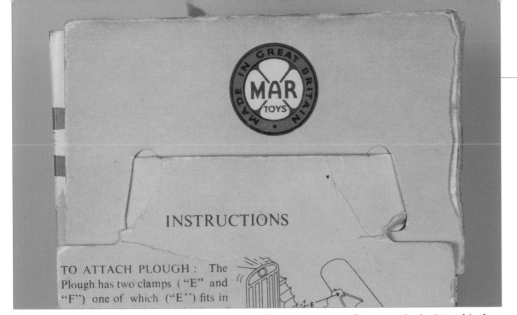

2. *The Marx logo on the end flap of the box for the Swansea-made tractor in (**19**). In this, less usual, version of the logo the* GREAT *of* GREAT BRITAIN *was not abbreviated to* GT. *but spelled out in full.*

3. *Marx Climbing Tractor, Van and Stake Truck set in a box illustrated in green, red and white with the lid marked* MADE IN ENGLAND BY LOUIS MARX & Co. LTD DUDLEY WORCS.

4. *Marx climbing tractor and van in (**3**). A metallic silver-and-red clockwork tinplate crawler on white rubber tracks with push-in driver and brake lever to the left of the driver. A green tinplate, four-wheeled van with white rubber tyres and* LEP TRANSPORT *transfers on both sides. The white tracks on the tractor have been borrowed from another toy for this picture, but we think they are the correct type (tractor 215 mm and van 215 mm).*

5. *Marx climbing tractor and stake truck in (**3**). The stake truck again with* LEP TRANSPORT *transfers on both sides on white rubber tyres (truck 200 mm).*

6. *Marx Climbing Tractor, Van and Stake Truck set similar to* (**3**) *to* (**5**) *but the tractor in bare metal and the van and stake truck without transfers and with black rubber wheels and a green drawbar. The tractor as illustrated is on replacement tracks and is missing a driver.*

7. *Marx tinplate clockwork Mechanical Horse with horse and small striated yellow-brown cart with a seat, but no driver, and a wheel either side of the nearside foreleg of the horse (240 mm).*

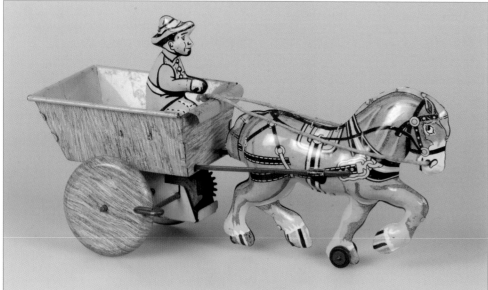

8. *Marx tinplate, clockwork horse and cart as* (**7**) *but with more yellow in the exterior striations and the interior all-yellow. The seat had a slot and a push-in driver.*

9. *Marx tinplate clockwork horse and cart with a large red cart and a highly decorated horse. There was a slot in the seat, but the push-in driver is missing from this example. MARX SUPER was written within the chequerboard pattern on the wheels (280 mm).*

10. Marx tinplate, clockwork Dual Horse & Cart in a colourfully illustrated end-opening box. The cart as (**9**) but dark blue and pulled by two horses with a driver as in (**8**) but more colourfully printed. An axle ran through the forelegs of the two horses with a wheel on each end (310 mm).

11. Marx tinplate clockwork Dual Horse & Cart as (**10**) but with a stylistically later design on the box.

12. Marx tinplate, clockwork Dual Horse & Cart as (**10**) but with a red cart.

13. Marx tinplate, clockwork tractor in red, black and yellow printed tinplate, with driver, rubber front wheels with silver metal hubs driven by the motor, black-and-white chequerboard tinplate rear wheels with MARX SUPER printed round the circumference, and a brake lever to the right of the driver (134 mm).

14. Marx tinplate, clockwork crawler tractor in red, black and yellow printed tinplate, with metal wheels with narrow black rubber tracks; the short-wheelbase version (125 mm).

17. Marx tinplate, clockwork crawler tractor, as (**16**) but with the key opening on the left-hand side. The key was of a smaller type.

15. Marx tinplate clockwork crawler tractor, as (**14**) but the long-wheelbase version (134 mm).

16. Marx tinplate, clockwork crawler tractor, as (**14**) but with plastic wheels and broad black rubber tracks. The key opening was on the right-hand side.

18. Marx Sparkling Climbing Tractor, tinplate, clockwork, sparking crawler tractor printed green, white and red with red plastic exhaust, black plastic wheels and black rubber tracks. The brake lever was to the left of the exhaust (215 mm).

19. *Marx tinplate, clockwork, Sparkling Climbing Tractor crawler tractor, as (**18**) but in yellow, red and black printed tinplate in a revised pattern not reflected on the illustrated, end-opening box.*

20. *Marx tinplate, Remote Controlled Electric Tractor. Tractor as (**19**) but complete with hand-held battery-operated control mechanism which allowed the tractor to climb, stop and reverse, in a new illustrated end-opening box (215 mm).*

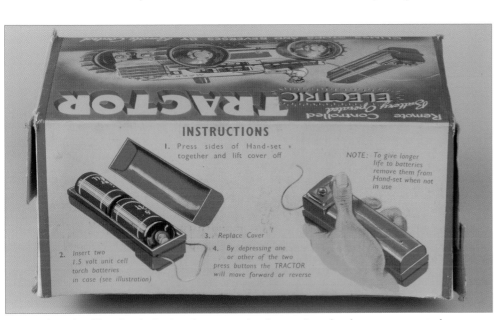

21. *Underside of box for (**20**), showing operating instructions for the remote control.*

22. *Marx Tricky Tommy "The Big Brain" Tractor with "MYSTERIOUS • AUTOMATIC • STEERING ACTION!" A battery-powered red, yellow and silver plastic tractor with black rubber front wheels and tinplate rear wheels, plastic driver, and switch for the battery on the bonnet. A highly illustrated and lettered box. (250 mm).*

24. *Marx Mechanically Operated Tractor and Trailer. Clockwork tractor with blue plastic body, printed tinplate chassis, black plastic wheels, yellow push-in driver and removable metal key, towing an orange plastic Jeep-type trailer with black plastic wheels in an end-opening illustrated box (tractor 155 mm and trailer 190 mm).*

23. *Marx Tricky Tommy tractor, as (22) but with green replacing the red.*

25. *Marx Mechanically Operated Tractor and Trailer as (24) with a yellow plastic trailer (driver missing in this example).*

26. *Marx logo incorporating the figure of a little man on the boxes for (24) and (25).*

Master Models (Kemlows)

Company History

The Master Models series of OO scale railway accessories was manufactured by Kemlows Die Casting Products and distributed by B.J. Ward Ltd. Kemlows was founded in 1946 by Charles Kempster and his brother-in-law William Lowe. A third partner, Ron Jackson, had worked as a toolmaker for Die Casting Machine Tools Ltd, but left the partnership around the end of 1948. Their premises were at 60 St John Street, Smithfield, London EC1. Later they were at 79a Westbury Avenue, Wood Green, London N22, and from 1958 moved to Potters Bar in Hertfordshire. They started making lead soldiers, but soon purchased a D.C.M.T. M55 diecasting machine to make their first diecast toy, an armoured car towing a field gun. Various other toys quickly followed, and Kemlows also undertook sub-contract casting, including making the hands for Pelham Puppets and wheels for the Kondor tractor (see the Kondor chapter). We know this because their account books from 1946 to 1952 have survived, and are a fascinating record not only of what was made, but of quantities produced and names of suppliers and customers.

A series of farm items was introduced from late 1948, unfortunately not including any vehicles. These were a gate, fence, stile, orchard ladder, bridge, swing gate, signboard, children on seat, windmill, rabbit hutch and hurdle (**1**).

H.R.H. The Duchess of Gloucester and Prince Richard inspecting "Mastermodels" on Messrs. B. J. Ward, Ltd.'s stand at the 1955 B.I.F. at Olympia.

Illustration from the 1955 Master Models catalogue showing the B.J. Ward Ltd stand at the 1955 British Industries Fair. The gentleman at the right of the picture with the round glasses is Bertie Ward. (Reproduced courtesy of West Wales Museum of Childhood.)

All were zinc diecastings, not lead. The signboard had a paper label, similar to the decoration used later on the Master Models, and we have noted the following: *Neale's Farm, Beware of the Bull, Beware of the Dog, Public Footpath, No Fishing, No Camping* and *Private.* The rabbit hutch in particular is often erroneously attributed to Crescent. The accounts record that the 'final lot' of farm pieces was sold in January 1951 to B.J. Ward Ltd, and in the previous month Ward had taken delivery of the first of the railway accessories.

B.J. Ward Ltd also started in business in 1946, and from 1951 was at 130 Westminster Bridge Road, London SE1. Bertie John Ward had previously worked for many years for the toy wholesalers S. Guiterman & Co. Ltd as a sales representative. The early railway accessories were not exclusive to Ward, but clearly he saw the potential for a series of models and made an exclusive arrangement with Kemlows, so that from the middle of 1951 he was the only customer for an expanding range of railway items.

The Master Models tractor and implements appeared around 1954, and although they were approximately OO scale, initially they were numbered in the 'K' series along with the Wardie garage equipment which was scaled for Dinky-size toys. The models were packaged in the following combinations:

- no.K47 Massey-Harris Tractor and Hay Wagon Trailer (this was the catalogue description);
- no.K49 Massey-Harris Tractor and Roller;
- no.K50 Massey-Harris Tractor and Rake;
- no.K63 Farm Implements Set containing Tractor, Hay Trailer, Roller and Rake.

The red-and-green boxes proclaimed *WEE WORLD SERIES* on the face. For nos.K49 and K50, the box for no.K48 Cement Mixer was used, with paper labels on the ends and in some cases on the sides to indicate the contents. In 1955, a new style of box was introduced for nos.K47, K49 and K50 in green, yellow and red, with *A WARDIE PRODUCT WEE WORLD SERIES* A *MASTER MODEL* on the face.

Around 1958, the individual items were re-numbered into the main Master Models OO scale series, and appeared in Master Models boxes. The Farm Implements Set was also catalogued but remained as no.K63. The Master Models numbers were:

- no.81 Massey-Harris Tractor and Roller;
- no.82 Massey-Harris Tractor and Rake;
- no.83 Massey-Harris Tractor and Hay Wagon Trailer.

Numbers 81 and 82 were still included in an advertisement in *Railway Modeller* in January 1961, but the Master Models range was by then much reduced. Competition from the similar series of railway accessories made in plastic by Merit (J. & L. Randall Ltd) caused the eventual demise of Master Models.

Models

No.K63 Farm Implements Set

This set contained the tractor, hay trailer, roller and rake, all packed in a lift-up lid display box. See below for details of the individual models. Curiously, no.K63 was not included in Master Models catalogues and lists in 1955–56, but it must have been issued at the same time or possibly earlier than the individual items, because the set often contained early versions of the models (**2** to **4**).

No.K47 (later no.83) Massey-Harris Tractor and Hay Wagon Trailer

Although described as a Massey-Harris in the Master Models catalogues, the tractor was a fairly approximate little model and was probably rather smaller than true OO scale, looking more like 1:90. It was always painted red, with a red or black baseplate, and unpainted or black wheels. It either had no lettering cast underneath, or *BRITISH MADE* was cast, or *BRITISH MADE MADE IN ENGLAND*. The driver was a push-fit onto the seat and he was not provided with a steering wheel! He was painted green (shades vary), blue or black. The hay trailer was always painted green, in various shades of medium or dark green, with unpainted, yellow or beige raves and unpainted or black wheels. *BRITISH MADE* was added underneath on later models. The first type of box was a long box in which the tractor and trailer were packed end-to-end. It was printed in red and green with *47* on the end flaps (**5**). This was superseded in 1955 by the green, yellow and red style of box with *TRACTOR & FARM CART K47* on the ends (**6**). Around 1958 the set was renumbered as no.83 in a *MASTER MODELS* cream card box with red, green and black printing (not illustrated).

No.K49 (later no.81) Massey-Harris Tractor and Roller

The tractor is described above under no.K47. The die for the roller frame and drawbar went through at least three stages. Early models were quite thin castings (this version was usually painted orange or blue) and perhaps there was some difficulty filling the die through the thin channels of the casting, because this was modified to a two-piece model with a separate roller frame riveted to the drawbar. We have seen this type in blue or yellow. The third version reverted to a one-piece casting, which had been thickened to give strength and aid the flow of metal, and can be found in medium or light blue, or dark green. The roller itself consisted either of five unpainted segments (on the first casting), or six narrower segments (on the two-piece second casting). The third casting usually had six segments, but can also be found with five.

No lettering was cast on the model (**7**). The first type of box was the box for no.K48 Cement Mixer, but with white paper labels on the ends to identify it as *TRACTOR AND ROLLER K49* (**8**). Another variation of this box had green labels on the box ends and also on two of the sides to cover up the illustrations of the cement mixer (**9**). This was followed by the green, yellow and red style of box (**10**) and finally by the *MASTER MODELS* cream card box in which the set was renumbered 81 (**11**).

No.K50 (later no.82) Massey-Harris Tractor and Rake

Details of the tractor are given above under no.K47. The rake was painted medium or light blue with red wheels, and came with or without *BRITISH MADE* cast underneath. The first type of box, as for no.K49, was the box for no.K48 with white paper labels on the ends reading *TRACTOR AND RAKE K50* (**12**). The green, yellow and red box followed in 1955 (**13**), and the *MASTER MODELS* no.82 box appeared around 1958 (**14**).

Model List

Serial No.	Model	Dates
K47	Massey-Harris Tractor and Hay Wagon Trailer	1954–57
K49	Massey-Harris Tractor and Roller	1954–57
K50	Massey-Harris Tractor and Rake	1954–57
K63	Farm Implements Set containing Tractor, Hay Trailer, Roller and Rake	1954–58
81	Massey-Harris Tractor and Roller (re-numbering of no.K49)	1958–61
82	Massey-Harris Tractor and Rake (re-numbering of no.K50)	1958–61
83	Massey-Harris Tractor and Hay Wagon Trailer (re-numbering of no.K47)	1958–60?

Further Reading

Brookes, Paul & Jennifer, 1998, *Master Models Listing of Models for Collectors*. Published by the authors.

Brookes, Paul, 2009, *The Illustrated Kemlows Story*. Published by the author.

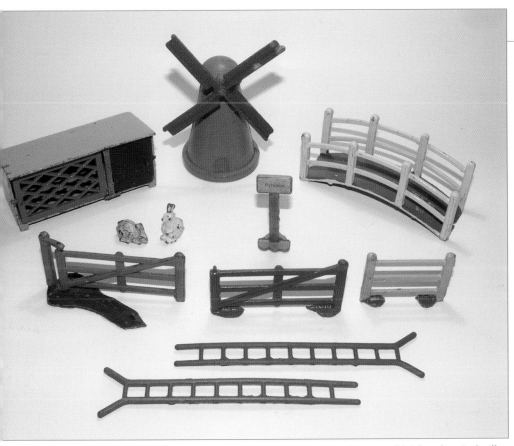

1. Some of the Kemlows farm items, made 1948 to 1950: (back row) rabbit hutch, windmill, signboard and bridge; (middle row) swing gate, gate and fence; (in front) two orchard ladders.

2. Master Models no.K63 Farm Implements Set (box).

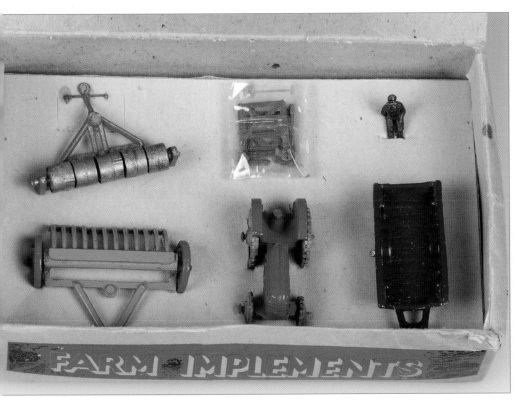

3. Master Models no.K63 Farm Implements Set containing Tractor, Hay Trailer, Roller and Rake.

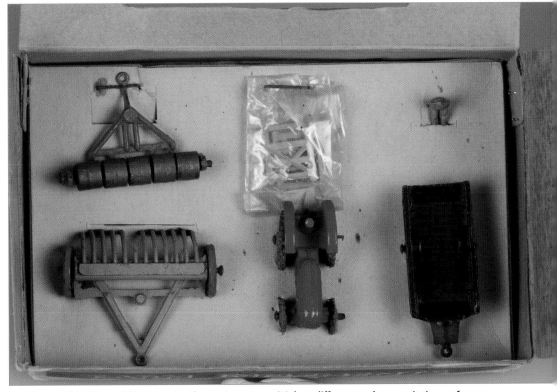

4. Master Models no.K63 Farm Implements Set, as (3) but different colour variations of some of the items.

5. Master Models no.K47 Massey-Harris Tractor and Hay Wagon Trailer, red tractor with black wheels and driver, green trailer with yellow raves, first type box (tractor 37 mm, trailer 47 mm).

6. Master Models no.K47 Massey-Harris Tractor and Hay Wagon Trailer, as (**5**) but second type box.

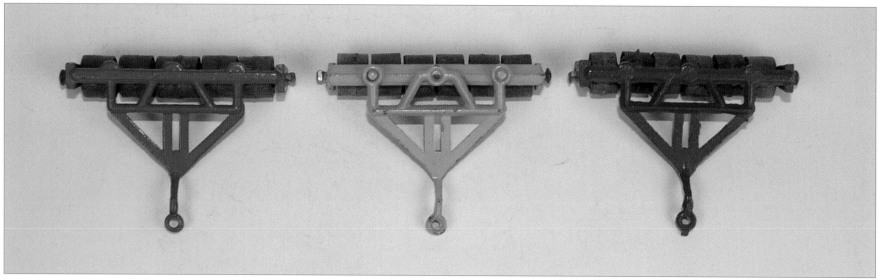

7. Master Models Rollers showing the three different castings: (left to right) first type thin one-piece casting in orange with five roller segments; second type two-piece casting in yellow with six roller segments; third type thicker one-piece casting in dark green with six roller segments.

8. *Master Models no.K49 Massey-Harris Tractor and Roller, red tractor with black wheels and driver, roller with blue frame, first casting with five roller segments, first type box (tractor 37 mm, roller 32 mm).*

9. *Master Models no.K49 Massey-Harris Tractor and Roller, as (**8**) but black base on tractor, roller with yellow frame, second casting (in two pieces) with six roller segments, first type box with a different style of labels and including labels on two of the sides as well as the ends.*

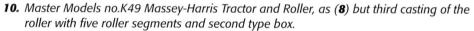

10. Master Models no.K49 Massey-Harris Tractor and Roller, as (**8**) but third casting of the roller with five roller segments and second type box.

11. Master Models no.81 Massey-Harris Tractor and Roller, as (**10**) but blue driver and dark green third casting of the roller with six roller segments, third type box.

12. Master Models no.K50 Massey-Harris Tractor and Hay Rake, red tractor with unpainted wheels and blue driver, blue rake with red wheels, first type box (tractor 37 mm, rake 39 mm).

13. *Master Models no.K50 Massey-Harris Tractor and Hay Rake, as (12) but black wheels and driver, second type box.*

14. *Master Models no.82 Massey-Harris Tractor and Hay Rake, as (13) but blue driver, third type box.*

Company History

The origins of Lesney Products & Co. Ltd, which was incorporated in March 1949, can be found in the Early Lesney Toys chapter in this Volume. Further details are also well presented by Kevin McGimpsey, Stuart Orr and others in *Collecting Matchbox Diecast Toys – The First Forty Years* (1989). This book is surely the bible for Matchbox collectors, and it includes useful sections on the 'Regular Wheels' Matchbox Series by Nigel Cooper and on gift sets by Philip Bowdidge. 'Regular Wheels' refers to models with conventional wheels and thicker axles, which were the norm for diecast vehicles until 1969. For clear illustrations showing the sequence of models it is best to turn to Charlie Mack's *Collecting Matchbox Regular Wheels, 1953-1969* (2001). His illustrations do include colour pictures of the early boxes as well as the models, which is a great help. He classifies the Regular Wheels boxes into six basic types, lettered A to F, and his classification has been used here. His *Big Book of Superfast Matchbox Toys 1969-2004 Vol.1* (2005) illustrates the later models, but without their boxes.

It was the remarkable success of the 1953 small Coronation Coach which provided Lesney with the capital and inspiration for a new miniature range which became known as the Matchbox Series. The small Coronation Coach sold over a million models and demonstrated that the buying public obviously had a liking for miniatures. The early factory at Shacklewell Lane, Dalston, London E8 was replaced by larger works at Barretts Grove, Stoke Newington in 1957 and again in 1959 at Eastway, Hackney Wick. By 1961 Lesney Products was employing 1,300 people, and further factories were opened in 1964 at Lee Conservancy Road, Hackney Wick and in 1969 at Rochford, Essex. The miniature Matchbox models were an extraordinary success. By the mid-1960s Lesney were producing 2,000 tons of zinc-alloy castings annually, 85 per cent of which were Matchbox vehicles, and the rest were industrial components. Lesney were making five million toys a week, 60 per cent of which were going to the U.S.A. It is small wonder that Matchbox collecting is now so popular in the United States.

In June 1953 an advertisement in *Games & Toys* announced the first four of the new Lesney miniature models to be distributed by J. Kohnstam Ltd, a highly successful toy wholesaler who operated under the trade name Moko. Moko was an abbreviation for the name of the founder of the company, Moses Kohnstam (see the Moko chapter). The first four Matchbox Series models were available at the end of November 1953 and were sold at the very affordable price of 1s. 6d. The models were displayed at the Harrogate Toy Fair in 1954 and again at the British

Industries Fair at Olympia in the spring of same year. After that the orders started to flow in from Britain and abroad, mainly from the U.S.A. and South Africa. J. Kohnstam Ltd proved adept at advertising the new series (**1**). In 1959 Lesney bought the Matchbox brand from J. Kohnstam Ltd and was able to drop the Moko name from the packaging.

Models

Matchbox 1–75 Series Regular Wheels

Only eight farm items were produced in the Matchbox Series in the Regular Wheels era (1953-69). Both the no.4 Massey-Harris Tractor and no.8 Caterpillar Tractor were scaled down from larger Lesney toys (see the Early Lesney Toys chapter).

Model numbering

Matchbox Series numbers started at no.1. By 1960 there were 75 models in the range, numbered from no.1 to no.75, and many of the early models had already been replaced with improved versions, using the same model numbers. After 1960 the new introductions each year were given the numbers of discontinued models, so that there were always 75 items in the series. Collectors usually add a suffix letter to indicate a change of casting or an entirely different model with the same number. These letters did not occur in contemporary literature.

Box Types

The study of boxes has been developed further by Philip Bowdidge in an important undated photocopied booklet published by the author entitled *Matchbox Boxes 1 to 75 1953-1962*. This demonstrates that the sequence is more complex than Mack's A to F classification, for there were many small changes, which add much interest to the subject. To the A to F sequence we can add a G for Superfast boxes.

Box A The concept of 'Matchbox' packaging came from Richard Kohnstam (of J. Kohnstam Ltd), who remembered little wooden dolls packaged in matchboxes that Moko used to sell. J. Kohnstam Ltd registered the Lesney miniatures under the trade name 'Matchbox' in 1954. As a result, the word *Moko* in script lettering followed by *LESNEY* in capitals is found in the scroll on the face of Type A boxes used for the first seven models from

1953 and 1954 (**2**). The boxes had dark blue ends, one side was blue, and the other was black representing the 'striker' side of the box.

Box B In late 1954 the word Moko was changed from script to capitals so that *MOKO* and *LESNEY* were given equal weight on the scroll (**4**). In 1957 a small black model number, which was often actually very difficult to see on a dark blue background, was added to each end flap to make models easier to identify when stacked on shelves (**5**). In 1959 these end numbers were put in a white circle (**14**). Late in 1959 the model name was also included on the end flaps within a white rectangular panel above the number (**15**). In 1960 the black lettering in the white panel was replaced with white lettering (**26**). All boxes had *REGD* in the bottom right-hand corner, with the exception of the very earliest boxes for Matchbox nos.8 and 9 (compare (**8**) and (**10**)).

Box C When, in 1959, Lesney bought out J. Kohnstam Ltd and acquired the Matchbox name, they took ownership of all the sales rights to the series. As a result, the name Moko was dropped from the scroll in 1961, and just the words *A LESNEY* remained. The model name was printed in white above the number on the end flap. Otherwise the 1961 boxes were the same as those in the previous style (**16**).

Box D In 1962 the words *"MATCHBOX" SERIES* in red which formed an arch over the illustration were changed to *"MATCHBOX" Series*. The illustration was made larger and printed in full colour. *A LESNEY PRODUCT* appeared along the lower edge, and the model numbers were printed without *No.* in front. On the end flap the model number appeared in white with the model name underneath (**19**).

Box E In 1964 the word *"MATCHBOX"* was enlarged to form a complete arch across the full width of the box, and the word *SERIES* was much smaller above a larger illustration (**22**). Until then the model number had been placed midway up either side of the illustration, but the number was now moved to the two bottom corners either side of *A LESNEY PRODUCT*. Later, a colour picture appeared occupying half of the end flaps. This was set within a yellow rectangle above an enlarged model number (**23**).

Box F In 1968 the whole design was changed again with the word *"MATCHBOX"* in a straight line along the top. The model number was put in a blue square in the top right-hand corner and the picture was made larger. There was a white-on-blue line drawing on the side, and the panel with the illustration on the end flap filled the whole area of the flap (**48**).

Box G This design showed Superfast elements without being labelled Superfast. Models with Regular Wheels which continued into the Superfast era were put into these boxes to suggest that the model could now move at great speed, even the Combine Harvester! The line drawing remained on the side (**49**).

Nos.4a & 4b Massey-Harris Tractor (1954–60)

• *No.4a* (1954-56) The Massey-Harris tractor was first produced with rear mudguards, a solid body and a tan (**2**) or light brown painted driver (**3**). It had spoked metal front wheels, and solid metal rear wheels with gold hubs and with *MADE IN ENGLAND BY LESNEY* around the circumference. The gap above the engine block could be either open (**2** and **3**) or filled (not illustrated). *MASSEY HARRIS* was cast in very small print on either side of the bonnet in front of the steering wheel, and there was hand-applied gold trim on the engine detail. There were four different combinations of *LESNEY* and *ENGLAND* cast inside the mudguards because the mudguards were interchangeable. Therefore, we have seen:

  • *LESNEY* on both;

  • *LESNEY* on the left and *ENGLAND* on the right;

  • *ENGLAND* on both;

  • *ENGLAND* on the left and *LESNEY* on the right.

The tractor was produced with a tow hook, but no implements were supplied to go with it. This version was in Types A and B boxes.

• *No.4b* (1957-60) The second version of the Massey-Harris tractor was without mudguards, and there was no gold trim on the engine (**4** to **6**). The engine block was no longer solid, and a cavity was visible from underneath, with *4 LESNEY ENGLAND* cast up inside the model. It first had solid metal wheels (**4** and **5**), with a small number *4* cast on one side of each front wheel and *No.4* cast inside the rear wheels. Later, the metal wheels were replaced by plastic (**6** and **7**). The plastic-wheeled version came with either crimped axle ends (**6**) or the axle ends were riveted, giving a domed appearance (**7**). The light brown driver was cast separately, and there were some very slight variations to the driver casting. This version was only in a Type B box.

It is worth examining the illustrations on the boxes in (**2** to **7**) because they were all slightly different!

Nos.8a, 8b, 8c & 8d Caterpillar Tractor (1954–66)

There were four castings of this crawler, although the difference between nos.8b and 8c was more a question of size than casting detail.

- **_No.8a_** (1954-57) The earliest version was yellow with yellow wheels and a red driver with a black hat (**8**). The Type B box for this version did not have *REGD.* in the bottom right-hand corner. The next was the same but with bare metal wheels (**9**), and then came a pale yellow version (**10**). This was followed by an orange tractor with a gold grille and a black hat on the driver (**11**), and finally a yellow tractor with a silver grille and again a black hat on the driver (**12**). In both (**11**) and (**12**) the driver was in the same colour as the tractor. Cast underneath was *LESNEY ENGLAND*. The model had four minor casting variations of the engine caused by part of the mould breaking during production. It was always in a Type B box.

- **_No.8b_** (1958-60) This was a yellow Caterpillar tractor with bare metal wheels, and *ENGLAND LESNEY No8 CATERPILLAR REG TRADE MK* cast underneath and *CAT D8* cast on the bonnet. The space between the two wheels was filled with yellow metalwork. This was in a Type B box (**13** to **15**).

- **_No.8c_** (1961-63) This was a larger Caterpillar tractor with wheels of bare metal (**16** and **20**), black plastic (**18**) or a rare silver plastic (**17** and **19**). The words *LESNEY ENGLAND* and *CATERPILLAR REG. TRADE MK* were again cast underneath as on no.8b, sometimes with *No.8* or *No.18* as well. This was in Types C and D boxes (**16** to **20**). Photo (**21**) shows how no.8b was significantly smaller than no.8c.

- **_No.8d_** (1964-66) This was a larger model than no.8c and an entirely different casting with a baseplate and large tow hook but no driver. It had black plastic wheels, tall or short exhaust and was with or without a hole in the baseplate. Cast in the baseplate was *No 8 "MATCHBOX" SERIES MADE IN ENGLAND BY LESNEY CATERPILLAR D8 REG. TRADE MK. TRACTOR.* This was only in a Type E box (**22** and **23**).

Note again the great variety in the detail on the line illustrations (**8**, **10** to **17**). From (**18**) onwards the illustrations more accurately represented the model.

No.72a Fordson Major Tractor (1959–66)

This blue Fordson tractor, advertised in *Games & Toys* as new in September 1959 (**24**) had just the one casting. *FORDSON MAJOR* was cast on both sides of the bonnet, *LESNEY ENGLAND* was cast under the front axle and *No.72* was cast under the back end near the tow hook. However, there were several wheel variations and combinations, and this was the first Matchbox model which had separate tyres on plastic hubs. The wheel variations were as follows:

1. Grey plastic front wheels, orange rear hubs with grey plastic tyres (**25** to **28**).
2. Black plastic front wheels with knobbly treads, orange rear hubs with black plastic tyres (**29**).
3. Black plastic front wheels with fine treads, orange rear hubs with black plastic tyres (**30**).

4. Orange front and rear hubs with grey plastic tyres, smooth treads on the front wheels, heavier treads and *LESNEY ENGLAND* moulded around the circumference on the rear wheels (**31**).
5. As 4 but with black tyres (**32**).
6. As 5 but with yellow hubs (**33**).

The Fordson was in Types B, C and D boxes. The illustrations on the boxes (**25** to **29**) again all showed minor variations. From (**30**) onwards the illustrations more accurately represented the tractor, although not the hub and tyre changes over the period.

No.G-4 Farm Gift Set (1960–1962)

The no.72a Fordson Major was included in the first Matchbox farm gift sets. The sets consisted of an inner box with compartments for the models and an outer sleeve illustrated with the box designs for the individual models. The underside of the sleeve listed the contents. The series was given 'G' prefixed numbers, and the models were grouped in themes, including no.G-4 for farm models. The first version was illustrated with Type B box designs and produced in 1960 (**34**), containing a cattle truck and five other vehicles, including the no.72a Fordson Major. Note that the box design for no.12 Land Rover appeared twice on the sleeve. The second version, with *G-4* on the sleeve, was illustrated with Type C box designs and produced in 1961 and 1962 (**35**). It contained the cattle truck with six other vehicles, the new addition being the no.50 Commer Pick-up.

No.50b John Deere-Lanz 700 Tractor (1964–68)

This green John Deere tractor had a yellow plastic steering wheel and hubs, a silver grille and either black (**38** and **39**) or, less frequently, grey (**36** and **37**) tyres. Cast on either side of the bonnet was *JOHN DEERE LANZ 700* and underneath the tractor was *"MATCHBOX" SERIES LESNEY ENGLAND No.50.* There were no known casting variations. It was put into a Type E box, which had four different end flaps: with *NEW MODEL 50 TRACTOR* (**36**), with *50 TRACTOR* (not illustrated), with *50 TRACTOR* under a small picture (not illustrated) and with *50 JOHN DEERE-LANZ TRACTOR* under a larger picture and *"MATCHBOX"* above (**37** and **38**). There was also a blister pack produced by the Fred Bronner Corporation, New York, for sales in the U.S.A. (**39**). A pre-production example of this model, coloured blue with grey mudguards and red hubs, was illustrated in McGimpsey and Orr (1989), page 96.

No.51b Tipping Trailer (1964–68)

The trailer, coloured green with yellow hubs to match the no.50 John Deere tractor, again had no casting variations (**40** to **43**). In a rare variation the plastic hubs were green (**41**). Cast underneath was *No.51 "MATCHBOX" SERIES TRAILER MADE IN ENGLAND BY LESNEY.* A load of three yellow plastic barrels joined on a yellow sprue was supplied with the trailer. Like the John Deere tractor, the trailer was fitted with either black (**41** and **43**) or, less frequently, grey (**40** and **42**) tyres. It was sold in a Type E box, which had three different end flaps: with *NEW MODEL 51 TRAILER* (**40** and **41**), with *51 TRAILER* (not illustrated) and *51 TIPPING TRAILER* under a picture with *"MATCHBOX"* above (**42** and **43**). There was also a pre-production model with a dark brown body and a yellow chassis, and with the words *"MATCHBOX" SERIES* omitted from the casting underneath (**44**).

No.39c Ford Tractor (1967–72)

This Ford tractor was dark blue and yellow (**45** and **46**) or light blue and yellow (**48** and **49**) with a black plastic steering wheel and yellow plastic hubs with *LESNEY ENGLAND* around the rear plastic tyres. Rare variations had a yellow (**46**) or translucent white (**47**) steering wheel. There was also a limited number of dark green models produced with black plastic hubs (**51**). *FORD* was cast over the grille and on the two sides of the bonnet, and underneath was *MATCHBOX SERIES No 39 LESNEY ENGLAND FORD TRACTOR*. The tow hook had two different configurations, and the exhaust was first 7 mm long (not illustrated) and then 5 mm (**45** to **47** and **50**) or 4 mm (**48**, **49** and **51**). It was in Types E, F and G boxes. The tall early exhaust is shown quite clearly in the illustration on the Type E boxes (**45** and **46**), but not on the later ones.

It is not difficult to buy single-colour blue or orange models in Types E, F and G boxes (**50**). However, the literature suggests that blue tractors were sold exclusively in the 1967 to 1972 King Size no.K-20-1 Tractor Transporter set (**109** to **113**) and the orange tractors were only in the no.BM-2 Big M-x Mechanised Tractor Plant and Winch Transporter (**117**). Perhaps this is an area for further research.

No.40c Hay Trailer (1967–71)

A four-wheeled hay trailer with a light blue (**52** and **54**) or dark blue (**53**) body, yellow plastic hubs and raves and black plastic tyres came out to be used with the no.39c Ford Tractor. Cast under the trailer body was *MADE IN ENGLAND BY LESNEY HAY TRAILER No 40 "MATCHBOX" SERIES* and a letter *A* near the back end (**55**). The trailer base is said to have had six minor variations, the trailer bed four, and the hay raves three. It was in Types E, F and G boxes. The tractor and trailer combination was actually promoted in a jigsaw by the Fred Bronner Corporation in the United States (**56**).

No.65c Claas Combine Harvester (1967–72)

A dark red combine harvester with a yellow plastic reel which rotated as the model was pushed along, black plastic rear wheels and yellow front hubs was produced unchanged from 1967 until 1972. The combine harvester was issued in Type E (**57**), Type F (**58**) and Type G (**59**) boxes. Cast on the baseplate was *"MATCHBOX" SERIES No 65 MADE IN*

ENGLAND BY LESNEY CLAAS COMBINE HARVESTER, and on a small circular knob near the front end of the baseplate there was a capital *A*. There were no casting variations.

No.G-3 Farm Gift Set (1968–69)

In 1968 the no.39c yellow-and-blue Ford Tractor, the blue no.40c Hay Trailer and the no.65c Claas Combine Harvester were put into the no.G-3 farm set in a window box, containing eight models altogether, covered with cellophane shrink wrap, which could only be opened by tearing off the cellophane (**60**). *"MATCHBOX"* was in large red letters across the top of the window next to *G-3*; below was *A LESNEY PRODUCT FARM SET*. To the right was an illustration of a combine harvester at work with a tractor and trailer behind. To the left were some ears of wheat. Covering the whole of the back of the box was a very busy farmyard scene featuring all the models in the set. They were also illustrated in a yellow panel on the end flap, reminiscent of the boxes for individual Regular Wheels models.

No.R-3 Fold Away Farm (1968–73)

For five years there was a folding card farmyard layout with farm buildings (**61**) described in the catalogue as:

'Another 1968 special – made especially for all your tractors, trailers and combine harvesters. Another great new "fold-away" playground printed in all the colours of a farmyard and incorporating many three-dimensional stand-up cut-outs giving added realism.'

Spread out, the card layout measured 890 mm x 580 mm, and there were four fold-up card buildings (which had to be glued together), a loading ramp, a tree and some bushes.

The Superfast Range

In 1968 the American Mattel toy company produced a new Hot Wheels series of low-friction, high-speed models. With heavy advertising, Hot Wheels sold in large numbers in the U.S.A. and around the world. The result was a rapid drop in demand for Lesney toys in the United States which, as a share of Lesney's production, went down from 60 per cent to 10 per cent.

The Regular Wheels models had strong steel axles of 1.6 mm diameter, but they did not run far or fast because of the amount of friction between wheels and axles. Lesney responded to Hot Wheels after a two-year re-tooling operation in 1969 and 1970 with their own low-friction Superfast range with wire axles of only 0.6 mm and much thinner wheels. The problem with these was that the wire axles were easily bent and the wire ends could be dangerous. Nevertheless, of the seventy-five Regular Wheels models, sixty were converted to Superfast by 1970, and only five Regular Wheels models were retained unchanged in Type G boxes, including the no.39c Ford Tractor, the no.40c Hay Trailer and the no.65c Claas Combine Harvester. The Hay Trailer was last illustrated in catalogues in 1971 and the Combine Harvester and Ford Tractor in 1972.

The best guide to the Superfast range is Charlie Mack's *Lesney's Matchbox Toys: The Superfast Years, 1969-1982* (1999), and his letters which follow the model numbers are used here. Some other authors use a different coding system. Matchbox Superfast models are usually indicated by the letters MB before the model number, a prefix adopted by Lesney in the late 1970s. Charlie Mack's coverage of Superfast boxes is minimal, and we are not aware of any literature which yet provides a clear dated sequence of Superfast packaging.

In the new Superfast series there were few new farm models, presumably because they did not lend themselves to speed. There was a very popular no.MB-25e Mod Tractor in 1973 (although hardly agricultural), then in 1978 came no.MB-46e Ford Tractor and Harrow and no.MB-51e Combine Harvester. The need for speed was otherwise a bit of a killer for farm vehicles!

No.MB-25e Mod Tractor (1973–78)

This highly idiosyncratic tractor with silver-painted plastic engine parts projecting in every direction, wide wheels and a high-backed seat caught the imagination of children and it sold in large numbers. It does indeed run fast a long way when pushed! While the model hardly relates to its farming counterparts, it did go into Two Packs with the hay trailer (**122**), so it is included here. Early versions had headlights cast on the rear mudguards (**62**), but after casting problems these were removed (**63** to **67**). Models were usually either metallic purple (**62**, **63**, **65** and **67**) or red (**64** and **66**). They all usually had a yellow seat and tow hook, but there was a rare version with a red plastic seat and tow hook (**65**). Cast on the baseplate was *MOD TRACTOR No25 LESNEY PRODS & Co Ltd © 1972 MADE IN ENGLAND*. We have also seen a pre-production version with an unpainted body and a white plastic steering wheel of a different design (**68**).

The Mod Tractors came first in Superfast boxes (**62** to **65**) and then in bubble packs with versions for the British (**66**) and U.S.A. (**67**) markets.

No.MB-46e Ford Tractor and Harrow (1978–84)

Having never before produced any implements for their tractors (other than trailers), in 1978 Lesney brought out a combination set of a Ford tractor in the correct Ford blue and a rather flimsy yellow plastic disc harrow. The yellow plastic varied in shade, and the harrow was sold attached to the back of the tractor with an elastic band (**69** to **72**). The model was sold boxed (**69** to **71**) or blister packed (**72**), with *"MATCHBOX" 75* branding on the packaging, indicating that the model

was part of the Matchbox 75 Series. The tractor colour was changed to metallic green from 1981. The cab interior was yellow plastic, the windows were unglazed and there was a minimal exhaust. The wheels were black plastic, although some rear hubs were yellow (**71** and **72**). The one-piece bare metal engine block, baseplate and tow hook had cast into it *No46 "MATCHBOX" FORD TRACTOR MADE IN ENGLAND © 1976 LESNEY PRODS & Co Ltd Superfast*. From 1985 the model was made in Macau and no longer had *MADE IN ENGLAND* cast underneath.

No.MB-51e Combine Harvester (1978–82)

After a gap of six years without a combine harvester, a new light red version came out in 1978, an entirely different casting to its Regular Wheels predecessor and with a swivelling yellow plastic grain elevator. It did, however, retain the same rotating yellow plastic reel. It had thick axles and black plastic wheels (**73** and **75**) or the front wheels had yellow hubs (**74** and **76**). It was issued in the new Matchbox 75 Series box (**73** and **74**) and in blister packs (**75** and **76**). Cast into the black baseplate was *No51 COMBINE HARVESTER MADE IN ENGLAND © 1977 LESNEY PRODS & Co LTD*, and cast under the cutter bar was *"MATCHBOX"*. Later models were produced without a baseplate, and had *MATCHBOX No51 COMBINE HARVESTER LESNEY ENGLAND 1977 ©* cast under the cutter bar (**76**). We have also seen a pre-production model in black with black plastic wheels (**77**). Although the model was deleted from the range in 1983, it was re-introduced in farm gift sets in 1987, made in Macau and no longer carrying *MADE IN ENGLAND* cast underneath.

Major Packs

No.M-5 Massey-Ferguson '780' Combine Harvester (1959–66)

Matchbox Major Packs were a larger-scale range of models featuring mainly commercial, road construction and agricultural vehicles, beginning with the Caterpillar DW20 Earth Scraper in 1958. The only true farming item was the no.M-5 Massey-Ferguson '780' Combine Harvester. It was produced at a scale of 1:58 for seven years from July 1959 (**78**). The last of the fifteen Major Pack items was brought out in 1965, and 1966 was the last year for the whole series, including the Combine Harvester, although some models were incorporated into the King Size range in 1967.

Box Types

There were six different boxes for this model, and the designs were similar to Types B to E boxes for Regular Wheels models. The comparable Regular Wheels box types are shown in brackets.

Type 1 (Type B) had a line-drawing illustration of the combine harvester within the central yellow panel of the box face which had blue panels to either side. *A MOKO LESNEY* was on the scroll under the illustration. In the bottom right-hand corner was *REGD* as on the Regular Wheels boxes. In the blue panels to either side was *No.5 MAJOR PACK*. The blue end flaps had *MASSEY-FERGUSON "780" COMBINE HARVESTER* in white lettering against the blue background, and in the middle was *M5* or *M-5* (with or without a hyphen) in black within a yellow circle (**79**).

Type 2 (Type C) had the same illustration but with *A LESNEY* on the scroll and *M-5* on the end flaps (**80**).

Type 3 (Type D), the rarest version, had a yellow panel occupying the full face of the box with a colour illustration showing a near side-on view of the combine harvester under an arched *"MATCHBOX" Series.* Model details were to either side of the illustration. The end flaps were mostly yellow forming a very large *M* with *M-5* in blue above *COMBINE HARVESTER* in red (**81**).

Type 4 (also Type D) was similar to Type 3, but with a larger colour illustration turned almost front-end to the viewer and occupying more of the space, reaching up to the words *"MATCHBOX" Series.* The end flaps were as in Type 3 (**82**).

Type 5 (Type E) had a still larger colour illustration showing a realistic farming scene which occupied most of the box face. The picture showed a man driving the machine in a corn field with trees behind. The one word *"MATCHBOX"* occupied the full width of the box face. *M-5* had dropped from the sides of the illustration to the bottom corners. The end flaps were mostly yellow, again forming a large *M*, with *MAJOR PACK M-5* in blue above *COMBINE HARVESTER* in red. This is the commonest box version and contains the widest range of colour variations of steering wheels and hubs (**83** to **89**).

Type 6 (also Type E) had an identical box face to Type 5 but there was also an illustration on half of the end flap above the model number and model description (**90**).

Although there is no way of being certain that dealers or previous collectors have not switched boxes in some instances, it appears that there were at least eighteen to twenty genuine combinations of box types and variations in hubs and steering wheels. To avoid too many pictures of this model, we have selected one of each box type, except for Type 5 for which we have shown the full range of model variations.

The Model

The combine harvester was a complex and ambitious piece of casting which was modified several times over the seven years production, mainly with strengthening underneath. The various modifications are described in detail in McGimpsey & Orr (1989, p.150). The most obvious were the addition of a heavier triangular brace above the front left axle and the thickening of the steering wheel column. The earliest models show the fine detailing of the lower sprocket wheel below the chain drive on the right-hand side, and this was soon replaced by a blank circle.

The model body was red, with white and black *MASSEY FERGUSON 780 SPECIAL* transfers on both sides. The reel varied from bright to mid-yellow and the driver was always tan-coloured. The steering wheels, usually open with spokes, were red metal (**79** and **80**), silver-painted metal (**81** to **83**, **85** and **86**) or yellow plastic (**84**, **88** and **89**), but a few were solid and bare metal (**87** and **90**). Red steering wheels were early and yellow plastic ones were late in the sequence, although there was considerable overlap between box types, wheel types and some variation in steering wheels. The early front wheels had silver plastic hubs (**79** to **81**), and then they were orange (**82** and **86** to **90**) or yellow (**83** to **85**). The rear wheels were usually solid black plastic (**79** to **84**, **86** to **88** and **90**), but could also have orange (**89**) or yellow (**85**) hubs to match the front wheels. The front tyres on the early models had fine treads (**79** to **82**) and no lettering, while later models had heavier treads and the words *LESNEY ENGLAND* moulded on both sides.

The disappearance of the detailing on the sprocket wheel, the thickening of the steering columns and the change to heavier tyre treads are the easiest ways to place a particular model in the sequence without reference to its box.

Cast on the baseplate was *MASSEY FERGUSON COMBINE HARVESTER MADE IN ENGLAND BY LESNEY* and cast under the arm supporting the cutter head was *MAJOR PACK No.5.*

King Size (1960–70) **and Super Kings** (from 1971)

The introduction of the King Size range in 1960 was an attempt to break into the traditional market occupied by Dinky, Corgi and Spot-On models, but it never did manage to match their achievements. The range was named after King Size cigarette packets to complement the Matchbox name. This was before people fully recognised the dangers of smoking. The larger scale allowed Lesney to add more detail than was possible on the miniatures or the Major Packs.

As with the Matchbox Series, new introductions were usually given the model numbers of items deleted from the range, so that there was always a continuous sequence of model numbers in the catalogue. We have used the convention of adding a hyphen and a number after Lesney's catalogue number to indicate successive models, for example models numbered K-1 would be shown as no.K-1-1 and no.K-1-2 etc.

In the 1971 catalogue the Super Kings were described as 'The big action models. Big in size – big in working parts, and that means lots more fun when you put them to work.'

No.K-4-1 McCormick International Tractor (1960–66)

The International Tractor was one of the first four models in the range to be launched in September 1960. It was illustrated in advance in the 1960 catalogue without the brand name on the bonnet and with yellow hubs, which were never used. By 1961 it was correctly branded and with green hubs.

Box Types

There were three quite different box designs, which equate roughly to the Regular Wheels box types shown in brackets:

Type 1 (Type C) The face had a black band horizontally through it, interrupted by the illustration, with *KING SIZE* in white in the black band. *"MATCHBOX" SERIES* formed an arch in red over the line-drawing illustration with a red crown to either side. It is interesting that the illustration showed nothing written on the bonnet of the tractor, as on the illustration in the 1960 catalogue. The scroll underneath had *BY LESNEY* with *No4* to either side. The end flaps were all-blue with a white circle containing *K-4* in red under a red crown. *INTERNATIONAL TRACTOR* was in white under the circle (**91**).

Type 2 (Type D) This had a much larger illustration of the tractor, correctly labelled on the bonnet. There was also a much larger *"MATCHBOX" Series* arched over the tractor with *A LESNEY PRODUCT* below and *KING K-4 SIZE* to either side. The end flap was mostly red forming a very large *K* with *K-4 INTERNATIONAL TRACTOR* in white (**92**).

Type 3 (Type E) The final box was reminiscent of the design of the Major Pack combine harvester with a large full-colour, more realistic illustration of the machine at work. *"MATCHBOX"* formed the arch, with *INTERNATIONAL TRACTOR* in small lettering underneath. Below that, in very small print, was *A LESNEY PRODUCT* with *K-4* either side. The design of the end flap was as in Type 2 with *KING SIZE K-4 INTERNATIONAL TRACTOR* in white (**93** and **94**).

The Model

The tractor was red throughout, with silver paint on the headlights on early models (soon omitted). Underneath was cast *No.4 MATCHBOX SERIES KING SIZE*. On both sides of the bonnet were transfers with *B-250* in red on a circle. The circle, together with *McCORMICK INTERNATIONAL* lettering was cream (**91** and **92**) or white (**93** and **94**). The only casting variation concerned the tow hook, which started small and was then made larger and more curved. The earliest hubs were dark green metal (**91**), then red plastic (**92** and **94**) or orange plastic (**93**), using the same red and orange hubs as on no.K-11-1 Fordson tractor produced from 1963 onwards. The small earlier tow hooks were only used on models with green hubs. It is reasonable to assume that this change from

green hubs took place in 1963 when the Fordson was introduced. We have also seen a rare colour variation with white-painted metal hubs (**95**).

No.K-8-1 Prime Mover Transporter for Caterpillar Tractor (1962–66)

This set, consisting of three separate pieces, was introduced in April 1962, prominently featuring the name of the John Laing construction company. It was 305 mm long.

Box Types

The box consisted of an open-ended sleeve with the same design on front and back. There were two versions of the sleeve, which again can be equated to Regular Wheels box types as shown in brackets.

Type 1 (Type D) The first box had *"MATCHBOX" Series* forming an arch over the illustration with *KING K-8 SIZE* on either side and with *A LESNEY PRODUCT* in fine lettering underneath. The end flap was mostly red with *K-8 TRACTOR and TRANSPORTER* in white (**96**).

Type 2 (Type E) The second box had a more realistic colour picture with background detail under an arch with *"MATCHBOX"* in red above. This time the picture showed the Caterpillar tractor being loaded. *K-8* was in the bottom corners and *PRIME MOVER AND TRANSPORTER WITH CATERPILLAR TRACTOR* and *A LESNEY PRODUCT* in smaller print were below the illustration. The end flap was again mostly red with *KING SIZE K-8 PRIME MOVER AND TRANSPORTER WITH CATERPILLAR TRACTOR* in white (**97**).

The Model

The transporter consisted of a tractor unit and a low loader, and the tractor unit was the first King Size model to be fitted with plastic windows. Both pieces were orange. There was silver trim on the radiator grille, and the wheel hubs were either bare metal (**96**), or later were red plastic (**97**). Transfers showing *LAING* in black on a yellow background featured on both the tractor and trailer, and the lettering was bolder on the earlier version. There was also a transfer on the cab door, reading *CIVIL ENGINEERING CONTRACTORS* in black on a white panel. The bare metal baseplate of the tractor was turned up at the rear end, and included a curved tow hook. Cast on the baseplate was *MATCHBOX SERIES KING SIZE No.8 SCAMMELL 6x6 TRACTOR MADE IN ENGLAND BY LESNEY*. The trailer had double wheels on all three axles, and *MATCHBOX SERIES KING SIZE No.8 TRAILER MADE IN ENGLAND BY LESNEY* was cast underneath.

The Caterpillar tractor was no.K-3-1 Bulldozer without the front blade, also used in no.G-3 Farming Set (**120** and **121**). It was yellow with a red engine block, bare metal or red plastic wheels and green rubber tracks. On the baseplate was *MATCHBOX SERIES KING SIZE No.8 D9 CATERPILLAR REGD. TRADE MK. TRACTOR MADE IN ENGLAND BY LESNEY.*

No.K-11-1 Fordson Tractor and Farm Trailer (1963–69)

This popular Fordson tractor and tipping trailer set lasted seven years and also appeared in the no.G-3 Farming Set (**120** and **121**).

Box Types

There were three box variations, two of which can be equated with the Regular Wheels box types shown in brackets.

Type 1 (Type D) was a long box with a face similar to the first box for the no.K-8 Prime Mover. The tractor and trailer occupied the centre of the face, and the trailer body was incorrectly shown as red. *"MATCHBOX" Series* in red formed the usual arch over the illustration with *KING K-11 SIZE* to either side. *A LESNEY PRODUCT* was in small print along the bottom. The end flap was mostly a large red *K* and had *K-11 FORDSON TRACTOR AND FARM TRAILER* in white (**98**).

Type 2 (Type E) had a realistic illustration showing a man smoking a pipe and driving the tractor and trailer loaded with what could be sand or grain with scenery behind. This scenery included another Fordson with a trailer tipping out a load. *"MATCHBOX"* in red formed an arch over the illustration; *K-11* was located to either side near the bottom, between them was *FORDSON TRACTOR AND FARM TRAILER* in red and underneath was *A LESNEY PRODUCT*. On the end flap was an illustration occupying half of the flap on a white background with the model details below (**99**).

Type 3 was a blue-and-yellow window box with *K-11 "MATCHBOX"* under the window; the same was on the top to the rear of the window, with *KING SIZE FORDSON TRACTOR AND FARM TRAILER* in smaller lettering. An illustration on a yellow background occupied most of the width of a wider end flap with the model details above it (**100** and **101**).

The Model

The tractor was blue with a red transfer having *FORDSON SUPER MAJOR* in white lettering on both sides of the bonnet. The grille and the headlights within it were a uniform silver. The early version had a silver-painted metal steering wheel and orange metal hubs (**98**). The only variations were that the steering wheel changed to blue (**99** to **101**) and the hubs changed from orange metal (**98**) to orange plastic (**99** and **100**) and then to red plastic (**101**). Cast underneath the tractor was *MADE IN ENGLAND BY LESNEY FORDSON TRACTOR KING SIZE No11.*

The tipping trailer was grey with a blue chassis and drawbar, with a white plastic loop on the end of the drawbar to fit over the curved tow hook on the tractor. A silver-painted metal ram fitted tightly into a plastic sleeve which kept the trailer body in the air even after hours of play. The hubs were either orange plastic or red plastic, matching the colours on the tractor hubs. Cast on the trailer body underneath was *WHITLOCK TRAILER KING SIZE No11 MADE IN ENGLAND BY LESNEY.*

No.K-9-2 Claas Combine Harvester (1967–72)

When the Major Pack no.M-5 Massey-Ferguson Combine Harvester was withdrawn at the end of 1966, this Claas combine harvester in a new window box took its place. In the same year Lesney also produced their miniature no.65c Claas Combine Harvester (**57**).

Box Types

There were three boxes used.

Type 1 was a blue-and-yellow window box with *K-9"MATCHBOX"* under the window, and the same on the top with *KING SIZE COMBINE HARVESTER*. Inside the box there was a panorama of a harvesting scene with the sea behind (**103**). An illustration filled most of the end flap with the model details above (**102** to **104**). On the back was a picture of a farmyard scene.

Type 2 was a hanging window box, similar to Type 1 but with a picture of the green, and later red combine at work below the suspension hole. Inside, the space was given up to line drawings demonstrating how the reel and the grain auger could turn and swivel. The end flaps had yellow panels with illustrations and model details of first green (**105** and **106**) and then red machines (**107**). There was the same farmyard scene on the back as Type 1.

Type 3 This 1971 Super Kings window box dispensed with the concept of a rural idyll and instead gave a rather false impression that the combine harvester was intended to be driven fast rather than have a harvesting use. The interior was purple (**108**).

The Model

The earlier models were green (**102** to **105**) and had a small plastic driver, which was later discontinued. The driver was subsequently re-introduced, and was first shown in the catalogue when the colour of the model changed from green to red in 1970 (**106** to **108**). There was a white-on-green circular *CLAAS* logo on the side (omitted in **105**) and the word *CLAAS* was repeated on the straw outlet at the rear. When the model was changed to red there was a time-lag when the green logos continued to be used (**106**) before they too were changed to red (**107** and **108**). On the baseplate was *CLAAS COMBINE HARVESTER "MATCHBOX" KING SIZE No K-9 MADE IN ENGLAND BY LESNEY.*

The model was a single casting with the addition of the rotating metal reel and the swivelling grain auger. On the green versions the reel was red, and on the red models the reel was yellow. The steering wheel was yellow plastic throughout. The plastic hubs were red on the green models and then yellow on the red models. The little plastic drivers were red (**102**) or white (**103**). After the period when the drivers were withdrawn (**104** and **105**) they were then supplied again in light brown (**106** to **108**).

In 1971 the model was re-packaged as Super Kings (**108**), but the casting remained the same until it was withdrawn after 1972.

No.K-20-1 Tractor Transporter with blue no.39c Ford Tractors
(1967–72)

Box Types

All the sets were in hanging window boxes with four variations.

Type 1 window box was clearly the first with *NEW* in large yellow letters over *KING SIZE TRACTOR TRANSPORTER* on the top left-hand corner of the backing card (which strangely did not have cut-out suspension holes, unlike later boxes). On the backing card were diagrams to show how to separate the tractor unit from the trailer and how to clip the tractors into the yellow plastic clips on the trailer. These diagrams were located under *"MATCHBOX"* in large red letters. The scene behind the model on the plinth was of a ship with a crane on the skyline (**109**).

Type 2 had a high backing card, but the diagrams were replaced by a loading/unloading scene beside a ship. In both Types 1 and 2 there were double end flaps, so the yellow illustration on the ends did not occupy the full depth of the box (**110**).

Type 3 had a lower backing card with a goods train standing behind the lorry, and the word *"MATCHBOX"* was under the window. The end flap with its larger illustration now occupied the full depth of the box (**111** and **112**).

Type 4, the final Super Kings version, had *NEW Super Kings* on the backing card and *ARTICULATED TRAILER WITH THREE DETAILED TRACTORS* behind the window. The interior was purple. Speed was now all-important, replacing the dockside theme (**113**).

The Model

The set was introduced in November 1967, and consisted of a Ford D tractor unit and articulated semi-trailer carrying three no.39c all-blue Ford Tractors, which were attached by a strip of yellow plastic hooks. It was a slightly strange arrangement because there was no ramp to load and unload the tractors, and to create an image of reality the illustrations on the boxes showed the tractors being loaded onto (or off) a ship by crane. The tractors were held in place by the yellow hooks which engaged with the tractor footplates. On the bare metal baseplate under the tractor unit was *MADE IN ENGLAND BY LESNEY "MATCHBOX" KING SIZE FORD TRACTOR PATENT No 1.049.164*. Cast under the trailer body was *MADE IN ENGLAND BY LESNEY "MATCHBOX" KING SIZE No K-20 TASKER TRANSPORTER*. On the Super Kings version, the lettering was amended, with just *FORD TRACTOR* underneath the tractor unit, and under the trailer was *MATCHBOX SUPER KINGS K-20 TASKER TRANSPORTER MADE IN ENGLAND © 1971 LESNEY PRODUCTS & Co Ltd.*

The tractor unit was painted red, with either a red (**109** to **111** and **113**) or yellow (**112**) fuel tank. The trailer was usually red (**109** to **112**) but later yellow (**113**). The tractor unit and trailer had either red plastic hubs with black tyres (**109** to **112**) or Superfast wheels with low-friction axles (**113**). The early tractors were dark blue with 5 mm exhausts (**109** to **112**), and those in the final Super Kings set were light blue with 4 mm exhausts (**113**). The base lettering on these tractors was the same as on the separately boxed no.39c tractors.

No.K-3-3 Massey-Ferguson Tractor and Trailer (1970–73)

Box Types

The set had two types of hanging window box.

Type 1 was similar in design to the no.K-20-1 Tractor Transporter box Type 3, with a low backing card illustrated, in this case, with a tractor and trailer loaded with large bales and another trailer tipped-up with an emptied green-coloured load. Inside the box was a diagram showing how the trailer tipped. The large yellow illustration on the end flap occupied the full depth of the box (**114**).

Type 2 The Super Kings version was without pictures of farm scenes, just a tractor and trailer travelling at speed on both the back and on the wide end flaps. On the top of the box behind the window under *NEW Super Kings* was *TOWING TRACTOR WITH TIPPING TRAILER*. The interior of the box was purple (**115**).

The Model

In 1970 this Massey-Ferguson 165 tractor and four-wheeled metal tipping trailer replaced the earlier very popular no.K-11-1 Fordson tractor and trailer, which had been withdrawn in 1969. In line with changes in farm practices the tractor had a cab and the trailer was larger and more suitable for heavier loads, with a double hydraulic ram (**114** and **115**). The tractor was red with a white grille with inset lights. The engine block and cab interior were all grey. The cab, except for the cab door, was glazed, and the windscreen had a clever mark on it depicting a wiper blade. It is rather surprising that the tractor had no exhaust. Both tractor and trailer had yellow plastic hubs. There was a paper stick-on label with *MASSEY-FERGUSON* in black lettering and *165* on a black circle on either side of the bonnet. Cast into the grey underside of the tractor was *K3* near the front axle

The red metal trailer had a yellow chassis and drawbar with a cream plastic loop to go over the tow hook on the tractor. The early trailers had smooth floors and the later ones were textured. Under the trailer body was *MADE IN ENGLAND BY LESNEY FARM TRAILER "MATCHBOX" KING SIZE NoK-3 2 © 1970 LESNEY PROD & Co Ltd.* The number *2* above the copyright date may have been a die number.

No.BM-2 Big M-x Mechanised Tractor Plant and Winch Transporter with orange no.39c tractors and No.BM-A Big M-x Power Activator (1972–73)

In 1972, Matchbox introduced the Big M-x series of King Size models with working features that could be made to operate using the battery-powered Power Activator. The tractor transporter was modified from no.K-20-1 and consisted of an articulated Ford lorry and trailer with a folding plastic loading ramp, three orange no.39c Ford Tractors, and buildings containing a conveyor belt, all in one large box (**116** and **117**). The Power Activator gun had to be purchased separately (**118**). The activator gun could engage with a winch on the back of the lorry cab to pull the tractors out of the buildings and up the loading ramp. There was a separate conveyor belt in the buildings, also operated by the Power Activator, to move the tractors in and out of the buildings. The 1973 catalogue read:

> 'Place the Power Activator into the Big 'M-x' drive points and the models are ready for some hard working action. Running along the track, lifting loads, erecting buildings – however big the task you and the "MATCHBOX" Big 'M-x' team are ready to take it on.'

> 'Drive the Transporter to the Tractor Factory where three brand new vehicles are awaiting collection. When the Transporter has been parked in front of the Factory connect the power winch to each tractor in turn and load them onto the back of the truck.'

The tractor transporter was one of six such Big M-x sets, however the whole concept was too cumbersome, and apart from one model they were all withdrawn after 1973. Children preferred just to handle their toys without the intrusion of these elaborate mechanisms, and it did not help that all the models were adapted from existing King Size vehicles, some of which were quite dated. They were also expensive - with the cost of the Power Activator, the tractor transporter was over three times the price of the no.K-20-1 from which it was derived.

The tractor unit of the lorry was metallic blue. The gold-coloured flatbed trailer had a red plastic folding loading ramp at the rear, hinged on the back of the yellow plastic strip. This strip had chocks, rather than clips, to hold the tractors in place. The lorry had Superfast wheels and a red light over the cab. The winch attached to the back of the cab was red plastic, as was the upright gantry through which the thread passed between the winch and the tractors. *FORD* was on the bare metal grille on the lorry, and on the baseplate under the cab was *FORD TRACTOR.* Under the trailer body was *"MATCHBOX" Super Kings K-20 TASKER TRANSPORTER MADE IN ENGLAND © 1971 LESNEY PRODUCTS & Co LTD.*

The box, labelled *"MATCHBOX" BIG M-X THE ACTION MOVERS WITH POWER CONTROL Mechanised Tractor Plant & Winch Transporter,* was a special one for this model and not a part of the usual Matchbox series. On one end it was recorded that the model was *PRODUCED UNDER LICENCE FROM MARVIN GLASS & ASSOCIATES.* On the lid illustration a boy, wearing a safety helmet, was shown activating the winch behind the cab to pull tractors out of the buildings and up the loading ramp onto the trailer, using the rear tow hook on the tractor.

No.K-3-4 Mod Tractor and Trailer (1974–79)

This was a King Size version of the Matchbox no.MB-25e Mod Tractor, complete with a streamlined trailer. We do not consider it to be a farming model, rather it is something from the world of drag racing, so we do not propose to describe and illustrate it.

No.K-35-2 Massey-Ferguson Tractor and Trailer (1979–81)

Some sanity returned in 1979, which was the last year the Mod Tractor was in the catalogue, and its replacement was introduced the same year, modelled on the Massey-Ferguson 595. The box was an all-blue Super Kings design, and *NEW! NEU! NOUVEAU!* screamed out from the backing card. The end flaps had a more realistic illustration than before on a white background (**119**).

The tractor was an entirely new casting, painted red with a plated plastic engine block, and a silver metal baseplate which formed the front weights, axles and tow hook. The white plastic steering wheel was the same as on no.K-3-4 Mod Tractor. The cab was white plastic without window glazing, and there was a large black plastic exhaust. The grille was black plastic with inset plated headlights. There was a white stick-on paper label on either side of the bonnet with *MF 595* lettering. On the baseplate was *MADE IN ENGLAND LESNEY PROD & Co LTD © 1978 K-35 MATCHBOX Super Kings MASSEY-FERGUSON.*

The red metal two-wheeled tipping trailer with Superfast wheels was based on the version in no.K-3-3 Tractor and Trailer set. It had rather flimsy brown plastic rails which fitted into slots inside of the trailer. The chassis and drawbar were silver metal with a cream loop to go over the tow hook on the tractor. A pair of bare metal rams fitted tightly into black plastic sleeves. Cast under the trailer body was *MADE IN ENGLAND BY LESNEY FARM TRAILER "MATCHBOX" SUPER KINGS K-35 © 1978 LESNEY PROD & Co LTD.* Under the chassis was *2 K32/3* and on the bar joining the two metal rams together was *K5 4.* The trailer carried six large plastic bales of straw coloured a rather dirty brown.

This set was followed in 1981 by no.K-87-1 Massey-Ferguson Tractor and Rotary Rake which used the same Massey-Ferguson 595 tractor and was withdrawn after 1982. It lies just outside the date range of this book.

No.G-3 Farming Set (1963–64)

Box Types

During the two years this set was in the catalogues it went through two box types:

Type 1 in 1963 with *"MATCHBOX" Series* forming an arch over the illustration (**120**).

Type 2 in 1964 with just *"MATCHBOX"* forming the arch and with *FARMING SET* in bold red letters underneath (**121**).

Models from the Major Packs and King Size series were combined into a farm gift set for two years. The Major Pack items were no.M-5 Massey Ferguson '780' Combine Harvester and no.M-7 Ford Thames Cattle Truck. The King Size items were a Caterpillar D9 Tractor from the no.K-8-1 Prime Mover set (the no.K-3-1 Caterpillar Bulldozer without the bulldozer blade) and no.K-11-1 Fordson Tractor and Farm Trailer. It was a slightly strange combination of scales which made the tractor and trailer bigger than the cattle lorry. Nevertheless, it was attractively produced in a presentation box with a stepped interior with the models resting on yellow card.

The contents of the set did not change, except that the Fordson tractor can be found in the Type 1 box with either a silver-painted (not illustrated) or blue (**120**) metal steering wheel.

"Two" Packs/900 Packs

Matchbox "Two" Packs were introduced in 1976, mostly consisting of two re-issued obsolete models in blister-packed pairs. The pairs often consisted of a towing vehicle and a trailer. The packs were re-named the '900' series in late 1978. There were two Two Packs which had a farm theme - no.TP-2-1 Tractor and Trailer and no.TP-11-2 Tractor and Hay Trailer.

No.TP-2-1 Tractor and Trailer (1976–78)

This pack made use of no.MB-25e Mod Tractor in red, and the old no.40c Hay Trailer which was painted yellow and modified to remove the hay raves (**122**). There was also a rare version of the trailer in orange (not illustrated). The trailer had one-piece black plastic wheels.

No.TP-11-2 Tractor and Hay Trailer (1979–81)

This set contained no.MB-46e Ford Tractor without the plastic harrow and no.40c Hay Trailer, modified again to allow for removable black plastic hay raves, which were different to those used previously. The first version, in 1979, had a blue tractor and a light blue trailer (**123**) followed later the same year by a blue tractor and a red trailer (**124**). In 1981 the colours were changed to a metallic green tractor with a beige trailer (**125**). The blue/red combination was re-issued as no.TP-108 in 1984, and from 1985 both the

tractor and trailer were produced by Matchbox in the Far East, using the same dies but with altered base lettering which no longer included *MADE IN ENGLAND.*

No.G-6 Farm Gift Set (1979–82)

Presented in a window box with an illustration to one side of a very happy driver were no.MB-46 Ford Tractor and Harrow, no.MB-51 Combine Harvester, no.MB-71 Cattle Truck and Trailer, no.MB-40 Horse Box, a plastic farm building and plastic hurdles (**126**).

'Live-n-Learn' range

No.LL-2700 Plastic Tractor (1976–79)

This plastic tractor and driver was issued in the 'Live-n-Learn' range of pre-school toys from 1976 to 1979 (**127**). It was also included in **no.LL-400 Tractor and Trailer Set** (not illustrated) which was shown as 'new' in the 1979 catalogue. These were the only farm items in the range, probably inspired by the Fisher Price Little People series, as mentioned in the Merit chapter. The tractor was all-plastic except for the axles. It was made in three parts: the white engine block, exhaust and chassis, the red bonnet and mudguards, and blue wheels. The driver, in orange, off-white and mauve, was removable and fitted over a large projection forming part of the chassis. Underneath the tractor was *"MATCHBOX" TRACTOR Live-n-Learn TM No.LL 2700 LESNEY PROD & Co LTD 1975 MADE IN ENGLAND.*

Models of Yesteryear

The Lesney Models of Yesteryear series started in 1957 with three models (**128**). The series was the brainchild of Jack Odell, who was Lesney's engineering genius, reflecting his interest in old vehicles and desire to make more intricate and detailed models. The delightful Yesteryear models appealed as much to adults as to children, and for many people Yesteryears were their introduction to the toy collecting hobby. At first the models were given sequential numbers (1, 2, 3 etc.) but a Y prefix was added from 1960, giving Y-1, Y-2, Y-3 etc.

No.Y-1 Allchin 7N.H.P. Traction Engine (1957–64)

The very first Yesteryear was also the only farm toy in the series, although it was not obviously a farm traction engine and may equally have represented a road engine. The reverse of the boxes gave a brief history of the real vehicle, as follows:

William Allchin of Northampton, established in 1847, was one of the pioneers among traction engine builders, which were the most economical form of transporting heavy loads for nearly a century. Built in 1925, this single cylinder engine weighing 9½ tons could pull up to 38 tons, and was capable of producing 32 brake H.P., it was the last traction engine made by William Allchin Ltd. Unfortunately very few are left working today, but enthusiasts are buying these fine examples of engineering for preservation and they may be seen from time to time at traction engine rallies.

The Lesney model was always painted dark green, with a black flywheel. Gold paint trim was applied in seven places on the vehicle, later reduced to five by omitting the trim on the lower chimney ring and the piston housing. *LESNEY No.1* was cast under the front axle support, and on the left side of the model was *MADE IN ENGLAND*. The smokebox door was a separate casting, on which was cast *WM ALLCHIN LTD NORTHAMPTON*, and it was usually given a plated finish, at first copper (**129**) and later gold-coloured (**132** and **133**). Less common was silver-coloured plating (**131**), and a rare version had the door painted dark green with the body (**130**). The wheels were painted dark red, and the sixteen-spoke rear wheels at first had the treads left unpainted (**129** to **131**), and later were dark red all over (**132**). There were minor variations of the rear wheel treads which we do not consider to be important, although Yesteryear collectors do like to catalogue every little difference. A short-run version was fitted with fourteen-spoke rear wheels intended for no.Y-11 Steam Roller, and this is easily recognised by the smooth unpainted rims (**133**).

The first type of box for the traction engine had a line drawing of the model on the front face and the number *1* in black on the end flaps (**129**). The end flaps were then altered to have the number *1* in black on a white circle (**130**), and subsequently to have *ALCHIN TRACTION ENGINE* in white and the number *Y-1* in red on a white circle (**131**). The line-drawing box was then replaced by one with a colour picture of the model on the front face (**132** and **133**). Note that Allchin was mis-spelled as *ALCHIN* on the end flaps of both types of box. The rear face of both types gave a history of the real vehicle, as mentioned above, and the colour-picture box had *Scale 80:1* added at the end of the text.

Model List

(in date order within each series)
Scales are taken from Matchbox catalogues and from Hammond (1972).

Serial No.	Scale	Model	Dates	Box types
Matchbox Series, Regular Wheels				
4a	1:75	Massey-Harris Tractor	1954–56	A and B
4b	1:75	Massey-Harris Tractor	1957–60	B
8a	1:137	Caterpillar Tractor	1954–57	B
8b	1:117	Caterpillar Tractor	1958–60	B
8c	1:102	Caterpillar Tractor	1961–63	C and D

Serial No.	Scale	Model	Dates	Box types
8d	1:102	Caterpillar Tractor	1964–66	E
72a	1:62	Fordson Major Tractor	1959–66	B, C and D
50b	1:61	John Deere-Lanz 700 Tractor	1964–68	E
51b	1:61	Tipping Trailer	1964–68	E
39c	1:56	Ford Tractor	1967–72	E, F and G
40c	1:56	Hay Trailer	1967–71	E, F and G
65c	1:106	Class Combine Harvester	1967–72	E, F and G
Matchbox Series, Superfast				
MB-25e	1:52	Mod Tractor	1973–78	
MB-46e		Ford Tractor and Harrow	1978–84	
MB-51e		Combine Harvester	1978–82	
Matchbox Major Packs				
M-5	1:58	Massey-Ferguson '780' Combine Harvester	1959–66	
Matchbox King Size and Super Kings				
K-4-1	1:37	McCormick International Tractor	1960–66	
K-8-1	1:70	Prime Mover Transporter for Caterpillar Tractor	1962–66	
K-11-1	1:42	Fordson Tractor and Farm Trailer	1963–69	
K-9-2	1:80	Claas Combine Harvester	1967–72	
K-20-1	1:62	Tractor Transporter (with blue no.39c Ford Tractors)	1967–72	
K-3-3	1:45	Massey-Ferguson Tractor and Trailer	1970–73	
BM-2	1:62	Big M-x Tractor Transporter (with orange no.39c Ford Tractors)	1972–73	
K-35-2		Massey-Ferguson 595 Tractor and Trailer	1979–81	

Serial No	Model		Dates
Gift Sets			
G-4	Farm Set first version with six Regular Wheels models		1960
G-4	Farm Set second version with seven Regular Wheels models		1961–62
G-3	Farming Set with King Size and Major Pack models		1963–64
G-3	Farm Set in window box with Regular Wheels models		1968–69
G-6	Farm Set		1979–82
"Two" Packs/900 Packs			
TP-2-1	Tractor and Trailer (with Mod Tractor)		1976–78
TP-11-2	Tractor and Hay Trailer (with Ford Tractor)		1979–81
Roadway Series			
R-3	Fold Away Farm layout		1968–73
'Live-n-Learn' Range			
LL-2700	Plastic Tractor		1976–79
LL-400	Tractor and Trailer Set		1979

Serial No.	Scale	Model	Dates
Models of Yesteryear			
Y-1	1:80	Allchin 7N.H.P. Traction Engine	1957–64

Further Reading

Bowdidge, Philip, undated, *Matchbox Boxes 1 to 75 1953-1962*. Published by the author.

Bowdidge, Philip, undated, *'Matchbox' Major and Accessory Pack Series 1957-1967. A Detailed Variation Catalogue*. Published by the author.

Bowdidge, Philip, undated, *'Matchbox' Miniatures 1 to 75 1953-1990*. Published by the author.

Bowdidge, Philip, undated, *Matchbox Twin Packs 1976-1991*. Published by the author.

Hammond, Maurice A., 1972, *Lesney Matchbox 1-75 Series Diecasts*. Reedminster Publications.

Leake, Geoffrey H.B., 2004, *A Concise Catalogue of 1-75 Series "Matchbox" Toys by Lesney* 4th Edition. Published by the author.

Leake, Geoffrey H.B., undated, *Concise Checklist of "Matchbox" Miniatures 1983-1989*. Published by the author.

McGimpsey, Kevin and Orr, Stewart, 1989, *Collecting Matchbox Diecast Toys – The First Forty Years*. Major Productions Ltd.

Mack, Charlie, 1999, *Lesney's Matchbox Toys: The Superfast Years, 1969-1982*. Schiffer Publishing Ltd.

Mack, Charlie, 2001, *Collecting Matchbox Regular Wheels 1953-1969*. Schiffer Publishing Ltd.

Mack, Charlie, 2002, *The Encyclopedia of Matchbox Toys* 3rd Edition. Schiffer Publishing Ltd.

Mack, Charlie, 2005, *The Big Book of Superfast Matchbox Toys 1969-2004, Vol. 1: Basic Models and Variation Lists*. Schiffer Publishing Ltd.

Scholl, Richard. J., 2002, *Matchbox Official 50th Anniversary Commemorative Edition*. Universe Publishing.

Stannard, Michael J., 1985, *"Matchbox" 1-75 Series 1953-1969 Collectors Catalogue*. Three M's Promotions.

Stoneback, Bruce and Diane, 1994, *Matchbox Toys: A Guide to Selecting, Collecting and Enjoying New and Vintage Models*. Apple Press.

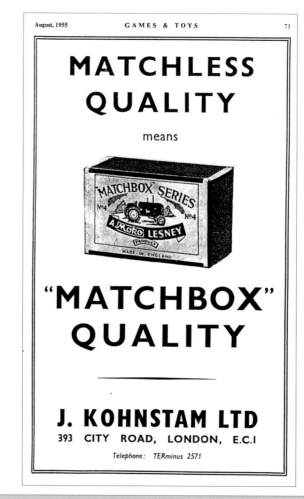

1. *Advertisement for the no.4 Massey-Harris Tractor in* Games & Toys, *August 1955 with the clever slogan 'MATCHLESS QUALITY means "MATCHBOX" QUALITY'.*

2. *Matchbox No.4a Massey Harris Tractor in a Type A box (with* Moko *in script). The boxes in (**2**) to (**6**) were all 58 mm long. The red solid-cast tractor had a tan driver between the mudguards. On this example, the exhaust must have broken off at the factory before painting, so that the model just has an air cleaner. The rear wheels were metal with gold hubs and the front wheels were metal with spokes (40 mm).*

3. *Matchbox No.4a Massey Harris Tractor in a Type A box as (**2**). Tractor as (**2**) but with an exhaust and a light brown driver.*

4. *Matchbox No.4b Massey Harris Tractor in a Type B box (with MOKO in block capitals). Red tractor without mudguards or gold trim and with a light brown driver. Metal rear wheels with gold hubs and metal front wheels without spokes.*

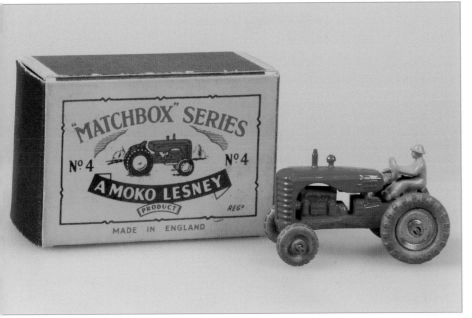

5. *Matchbox No.4b Massey Harris Tractor in a Type B box with the vegetation at either end of the tractor depicted a little differently to (**4**) and with a large black number 4 on the end flap (barely visible). Tractor as (**4**).*

6. *Matchbox No.4b Massey Harris Tractor in a Type B box as (**4**) but with a smaller black number 4 on the end flap. Tractor as (**5**) but with all-grey plastic wheels and crimped axle ends.*

7. *Matchbox No.4b Massey Harris Tractor in a later Type B box, 68 mm long, with a new picture including a driver. On the end flaps Tractor was in black on a white panel, above the model number in black on a white circle. Tractor as (6) but with riveted axle ends.*

9. *Matchbox unboxed No.8a Caterpillar Tractor, as (8) but with bare metal wheels. The driver originally had a black hat, but most of the black paint has chipped away on this example*

8. *Matchbox No.8a Caterpillar Tractor in a very early Type B box without REGD. in the bottom right corner. Yellow tractor with a yellow grille, red driver with a black hat, rare yellow wheels and green rubber tracks (41 mm).*

10. *Matchbox No.8a Caterpillar Tractor in a Type B box as (**8**) but with* REGD. *at the bottom right-hand corner of the box face. Tractor as (**9**) but pale yellow.*

11. *Matchbox No.8a Caterpillar Tractor in a Type B box as (**10**). Orange tractor with a gold grille, orange driver with a black hat and bare metal wheels with green rubber tracks.*

12. *Matchbox No.8a Caterpillar Tractor in a Type B box as (**10**). Tractor as (**10**) but yellow with a silver grille and yellow driver with a black hat.*

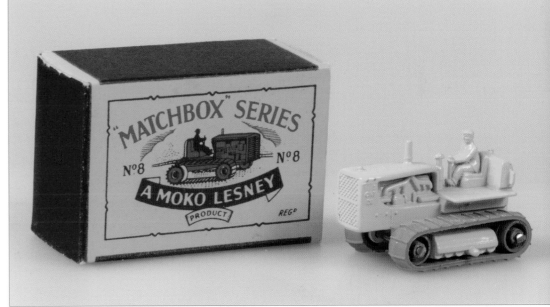

13. *Matchbox No.8b Caterpillar Tractor in a Type B box; in contrast to the preceding boxes (**8**) to (**12**), the illustration showed a tractor with a taller engine, the driver had a hand on a control lever, the top track was coloured red with the tractor and there was minimal vegetation at either end of the tractor; the reader can soon spot other minor differences. There was a black number 8 on the end flaps. Yellow tractor with a yellow grille, an all-yellow driver and bare metal wheels with CAT D8 cast near the front of the bonnet (41 mm).*

14. *Matchbox No.8b Caterpillar Tractor in a Type B box as (13) except that a black number 8 in a white circle was printed on the end flaps. Tractor as (13).*

15. *Matchbox No.8b Caterpillar Tractor in a Type B box but with more vegetation at either end of the tractor than is visible on (14). Caterpillar Tractor appeared in black within a white rectangle above the model number on the end flaps. Tractor as (13).*

16. *Matchbox No.8c Caterpillar Tractor in a Type C box similar to (15) but deeper. Caterpillar Tractor was printed in white above the model number on the end flap. In the bottom left-hand corner was REGD. U.S. PAT. OFF. Pale yellow tractor with bare metal wheels (46 mm).*

17. *Matchbox No.8c Caterpillar Tractor in a Type C box as (16) but with an entirely new illustration. Mid-yellow tractor with silver plastic wheels and without No.8 cast underneath.*

18. *Matchbox No.8c Caterpillar Tractor in a Type D box. Pale yellow tractor as (16) but with black plastic wheels and without No.8 cast underneath.*

19. *Matchbox No.8c Caterpillar Tractor in a Type D box. Tractor as (**17**).*

20. *Matchbox No.8c Caterpillar Tractor in a Type D box. Mid-yellow tractor as (**17**) but with bare metal wheels and No18 cast underneath.*

21. *Matchbox models No.8b as (**15**) and No.8c as (**16**) for a comparison of size.*

22. *Matchbox No.8d Caterpillar Tractor in a Type E box. Mid-yellow tractor without a driver and with black plastic wheels (50 mm).*

23. *Matchbox No.8d Caterpillar Tractor in a Type E box with an illustration in a yellow panel on the end flap. Mid-yellow tractor as (**22**).*

25. *Matchbox No.72a Fordson Tractor in a Type B box with the illustration showing a red tractor with red hubs, spokes on the front wheels and uncoloured hub centres to the rear. FORDSON TRACTOR in black on a white panel above the model number on the end flaps. Tractor with grey plastic front wheels, orange rear hubs with grey plastic tyres (49 mm).*

26. *Matchbox No.72a Fordson Tractor in a Type B box with an illustration as (25) but showing front hubs with red details and uncoloured rear hubs but with red centres. Fordson Tractor in white lettering on the end flaps. Tractor as (25).*

24. Advertisement in Games & Toys, September 1959 announcing the No.72 Fordson Major Tractor as 'NEW FOR OCTOBER'.

27. *Matchbox No.72a Fordson Tractor in a Type B box, with an illustration as (25) but with most of the engine not coloured in red, and end flaps as (26). Tractor as (25).*

28. *Matchbox No.72a Fordson Tractor in a Type B box with an illustration showing all-red hubs and engine, and largely uncoloured tyres. End flaps as (**26**). Tractor as (**25**).*

29. *Matchbox No.72a Fordson Tractor in a Type C box with illustration similar to (**28**). Tractor as (**25**) but with black plastic front wheels with knobbly treads, and black plastic rear tyres.*

30. *Matchbox No.72a Fordson Tractor in a Type D box with a colour illustration in the correct Fordson blue. Tractor as (**29**) but with fine treads on the front wheels.*

31. *Matchbox No.72a Fordson Tractor in a Type D box as (**30**). Tractor with orange hubs and grey tyres front and rear, with smooth treads on the front and heavier treads on the rear with* LESNEY ENGLAND *moulded around the rear tyres.*

32. *Matchbox No.72a Fordson Tractor in a Type D box as (**30**). Tractor as (**31**) but with black tyres.*

33. *Matchbox No.72a Fordson Tractor in a Type D box as (**30**). Tractor as (**32**) but with yellow hubs.*

34. *Matchbox No.G-4 Farm Set, issued in 1960. The sleeve was illustrated with Type B box designs for nos.12 (twice), 23, 31, 35 and 72. On the back of the sleeve the contents were listed as no.31 Ford Station Wagon, no.12 Land Rover, no.72 Fordson Tractor, no.23 Caravan, no.35 Horse Box and Major Pack 7 Cattle Truck.*

35. *Matchbox No.G-4 Farm Set, issued 1961-62. This revised set had G-4 on the sleeve and illustrations of Type C box designs for nos.12, 23, 31, 35, 50 and 72. On the back of the sleeve the contents were listed as no.12 Land Rover, no.23 Caravan, no.31 Ford Station Wagon, no.35 Horse Box, no.50 Commer Pick-up, no.72 Fordson Tractor and Major Pack 7 Cattle Truck.*

36. *Matchbox No.50b John Deere-Lanz 700 Tractor in a Type E box with* NEW MODEL TRACTOR *in white above and below the model number on the end flaps. Green tractor with a silver grille, yellow plastic steering wheel, yellow plastic hubs and grey plastic tyres (51 mm).*

37. *Matchbox No.50b John Deere-Lanz 700 Tractor in a Type E box with a picture of the tractor in a yellow panel on the blue end flaps. Below the picture was the model number in yellow and* JOHN DEERE-LANZ TRACTOR *in white. Tractor as (36).*

38. *Matchbox No.50b John Deere-Lanz 700 Tractor in a Type E box as (37). Tractor as (36) but with black plastic tyres.*

39. *Matchbox No.50b John Deere-Lanz 700 Tractor in a Type E box enclosed in a Fred Bronner Corporation, New York blister pack for the U.S.A. market. Tractor as (38).*

40. *Matchbox No.51b Tipping Trailer in a Type E box with NEW MODEL TRAILER in white above and below the model number on the end flaps. Green trailer with yellow plastic hubs and grey plastic tyres (63 mm).*

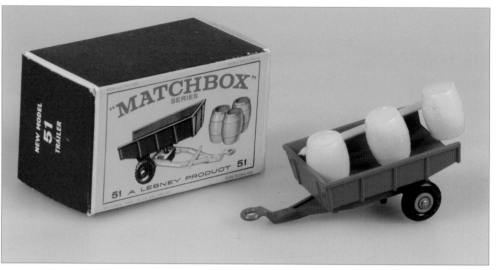

41. *Matchbox No.51b Tipping Trailer in a Type E box as (**40**). Trailer as (**40**) but with green plastic hubs and black plastic tyres.*

42. *Matchbox No.51b Tipping Trailer in a Type E box with a picture of the trailer in a yellow panel above the model number on the end flap. Trailer as (**40**).*

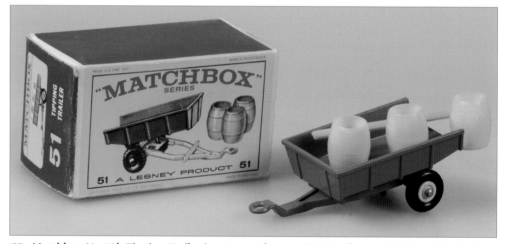

43. *Matchbox No.51b Tipping Trailer in a Type E box as (**42**). Trailer as (**40**) but with black plastic tyres.*

44. *Matchbox No.51b Tipping Trailer unboxed pre-production prototype with a dark brown body and yellow chassis.*

45. *Matchbox No.39c Ford Tractor in a Type E box with a picture of the model in a yellow panel on the end flaps above the model number. A dark blue and yellow tractor with a black plastic steering wheel, yellow plastic hubs with black plastic tyres and 5 mm exhaust (52 mm).*

46. *Matchbox No.39c Ford Tractor in a Type E box as (45). Tractor as (45) but with a yellow plastic steering wheel.*

47. *Matchbox No.39c Ford Tractor. Tractor as (45) but with a translucent white plastic steering wheel.*

48. *Matchbox No.39c Ford Tractor in a Type F box with an illustration on the end flaps in a yellow panel occupying the whole flap. Tractor as (45) but light blue and yellow with a 4 mm exhaust.*

49. *Matchbox No.39c Ford Tractor in a Type G box. Tractor as (48).*

50. *Matchbox No.39c Ford Tractor in a Type G box as (49). Tractor as (45) but single colour dark blue.*

51. Matchbox No.39c Ford Tractor. Tractor as (**48**) but dark green with black plastic hubs. This was not a pre-production model, as it had the later 4 mm exhaust. It was probably a factory colour trial.

52. Matchbox No.40c Hay Trailer in a Type E box with an illustration in a yellow panel above the model number on the end flaps. Light blue trailer with yellow plastic hubs, black plastic tyres and yellow plastic raves (86 mm).

53. Matchbox No.40c Hay Trailer in a Type F box. Trailer as (**52**), but dark blue.

54. Matchbox No.40c Hay Trailer in a Type G box. Trailer as (**52**)

55. Matchbox No.40c Hay Trailer showing examples of minor casting variations under the rear end of the trailer body.

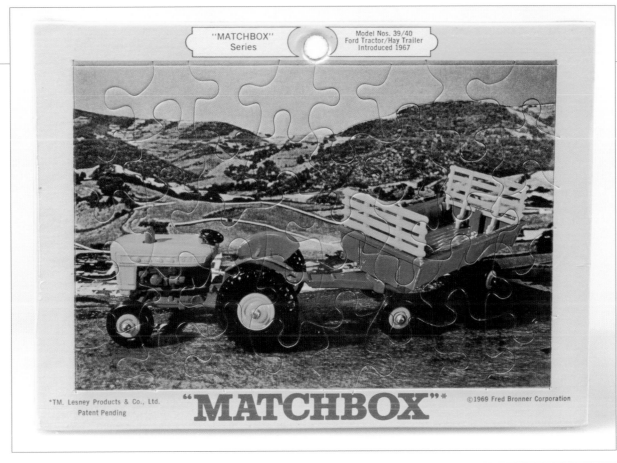

56. *Matchbox Jigsaw Puzzle marked "MATCHBOX" Series Model Nos.39/40 Ford Tractor/Hay Trailer Introduced 1967. *TM. Lesney Products & Co., Ltd. Patent Pending © 1969 Fred Bronner Corporation (178x127 mm).*

57. *Matchbox No.65c Claas Combine Harvester in a Type E box with an illustration of the model over the model number on the end flaps. Red combine harvester with yellow plastic reel and front hubs and black plastic rear wheels (78 mm).*

58. *Matchbox No.65c Claas Combine Harvester in a Type F box with an illustration of the model in a yellow panel occupying the whole of the end flaps. Combine harvester as (57).*

59. *Matchbox No.65c Claas Combine Harvester in a Type G box. Combine harvester as (57).*

60. Matchbox No.G-3 Farm Set in a window box shrink-wrapped in cellophane with six models in compartments, including the no.39c Ford Tractor, the no.40c Hay Trailer and the no.65c Claas Combine Harvester.

61. Matchbox Roadway Series No.R-3 Fold Away Farm layout with fold-up card farm buildings.

62. Matchbox No.MB-25e Mod Tractor in a Superfast box. Metallic purple body with headlights on the mudguards, silver plastic engine, yellow plastic seat and tow hook (57 mm).

63. Matchbox No.MB-25e Mod Tractor in a box as (**62**) but with NEW above the model number. Metallic purple tractor as (**62**) but with V-shaped marks on the mudguards in place of headlights.

64. *Matchbox No.MB-25e Mod Tractor in a box as (**63**). Tractor as (**63**) but red.*

65. *Matchbox No.MB-25e Mod Tractor in a box as (**63**). Metallic purple tractor as (**63**) but with red plastic seat and tow hook.*

68. *Matchbox No.MB-25e Mod Tractor pre-production model with unpainted body, and a white plastic steering wheel different in design to that on the issued model.*

66. *Matchbox No.MB-25e Mod Tractor in a blister pack mounted on a Superfast card. On the back of the card were details of five other Matchbox models. Tractor as (**64**).*

67. *Matchbox No.MB-25e Mod Tractor in a blister pack mounted on a yellow Superfast card. On the back are membership details of the U.S.A. "Matchbox" Collectors Club with a New Jersey address, so this pack was clearly for the American market. Tractor as (**63**).*

69. Matchbox No.MB-46e Ford Tractor and Harrow in a "MATCHBOX" 75 box. Dark blue tractor with black plastic wheels, silver metal engine block and yellow plastic cab interior. Dark yellow plastic disc harrow fixed to the cab with an elastic band (length with harrow on the tow hook 90 mm).

70. Matchbox No.MB-46e Ford Tractor and Harrow in a box as (**69**). Tractor as (**69**) but with a light yellow plastic disc harrow.

71. Matchbox No.MB-46e Ford Tractor and Harrow in a box as (**70**). Tractor and harrow as (**70**) but with yellow rear hubs on the tractor.

72. Matchbox No.MB-46e Ford Tractor and Harrow in a blister pack mounted on a "MATCHBOX" 75 Superfast card with illustrations on the back of six other models in the series. Tractor and harrow as (**71**).

73. *Matchbox No.MB-51e Combine Harvester in a "MATCHBOX" 75 box with two identical illustrations in white panels showing the moving parts. Red combine harvester with black wheels, rotating yellow plastic reel and swivelling yellow plastic grain unloading auger (78 mm).*

74. *Matchbox No.MB-51e Combine Harvester in a "MATCHBOX" 75 box with two panels to illustrate working parts, but one blank. Combine harvester as (73) but with yellow front hubs.*

77. *Matchbox No.MB-51e Combine Harvester pre-production version in black with black plastic wheels.*

75. *Matchbox No.MB-51e Combine Harvester in a blister pack mounted on a yellow Superfast card. On the back were membership details of the U.S.A. "Matchbox" Collectors Club with a New Jersey address. A pack for the American market. Combine harvester as (73).*

76. *Matchbox No.MB-51e Combine Harvester in a blister pack mounted on a blue-and-white card. On the back of the card were details of eight other models. Combine harvester as (74), but without a baseplate.*

78. Advertisement in Games & Toys, July 1959 announcing the new Major Pack No.5 Massey-Ferguson Combine Harvester.

79. Matchbox Major Pack No.M-5 Massey-Ferguson Combine Harvester in a Type 1 box. Red combine harvester with yellow reel, silver plastic front hubs, black plastic rear wheels and red metal steering wheel. Fine treads on front tyres.

80. Matchbox Major Pack No.M-5 Massey Ferguson Combine Harvester in a Type 2 box. Combine harvester as (79).

81. Matchbox Major Pack No.M-5 Massey Ferguson Combine Harvester in a Type 3 box. Combine harvester as (79) but with silver-painted metal open steering wheel.

82. *Matchbox Major Pack No.M-5 Massey Ferguson Combine Harvester in a Type 4 box. Combine harvester as (**81**) but orange plastic front hubs.*

83. *Matchbox Major Pack No.M-5 Massey Ferguson Combine Harvester in a Type 5 box. Combine harvester as (**82**) but yellow plastic front hubs with heavier treads on the tyres.*

84. *Matchbox Major Pack No.M-5 Massey Ferguson Combine Harvester in a Type 5 box. Combine harvester as (**83**) but yellow plastic steering wheel.*

85. *Matchbox Major Pack No.M-5 Massey Ferguson Combine Harvester in a Type 5 box. Combine harvester as (**83**) but with yellow plastic rear hubs.*

86. Matchbox Major Pack No.M-5 Massey Ferguson Combine Harvester in a Type 5 box. Combine harvester as (**83**) but with orange plastic front hubs.

87. Matchbox Major Pack No.M-5 Massey Ferguson Combine Harvester in a Type 5 box. Combine harvester as (**86**) but with dull bare metal solid steering wheel.

88. Matchbox Major Pack No.M-5 Massey Ferguson Combine Harvester in a Type 5 box. Combine harvester as (**87**) but with yellow plastic steering wheel.

89. Matchbox Major Pack No.M-5 Massey Ferguson Combine Harvester in a Type 5 box. Combine harvester as (**88**) but with orange plastic rear hubs.

90. Matchbox Major Pack No.M-5 Massey Ferguson Combine Harvester in a Type 6 box. Combine Harvester as (**87**).

91. *Matchbox King Size No.K-4-1 McCormick International Tractor in a Type 1 box. Tractor with cream lettering on bonnet, green metal hubs and a small tow hook (73 mm).*

92. *Matchbox King Size No.K-4-1 McCormick International Tractor in a Type 2 box. Tractor as (**91**) but with red plastic hubs and a larger tow hook.*

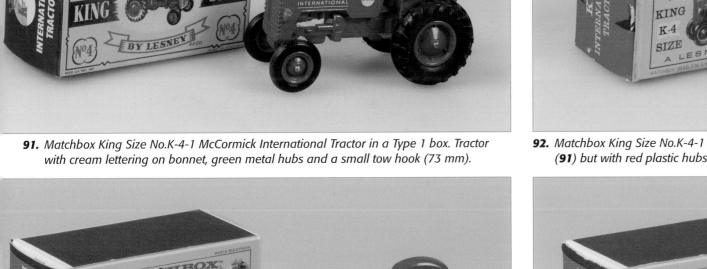

93. *Matchbox King Size No.K-4-1 McCormick International Tractor in a Type 3 box. Tractor as (**92**) but with white lettering on bonnet and with orange plastic hubs.*

94. *Matchbox King Size No.K-4-1 McCormick International Tractor in a Type 3 box. Tractor as (**93**) but with red plastic hubs.*

95. *Matchbox King Size No.K-4-1 McCormick International Tractor as (**93**) but with white-painted metal hubs.*

96. *Matchbox King Size No.K-8-1 Prime Mover for Caterpillar Tractor in a Type 1 box. Transporter consisting of orange tractor unit and low loader with bare metal hubs. Yellow D9 Caterpillar Tractor with a red engine block and bare metal wheels (305 mm).*

97. *Matchbox King Size No.K-8-1 Prime Mover for Caterpillar Tractor in a Type 2 box. Transporter as (96) but with red plastic hubs and smaller lettering on the transfers. Tractor as (96) but with red plastic wheels.*

98. *Matchbox King Size No.K-11-1 Fordson Major and Farm Trailer in a Type 1 box. Tractor with silver steering wheel and orange metal hubs; trailer with orange plastic hubs (170 mm).*

99. *Matchbox King Size No.K-11-1 Fordson Major and Farm Trailer in a Type 2 box. Tractor as (98) but with blue steering wheel and orange plastic hubs. Trailer as (98).*

100. *Matchbox King Size No.K-11-1 Fordson Major and Farm Trailer in a Type 3 box. Tractor and trailer as (**99**).*

101. *Matchbox King Size No.K-11-1 Fordson Major and Farm Trailer in a Type 3 box. Tractor and trailer as (**99**) but with red plastic hubs.*

103. *Matchbox King Size No.K-9-2 Claas Combine Harvester in a Type 1 box. Combine harvester as (**102**) but with a white driver.*

102. *Matchbox King Size No.K-9-2 Claas Combine Harvester in a Type 1 box. Green combine harvester with red reel, red plastic hubs and red driver (142 mm).*

104. *Matchbox King Size No.K-9-2 Claas Combine Harvester in a Type 1 box. Combine harvester as (**102**) but without a driver.*

105. *Matchbox King Size No.K-9-2 Claas Combine Harvester in a Type 2 box. Combine harvester as (**104**) but the usual circular logo is missing.*

106. *Matchbox King Size No.K-9-2 Claas Combine Harvester in a Type 2 box. Red combine harvester with a green background on the Claas labels, orange-yellow reel, yellow plastic hubs and a light tan driver.*

107. *Matchbox King Size No.K-9-2 Claas Combine Harvester in a Type 2 box. Combine harvester as (**106**) but with pale yellow reel and a red background on the Claas labels.*

108. *Matchbox Super Kings No.K-9-2 Claas Combine Harvester in a Type 3 box. Combine harvester as (**107**).*

109. *Matchbox No.K-20-1 Tractor Transporter with blue no.39c tractors in a Type 1 box. Red lorry with red plastic hubs, red fuel tank on the tractor unit. Dark blue tractors with 5 mm exhausts (233 mm).*

110. *Matchbox No.K-20-1 Tractor Transporter with blue no.39c tractors in a Type 2 box. Model as (109).*

111. *Matchbox No.K-20-1 Tractor Transporter with blue no.39c tractors in a Type 3 box. Model as (109).*

112. *Matchbox No.K-20-1 Tractor Transporter with blue no.39c tractors in a Type 3 box. Model as (109) but with a yellow fuel tank on the tractor unit.*

113. *Matchbox Super Kings No.K-20-1 Tractor Transporter with blue no.39c tractors in a Type 4 box. Lorry as (109) but with Superfast wheels and yellow trailer. Tractors light blue with 4 mm exhausts.*

114. *Matchbox No.K-3-3 Massey Ferguson Tractor and Trailer in a Type 1 box. Red tractor with white grille and grey metal engine block, red four-wheeled metal trailer with yellow chassis, both with yellow plastic hubs (205 mm).*

115. *Matchbox Super Kings No.K-3-3 Massey Ferguson Tractor and Trailer in a Type 2 box. Tractor and trailer as (114).*

116. *Matchbox No.BM-2 Big M-x Mechanised Tractor Plant and Winch Transporter set, box exterior.*

117. *Matchbox No.BM-2 Big M-x Mechanised Tractor Plant and Winch Transporter set, metallic blue tractor with gold trailer and three no.39c orange Ford Tractors. Black plastic roadway and blue and light brown plastic buildings. The factory roof is upside down as packed in the box.*

118. *Matchbox No.BM-A Big M-x Power Activator, blue plastic activator and battery case.*

119. *Matchbox Super Kings No.K-35-2 Massey Ferguson 595 Tractor and Trailer. Red tractor with black grille, plated lights in grille, plated engine block, silver metal front weights, baseplate and tow hook all in one, black plastic exhaust and white plastic cab. Red metal two-wheeled tipping trailer with silver metal chassis, brown plastic rails and containing six brown plastic bales (222 mm).*

120. *Matchbox No.G-3 Farming Set in a Type 1 box containing Major Packs Massey-Ferguson '780' Combine Harvester and Ford Thames Cattle Truck, and King Size Caterpillar Tractor and Fordson Tractor and Trailer.*

121. *Matchbox No.G-3 Farming Set in a Type 2 box with models as (120).*

122. *Matchbox "Two" Pack No.TP-2-1 Tractor and Trailer. Red Mod Tractor and yellow hay trailer with black wheels.*

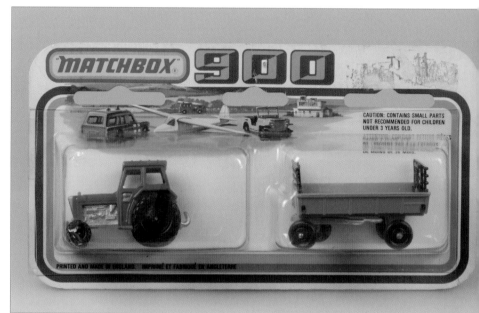

123. *Matchbox 900 Pack No.TP-11-2 Tractor and Hay Trailer. Blue Ford tractor and light blue hay trailer with black wheels and black plastic raves.*

124. *Matchbox 900 Pack No.TP-11-2 Tractor and Hay Trailer, as (123) but with red trailer.*

125. *Matchbox 900 Pack No.TP-11-2 Tractor and Hay Trailer, as (123) but with metallic green Ford tractor and beige trailer.*

126. *Matchbox No.G-6 Farm Set with Ford Tractor, plastic Harrow, Combine Harvester, Horse Box, Cattle Truck and Trailer, a plastic building and plastic hurdles.*

127. *Matchbox No.LL-2700 'Live-n-Learn' plastic tractor (99 mm).*

Introducing

MODELS OF YESTERYEAR
by LESNEY

No. 1 "ALLCHIN" TRACTION ENGINE

No. 2 "B" TYPE BUS

No. 3 "E" CLASS TRAMCAR

Sole Distributors for U.K.

J. KOHNSTAM LTD.
184/186   GOSWELL   ROAD,   LONDON,   E.C.1
Tel.: CLErkenwell 0575/6.

Export Enquiries to:
LESNEY PRODUCTS & CO. LTD., 38 Barretts Grove, London, N.16.

128. *Advertisement from Games & Toys July 1957 by the distributor J. Kohnstam Ltd introducing the first three Lesney Models of Yesteryear, no.1 Allchin Traction Engine, no.2 'B' type Bus and no.3 'E' class Tramcar.*

129. *Matchbox Models of Yesteryear No.Y-1 Allchin 7N.H.P. Traction Engine, dark green with black flywheel, dark red front wheels, dark red rear wheels with unpainted treads, copper-plated smokebox door, line-drawing box with 1 in black on the end flaps (66 mm).*

130. *Matchbox Models of Yesteryear No.Y-1 Allchin 7N.H.P. Traction Engine, as (129) but green-painted smokebox door, line-drawing box with 1 in black on a white circle on the end flaps* (photo by Vectis Auctions)**.**

131. *Matchbox Models of Yesteryear No.Y-1 Allchin 7N.H.P. Traction Engine, as (130) but silver-plated smokebox door, reduced amount of gold trim on the model, line-drawing box with model name on the end flaps.*

132. *Matchbox Models of Yesteryear No.Y-1 Allchin 7N.H.P. Traction Engine, as (131) but gold-plated smokebox door, rear wheels painted all over including the treads, colour-picture box.*

133. *Matchbox Models of Yesteryear No.Y-1 Allchin 7N.H.P. Traction Engine, as (132) but fitted with 14-spoke rear wheels intended for no.Y-11 Steam Roller. Rear wheels dark red with smooth, unpainted rims.*

Maylow

Company History

Where a model has a brand name, either on the model or the packaging, it is usually possible to find out something about the manufacturer, but unfortunately Maylow is an exception to this rule. Their tractor had *MAYLOW PRODUCTS MADE IN ENGLAND* cast underneath, and the box labels carried the same information, but without any address. Searches of phone directories, London street directories, company records and trade magazines have proved fruitless. The Maylow tractor is very much in the style of diecast toys made in north London in the late 1940s, but there is really no evidence for London manufacture any more than somewhere else in England.

We can be more certain about the date, because the tractor and binder (or reaper) are fairly close copies of the Charbens set (no.19). It is likely that the Maylow models were introduced in the late 1940s and disappeared when the zinc ban halted production of diecast toys in 1951 (see the Introduction in Volume 1). It is not known whether the tractor was sold separately or was only available in the three sets illustrated – with timber wagon, binder, and set of implements – and we are not aware of any other Maylow products apart from the farm items.

Models

Tractor

The first version of the tractor had a triangular drawbar at the rear (**1**), suitable for towing the timber wagon or the four small farm implements. The wire drawbars on the wagon and implements would simply hook into the triangle on the tractor. When the binder was introduced, a vertical pintle was added at the apex of the triangle, over which the towing eye of the binder could be fitted (**2**). *MAYLOW PRODUCTS MADE IN ENGLAND* was cast underneath the tractor. Colours of the model were:

• Dark blue tractor with orange hubs (**3**);

• Red tractor with dark green hubs (**4**, **5** and **7**);

• Dark green tractor with red hubs (**8**).

All Maylow tractors had diecast wheels with unpainted tyres, an unpainted steering wheel, and a driver painted in shades of brown which varied from light brown to medium brown to tan, always with dark brown trim on his cap.

Much scarcer than the normal tractor is a version with the addition of a clockwork motor (**9** to **11**). The die was modified to give space inside the

casting for the motor, resulting in less space for the driver's legs, so that when he was perched on top he could no longer reach the steering wheel! The die alteration removed the lettering cast underneath. Keyhole slots were provided on both sides of the model, and the clockwork motor was secured by two steel pins. The motor powered the rear axle, which was a thicker gauge of wire than on the earlier tractors. The wheels were driven tightly on to the axle so that they could not turn independently but only when turned by the motor. The same thicker wire with wheels fixed on the axle was also used at the front of the model, although not necessary for traction. The clockwork model was painted red with yellow hubs.

For the final version of the model, the clockwork motor was omitted and the die was modified to reinstate the rear axle supports, but the tell-tale keyhole slots and space for the motor remained. The wheels also continued to be fixed on the thicker axles. This version was painted dark green with yellow hubs (not illustrated).

Tractor and Log Carrier

Details of the tractor are given above. The chassis of the timber wagon was made from a rod which slotted into diecast bolsters front and rear. At first, the rod was a piece of aluminium tube, and later a solid bakelite rod. The drawbar was a single piece of bent wire, attached to the front axle. At first this was bent down at the end to fit into the triangular drawbar on the tractor, later the wire formed a loop to fit over the pintle. Chains were provided to secure the real timber load. Wheels were the same as the tractor's rear wheels, and had unpainted tyres. The version with the aluminium chassis was painted dark blue with orange hubs, or dark green with red hubs. The model with the bakelite chassis was painted dark green with yellow hubs and came with a clockwork tractor. The set of the tractor and log carrier was packed in a plain card lift-off lid box with a label on the lid (**4** to **6**).

Tractor and Binder

Details of the tractor are given above. Charbens used the word 'reaper' but Maylow used 'binder' to describe their harvesting machines. In the real world this sort of equipment would both cut and bind the crop, so both terms are correct. The Maylow model had *MADE IN ENGLAND* cast underneath and *MAYLOW* prominently cast in a large rectangle on the top of the casting, so there is no risk of confusion with the more common Charbens model. The sails were driven by a rubber band, and the seated operator was the same figure as the tractor driver. Colours were red or dark green body, with yellow wheels and sails. The tractor and binder set was packed in plain card lift-off lid box with a label on the lid, similar to the box for the timber wagon (**7**, **8** and **11**).

Tractor and Farming Implements Set

This set contained the tractor (with driver) and four implements – a hay rake, seed drill, plough and roller. All but the roller were tinplate pressings with diecast wheels. The set was packed in a lift-off lid box covered in red paper, which looked rather smarter than the boxes for the other items (**12** and **13**).

Hay Rake

The wheels were the same as the tractor's front wheels, with unpainted tyres and metallic dark green hubs. The tinplate rake was painted red and the wire drawbar was metallic dark green (**14**).

Seed Drill

The wheels were either the same as the tractor's front wheels, or the same as fitted to the binder, with a cast-in cross pattern. Both types had unpainted tyres and metallic dark green centres. The tinplate seed trough was red, and the wire drawbar was dark green or metallic dark green (**15**).

Plough

Like the hay rake, the wheels were the same as the tractor's front wheels, with unpainted tyres and metallic dark green centres. The tinplate plough was red and the wire drawbar was metallic dark green (**13**).

Roller

The roller was made from aluminium rod, painted red, with a metallic dark green wire drawbar (**16**).

Model List

Model	Dates
Tractor and Log Carrier set	late 1940s
Tractor and Binder set	late 1940s
Tractor and Farming Implements set	late 1940s

1. Maylow Tractor showing the early drawbar without pintle.

2. Maylow Tractor showing the later drawbar with pintle added.

3. Maylow Tractor, dark blue with orange hubs (88 mm).

4. *Maylow Tractor and Log Carrier, red tractor with dark green hubs, dark green trailer with red hubs, aluminium tube chassis on the trailer (tractor 88 mm, trailer 192 mm).*

5. *Maylow Tractor and Log Carrier, as (4) with original box.*

6. *Maylow Log Carrier with bakelite chassis, dark green with yellow hubs. The chassis is broken leaving only a small piece of bakelite fixed to the front bolster with a nail.*

577

7. Maylow Tractor and Binder with box, red tractor with dark green hubs, red binder with yellow sails and wheels (tractor 88 mm, binder 92 mm).

8. Maylow Tractor and Binder with box, as (**7**) but dark green tractor with red hubs.

9. Maylow Tractor with clockwork motor, red with yellow hubs (88 mm) (photo by Ray Crilley).

10. Maylow Tractor with clockwork motor, as (**9**), underneath view (photo by Ray Crilley).

11. Maylow Tractor (with clockwork motor) and Binder, red tractor with yellow hubs, dark green binder with yellow sails and wheels (tractor 88 mm, binder 92 mm) (photo by Ray Crilley).

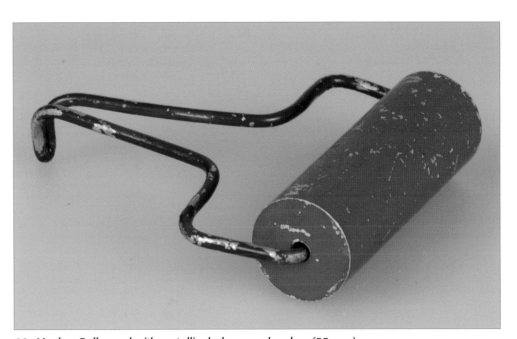

12. Maylow Tractor and Farming Implements set, containing dark green tractor with red hubs (steering wheel missing and incorrect driver), hay rake, seed drill, plough and roller.

13. Maylow Tractor and Farming Implements set, as (**12**).

14. Maylow Hay Rake, red with metallic dark green drawbar and hubs (85 mm).

16. Maylow Roller, red with metallic dark green drawbar (55 mm).

15. Maylow Seed Drill, red with metallic dark green drawbar and hubs (60 mm).

Company History

Mears was one of many companies that made a few toys in the years immediately after the Second World War. Some of those companies persisted in the toy trade and became household names, but Mears was one that fell by the wayside. Like many other diecasters they were in north London.

In the 1947 and 1948 street directories there were entries for 'Mafwo Products (Frederick Mears & William Freeman), woodware manufacturers' and also in 1948 there was a separate entry for 'Frederick C. Mears & Son, Mam Products, die-casters'. In 1949 this was shortened to just 'Frederick C. Mears & Son, die-casters'. All these directory entries gave the same address: 211 Northumberland Park, Tottenham, London N17.

The Mafwo name was no doubt derived from **M**ears **A**nd **F**reeman **WO**odware, and there were two diecast toys (a roadster and a racing car) which had *MAFWO PRODUCTS* cast underneath. Mam Products (possibly **M**ears **A**nd **M**ears?) did not appear on any diecast toys as far as we know. The Mears tractor had *MEARS & SONS LONDON ENGLAND* cast underneath and presumably dates from 1949, once the Mafwo and Mam brands had been abandoned and Mr Freeman was no longer involved. Also note that the directory entries were for Mears & Son (singular) while the lettering on the model was *MEARS & SONS* (plural). Perhaps a second son joined the business?

The prohibition on the use of zinc for toy production from March 1951 (see the Introduction in Volume 1) probably put an end to Frederick Mears' diecasting venture. This is confirmed by an entry in the account books of Kemlows Die Casting Products (see the Master Models chapter) - in November 1951 Kemlows bought a diecasting machine from Mears for £50. A new machine had cost Kemlows £130 in 1946, so it looks like this was a second-hand item, and it was probably Mears' one and only diecasting machine.

Model

Tractor

As mentioned above, the tractor probably dates from 1949. It was an unusually large model compared with most diecast toys of the time, and was a far more ambitious project than the earlier Mafwo cars. It must have been an expensive toy, making it a rare model to find today. It is a shame that it was not a model of any real tractor. It was assembled from a considerable number of components, probably because a small

D.C.M.T. diecasting machine would not have been large enough to make the tractor body as a single casting. The distinctive perforated wheels were each cast in two halves, and the front wheels were steerable from the steering wheel, linked by a single wire on the right side of the model. The driver's seat was a separate removable casting and is easily lost. The model was painted red, and there were two variations – either the wheels, steering wheel and driver's seat were unpainted (**1** and **2**), or the steering wheel was painted red and the wheel hubs were painted green on the outer faces with the smaller holes in the rear hubs omitted (**3** and **4**). *MEARS & SONS LONDON ENGLAND* was cast underneath. The tractor was individually boxed in a plain card lift-off lid box. There was a paper label on the lid showing the tractor in a farmyard scene, with *MEARS & SONS TOTTENHAM* printed at the bottom of the illustration.

1. Mears & Sons Tractor, red with unpainted wheels, steering wheel and seat (170 mm).

2. *Mears & Sons Tractor, rear view of the model in (**2**).*

3. *Mears & Sons Tractor, as (**2**) but green wheel hubs and red steering wheel, plain card box with paper label on the lid.*

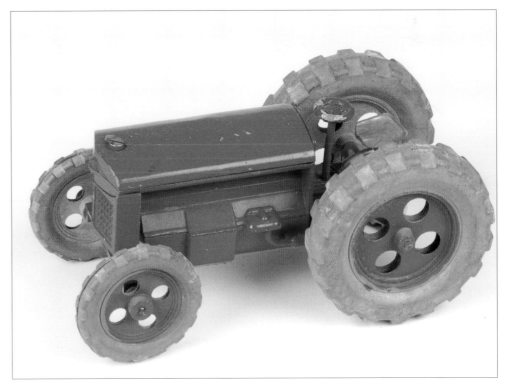

4. *Mears & Sons Tractor, as (**3**).*

Merit

Company History

Jack and Lena Randall were the founders of J. & L. Randall Ltd, manufacturers of the Merit brand of toys. Jack Randall was born in Warsaw in 1906 and came to Britain at the age of three. His first job in the toy trade was as a salesman for the wholesalers Kay (Sports & Games) Ltd of Muswell Hill, north London, in 1934. In 1936 he left Kay, and with two other partners founded his own wholesale company, Bell (Toys & Games) Ltd. Then in 1940 he resigned from Bell and started J. Randall (Toys & Games) Ltd, using the Merit brand, with his wife Lena. Her involvement was recognised by a change of the company name to J. & L. Randall Ltd around 1943. After just a few months trading, their premises in the City of London were bombed, but the Randalls started again in Potters Bar as manufacturers of signalling equipment for the Ministry of Defence, for which they incorporated another company, Signalling Equipment Ltd. They made items such as morse keys and buzzer units. According to the telephone directories, from 1941 the factory was at 163 High Street, Potters Bar, and the offices were at 16 Southgate Road, known from 1943 as Merit House.

It is interesting that Kay, Bell and J. Randall all sold sets containing lead figures and vehicles but with their own company name on the sets. Kay used Taylor & Barrett vehicles and figures in their sets, while Bell and J. Randall used Johillco vehicles and figures. The Bell sets were marked *BELL – LONDON* while the Randall sets (which must date from 1940) were marked *J. RANDALL (TOYS & GAMES) LTD.* and had *MERIT* in a scroll. We have not seen any farm vehicles in these sets, although there is a possibility that the Johillco tractor was included in some of them (see the Johillco chapter).

After the war, Merit became toy manufacturers in their own right, rather than wholesalers of other firms' products. Signalling Equipment Ltd produced scientific toys, such as microscopes and model steam engines, while a diverse range of general toys was produced by J. & L. Randall Ltd under the Merit brand. All the lines were included in one catalogue and marketed together. Around 1948 production moved to a new purpose-built factory, again called Merit House, at Cranborne Road, Potters Bar.

Merit toys included boxed games and puzzles, dressing-up sets, craft sets and pre-school toys. The latter were made in a high-gloss polythene developed for Merit by I.C.I. In 1958 Merit introduced a series of plastic OO scale model railway accessories which successfully contested the market for the similar range of Master Models diecast accessories made by Kemlows Die Casting Products, who coincidentally were also based in Potters Bar (see the Master Models chapter). However, unlike Master Models, Merit never produced a tractor in this series, and Merit never entered the market for toy vehicles, perhaps wisely given the strength of competition in that area.

In 1978 the Randalls sold the company to Letraset, who in turn sold it to a management buy-out in 1982. Jack and Lena Randall retired to Monte Carlo, where they died within a short time of each other in 1984.

Models

No.4470 David Brown Tractor Three-Dimensional Puzzle

Merit produced a series of ten little take-apart puzzles, numbered 4461 to 4470, some of which had a loop for a key chain. The date of introduction is unknown, but from the subjects included we think they started in the early 1950s. They were still available in 1974, which is the latest year for which we have located a Merit catalogue (**1**). The puzzles were made of a hard, glossy plastic, possibly cellulose acetate. Similar puzzles were sold by other firms, including Peter Pan, and Jack Randall's former company Bell. The tractor was one of the nicest items in the series, modelled on the David Brown Cropmaster of 1947. It could be dis-assembled into six pieces, starting by pulling up the seat. *MADE IN ENGLAND* was moulded underneath the bonnet. Each piece of the tractor was usually a different colour, and we have seen pale blue, dark blue, mauve, red, yellow, green and pink used for the components, although the main body of the tractor was usually pale blue or green. The tractor was secured for sale by an elastic band to a printed card, which identified it as a David Brown and included the *Merit* scroll logo. Later cards had the catalogue number added, and a diagrammatic solution for the puzzle was printed on the reverse of the card in black or red (**2** to **6**).

Around 1959 or 1960, copies of the Merit puzzles were given away with Kellogg's Corn Flakes. The series was called 'Jigtoys' and they were advertised in contemporary children's comics (**7**). The models were produced in soft polythene rather than hard plastic, and each item was a single colour rather than having multi-coloured parts. Although the tractor was not shown in the advertisement, there was a polythene version produced in red, blue, green or yellow. Compared with the Merit version, the Jigtoy tractor had a lengthened base moulding which engaged with a pillar extending down from the top of the bonnet, helping to keep the puzzle together, and also the *MADE IN ENGLAND* lettering was omitted. We think that the Jigtoy tractor was a copy of the Merit (rather than from the same mould) because there were small differences in the models, particularly in the patterns on the wheel hubs. The Jigtoys were probably made in Hong Kong – although the tractor was not marked as such, the polythene version of the Fire Engine did have *MADE IN HONG KONG* moulded on it – but as we are uncertain on this point, we have included the tractor here (**8** to **10**).

No.7667 Tractor

Although this was a generic tractor, it was a pleasing design and made a large colourful toy for small children. It was described on the end flaps of the first type of box as a *PULL-ALONG TAKE-APART TRACTOR*, and *A MERIT SAFE-PLAY NURSERY TOY* and *PULL-ALONG TAKE-APART NURSERY TOY* appeared on the box sides. The box also helps with the date of introduction, because © *by J & L RANDALL LTD 1960* was shown on the end flaps and *PATENT APPL. NO. 34357/60* was printed on two of the sides, indicating a patent application date of 1960. The tractor was still shown in the catalogue for 1974. It was made of the glossy polythene referred to in the Company History, although this material did not have quite such a reflective glossy appearance as the cellulose acetate of the puzzle tractor. The toy was held together by a threaded plastic bolt passing through the bonnet, which had a knob at the top end by which it could be unscrewed. This allowed the bonnet and engine to be dismantled from the base, and the other parts were just a push-fit. A string of twisted red and yellow cords was tied to the pivoting front axle, and had a plastic ball at the other end by which the model could be pulled along. Moulded underneath the chassis was *A MERIT SAFE PLAY NURSERY TOY* and *J & L R LTD MADE IN ENGLAND*. Earlier models were in one of two colour schemes (**11** and **12**). The example in (**11**) used red plastic for the bonnet, front hubs and driver's legs, yellow for the chassis (including mudguards) and steering wheel, pink for the engine, front axle, threaded bolt, gear lever and driver's body, light blue for the rear hubs, seat, exhaust pipe, front tyres, engine cables and driver's hat, and green for the rear tyres. The example in (**12**) had the yellow and red colours reversed and also the pink and light blue reversed, with only the green rear tyres unchanged. These models had the earlier type of box with the 1960 copyright date and pictures of a model with a red bonnet and yellow chassis. Sometimes various foreign language descriptions were shown on the inner end flaps. A later style of box had illustrations of two small children playing with the toy. The colours of the model in this case were the same as the box pictures, with light blue plastic for the bonnet, front hubs and driver's

legs, pink for the chassis (including mudguards) and steering wheel, yellow for the engine, front axle, threaded bolt, gear lever and driver's body, red for the rear hubs, seat, exhaust pipe, front tyres, engine cables and driver's hat, and green for the rear tyres (**13**). We know that this version was issued before 1974 because a leaflet was included in the box illustrating the full range of Merit Nursery Toys, and including some items that the 1974 catalogue indicated were discontinued. There was also an unusual version of the toy entirely made in light blue plastic, except for an off-white driver's hat and green rear tyres (**14**). It is not clear where this should be placed in the sequence of variations.

No.7451 Little People on the Farm

The 1974 catalogue illustrated this tractor and trailer set (**15**), but unfortunately we have not been able to locate an example to give a better description. The tractor and trailer were copies of the Fisher-Price Little People toys (made in the U.S.A.), and it would be interesting to find an example of the Merit products in their packaging which might indicate, for example, that they were made under licence from Fisher-Price, or that they were made by Fisher-Price outside the U.K. and imported and packaged by Merit.

Further Reading

Obituary of Jack and Lena Randall, *Toys International & The Retailer,* August 1984.

'Merit Moves Next Door', *Toys International & The Retailer*, October 1984.

CAT. NO. 4461/70
THREE-DIMENSIONAL TAKE-APART PUZZLES

Three-dimensional take-apart Puzzles – each item breaking down into several sections. Each one carded, showing easy to follow solution.

Assorted different puzzles as illustrated, not available separately.

1. Merit no.4461/70 Three-Dimensional Take-Apart Puzzles, shown in the 1974 catalogue.

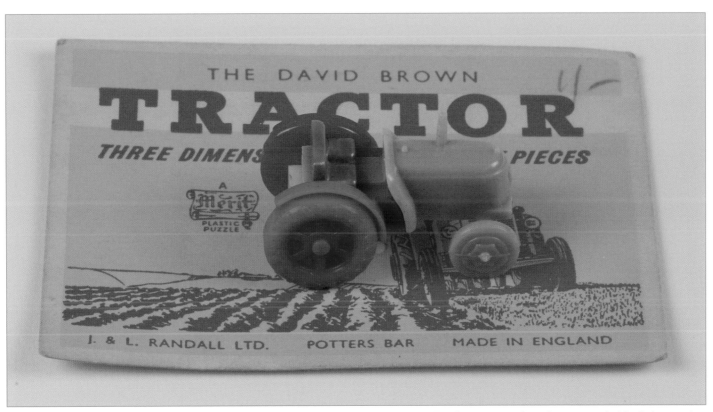

2. *Merit no.4470 David Brown Tractor Three-Dimensional Puzzle with original card, solution printed on the reverse in black (46 mm).*

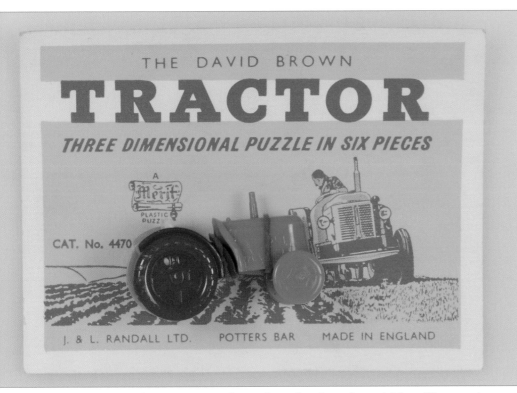

3. *Merit no.4470 David Brown Tractor Three-Dimensional Puzzle, as (2) but different colours, CAT. No. 4470 added to the card.*

4. *Merit no.4470 David Brown Tractor Three-Dimensional Puzzle, reverse of the card in (3) with printing in red.*

5. *Merit no.4470 David Brown Tractor Three-Dimensional Puzzle, as (**3**) but different colours.*

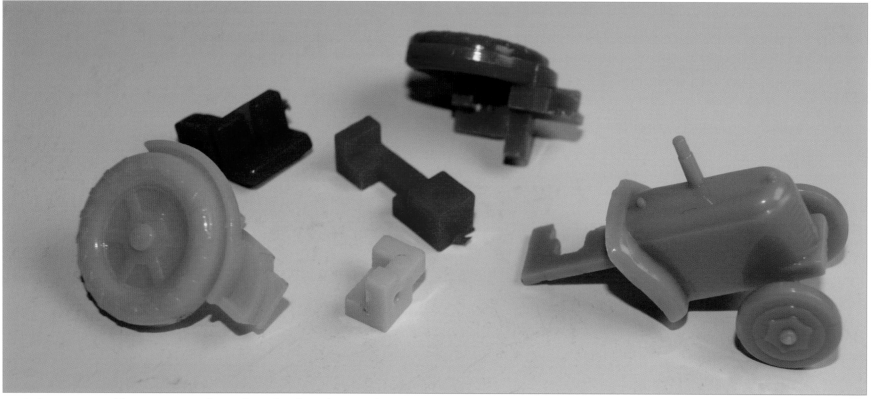

6. *Merit no.4470 David Brown Tractor Three-Dimensional Puzzle, as (**5**) but dis-assembled.*

7. *Kellogg's Corn Flakes advertisement from around 1959 showing the Jigtoy series of puzzles.*

8. *Jigtoy David Brown Tractor Three-Dimensional Puzzle, a polythene copy of the Merit puzzle, probably made in Hong Kong, red (46 mm).*

9. *Jigtoy David Brown Tractor Three-Dimensional Puzzle, as (**9**) but blue.*

10. *Jigtoy David Brown Tractor Three-Dimensional Puzzle, as (**9**) but dis-assembled.*

12. *Merit no.7667 Tractor, as (**11**) but yellow and red colours reversed, and pink and light blue colours reversed.*

11. *Merit no.7667 Tractor, early version with box showing a copyright date of 1960 (199 mm).*

14. *Merit no.7667 Tractor, unusual model entirely in light blue plastic, except for off-white driver's hat and green rear tyres (199 mm).*

13. *Merit no.7667 Tractor, later version with box showing two children (199 mm).*

CAT. NO. 7453

CAT. NO. 7452 CAT. NO. 7454 CAT. NO. 7451

CAT. NO. 7450C CAT. NO. 0450B CAT. NO. 7450A

'LITTLE PEOPLE' TOYS
CAT. NO. 7450A B C SAFETY PATROL
(Assorted 3 types)
CAT. NO. 7451 ON THE FARM
CAT. NO. 7452 BOAT FLEET
CAT. NO. 7453 ON THE ROAD
CAT. NO. 7454 AT SCHOOL

This most attractive range of junior activity toys is made to quality standards, each set attractively presented on colourful backing cards punched for wire rack display.

15. *Merit no.7451 Little People on the Farm tractor and trailer set, pictured in the 1974 Merit catalogue.*

Mettoy

Company History

The Mettoy Company, named after a conflation of the words Metal and Toy, was renowned for its tinplate toys made from the mid-1930s to the late 1950s. These were manufactured from printed metal sheets and were usually fitted with clockwork motors. Mettoy's tinplate range ceased soon after the company launched its highly successful diecast Corgi Toys in 1956.

The main sources of information for Mettoy are Marcel R. Van Cleemput's *The Great Book of Corgi* (1989), pp.8-27, David Cooke's *Corgi Toys* (2008), pp.7-11 and contemporary Mettoy catalogues. Catalogues do, however, have to be used with care because the names given to toys in the catalogues do not always correspond with the names on the boxes.

The company was founded by a German refugee and experienced toy manufacturer, Philipp Ullman, who had run Tipp & Co, a well-known toy company, founded in Nuremburg in 1912 and famed for their tinplate toy cars. Germany had a long history of producing fine tinplate toys, and no doubt he brought that expertise with him. Companies House records show that The Mettoy Co. Ltd was incorporated on 31 August 1932, which suggests that Ullman arrived in Britain before 1933, the date given by Van Cleemput. He was joined by his colleague, Arthur Katz, in July 1934, and this highly successful partnership lasted until Ullman retired in 1969. He died in 1971. Katz retired in 1980 and died in 1999.

Telephone directories record The Mettoy Co. Ltd from 1934 to 1947 at 70 Finsbury Pavement, London EC2 and from 1949 at 120 Moorgate, London EC2. These addresses were presumably their London offices. The two men developed their factory from 1935 to 1937 in premises owned by the model engineers Winteringham Ltd at Stimpson Avenue, Northampton. In 1938 they moved to their own site at Harlestone Road, Northampton, which remained the centre of their operations for the next forty years.

Mettoy's black-and-white catalogue for 1936 boasted a wide range of tinplate items, including cars, a caravan, fire engines, lorries, a motor bike and a road roller, most of which were exported to the U.S.A. After war broke out, like so many British toy companies, they switched to munitions production. The company expanded into a new factory in Swansea, possibly in 1944. From 1946 Mettoy was recorded in telephone directories on the Trading Estate at Carmarthen Road, Fforestfach in Swansea.

Toy production resumed in 1945, and the company had a full-colour catalogue for their Mettoy Playthings out by 1946. Their first tinplate tractor with its implements appeared in 1948 and, judging by the number of items still circulating among collectors today, their tinplate farm toys must have sold in large numbers, although good boxed examples are hard to find. In the same year the company also released the first six of their diecast 'Heavy Cast Mechanical Toys Perfect Scale Models', later known as Castoys, made for Mettoy at the Birmingham Aluminium Casting Co. Ltd at Smethwick. A Ferguson tractor, which they just called a 'Farm Tractor', was added to this range in 1949. These cast items were relatively large, unlike the smaller and more successful Dinky Toys produced by Meccano at 1:48 scale. Castoys did not sell well and are rare today.

A second range of smaller diecast 'Miniature Clockwork Models' was launched in 1950, including a tractor and trailer, but these too failed to capture the market, probably because Dinky were so dominant at the time and were constantly adding new models to their own range. From 1952, the miniature tractor and trailer were produced in plastic. Throughout the 1950s, plastic items increasingly replaced Mettoy's tinplate toys.

It was only when Mettoy started their superbly detailed and packaged Corgi Toys in 1956 that the company managed to break into the diecast market in a big way. The story of Mettoy's transformation from tinplate to diecast production was one of trial and error in which they eventually succeeded.

Corgi Toys are covered in the Corgi chapter in Volume 1.

Models

Tractors based on the Fordson E27N and their implements

No.3262 Farm Tractor or Heavy Tractor (1948–54)

Mettoy's first tinplate tractor, released in 1948, was a dark green clockwork steerable tractor with a silver grille, a long bare metal air cleaner and exhaust, a bare metal steering wheel, red seat, grey-and-yellow hubs and grey tyres. It looked like a Fordson E27N (**1**). The spring was in the rear left-hand wheel and the tractor was in an illustrated blue box with yellow and red strips. Printed on the box was *METTOY PLAYTHINGS FARM TRACTOR HEAVY STEEL MADE FOR HARD WEAR PERFECT MECHANICAL TOYS MADE IN GREAT BRITAIN.* The illustration on the box showed a model with green hubs.

In 1951 the tractor was re-issued in light green with a yellow grille and a black seat. *METTOY PLAYTHINGS* was written across the bonnet. A blue-and-red plastic driver was added in 1952, when the model number was extended to no.3262/1 (**2**). The box had an entirely new design with a realistic illustration and on the end flap was *This Tractor can pull the following Farm Implements 1-Mowing Machine, 2-Rake, 3-Disc Harrows, 4-Farm Trailer.*

Released with the tractor in 1948 were the Mowing Machine, Rake, Disc Harrows and Farm Trailer. Each was boxed and sold separately.

No.3270 Mowing Machine or Tractor-Mower (1948–51)

The Mowing Machine, called Tractor-Mower on the box, had an upper metal shield and rotary blades below. The box incorrectly illustrated the model with the axle above the metal shield whereas it was actually underneath. The illustration also showed a yellow tow bar and blue blades, but these colours were never used on this model, which was first coloured orange with green blades and had large bare metal spoked wheels (**3**). The orange was later changed to red (not illustrated).

The blue, yellow and red colour scheme on the early tractor box (**1**) was also used on the implement boxes (**3**), (**7**), (**9**) and (**17**). On the implement boxes, where the tractor was not a part of the contents, it was always coloured red, but this colour was never used on the tractors.

No. 3274 Rake (1948–51)

We have not yet seen a boxed version of the rake, but the Mettoy 1948 catalogue showed it with an orange tow bar and yellow tines with bare metal wheels as in (**4**). The tines were later green (**5**). On this version the rake had bare metal spoked wheels which were later replaced in brown plastic (**6**). The rake continued to be sold in set no.3262/2468 until 1954.

No.3272 Disc Harrows (1948–51)

The harrows had a yellow frame and drawbar and two rows of thin red tinplate discs underneath, correctly shown on the blue, yellow and red box (**7**). The frame was later changed to green (**8**) and the model continued to be available in sets until 1958.

No.3276 Farm Trailer, Tractor Trailer or Farm Tip Trailer (1948–57)

The tipping trailer was the longest-lived of the group. The first ones were deep red with green wire raves. Their wheels were similar in colour to the tractors. The first box, with end flaps, was the same design as used for the mower and disc harrows (**9**). Later, the box design was changed to plain red with a lift-off lid (**10**). At the very end of the sequence the trailers were given yellow end boards and black raves (**11**). The trailer continued for a further year, until 1958, in sets.

No.3262/76 Mechanical Tractor and Trailer Set (1949–54)

From 1949 the tractor and trailer were sold as no.3262/76 Mechanical Tractor and Trailer Set (**12**) with a driver added in 1952 (**13**). From 1952 the original range of farm implements was available only in the mechanical tractor sets, except for the farm tipping trailer which was always available separately.

No.3262/0246 Mechanical Farm Tractor Set (1949–50)

In 1949 the tractor with its four implements was released as a boxed set, the no.3262/0246 Mechanical Farm Tractor Set (**14**).

No.3264 Mechanical Farm Tractor or Giant Tractor (1951–56)

In 1951 a Mechanical Farm Tractor or Giant Tractor appeared, with a rubber skin over the rear tinplate tyres. This was a large red-and-blue clockwork steerable tractor with black seat, bare metal steering wheel, long exhaust and air cleaner, wind-up left-hand rear wheel and starting handle. Turning the starting handle created a clicking noise. A plastic driver was added in 1952 when the tractor became no.3264/1 (**15**). In 1956 the colours were changed to green and red (**26**), but for one year only.

The box, showing a green tractor, had four farming scenes representing farming in different parts of the world, one on each side, captioned *IN THE GREAT CANADIAN WHEATFIELDS, MIDST THE DATE PALMS IN THE TROPICS, HARVEST TIME IN SOUTHERN ENGLAND* and *BENEATH THE SOUTHERN ALPS, NEW ZEALAND.* On the end flaps was *METTOY PLAYTHINGS Mechanical FARM TRACTOR MADE IN GREAT BRITAIN.* The implements shown with the four farming scenes, a self-binder, a plough, a four-wheeled trailer and a two-wheeled trailer, were not those issued with the tractor, which actually pulled the implements already available for the smaller green no.3262 tractor.

No.3262/2468 Farm Tractor and Implement Set (1951–54)

In 1951 the tractor and implement set with the smaller tractor was re-released with the original mowing machine replaced by a high speed hay cutter (**16**) which was also sold separately. A driver was added from 1952. The large boxed sets must have been popular because they are still seen quite frequently. In 1952 the large set sold for 37/6d and the tractor and trailer set for 22/6d.

No.3278 High Speed Hay Cutter (1951)

The High Speed Hay Cutter with a reciprocating blade first came out with bare metal spoked wheels in a box similar to the one used for the other implements (**17**). This was followed by a cutter with brown plastic spoked wheels (**19**) or red plastic spoked wheels (not illustrated). There was also a variation of the rather clever single (**17**) or double (**18**) notched plastic wheel on the axle which caused the cutter blade to reciprocate. The body of the cutter was always green and the blades red and yellow. The illustration on the box showed a blue body, but no such body has been found. The Hay Cutter was only available individually boxed for one year, but continued in sets until 1958.

No.3263 Farm Tractor (1955–58)

In 1955 the original green no.3262 tractor was replaced with a new orange, yellow and grey clockwork steerable Farm Tractor or Mechanical Tractor with bare metal steering wheel and air cleaner and a black seat (**20**). Although there was no exhaust, this still appears to have been based on the E27N. The spring was again in the left-hand rear wheel, but unlike no.3262, no driver was provided. An innovation was a grey plastic engine block inserted into the tinplate body. There was a new design for the box which had *Mechanical TRACTOR* on all faces, two of which showed the model and two the tractor at work pulling disc harrows.

No.3263/76 Tractor and Trailer Set (1955–58)

Also released in 1955 was the Tractor and Trailer Set for this tractor using the same tipping trailer as before (**21**). On the box was *Mechanical TRACTOR and TRAILER in heavy pressed steel,* and on the two sides was an illustration showing the tractor and trailer in profile. On the lid was a realistic scene of the tractor towing a load of straw. The two illustrations were both set in white ovals.

No.3273 Farm Plough (1955–57)

Also in 1955 came the most sophisticated farm implement produced by Mettoy, the Farm Plough, a trailed two-furrow plough with an elaborate diecast frame, mouldboards and two adjustment levers to control the height of the two spoked wheels. There were two versions, a red frame with yellow wheels (**22**) and an orange frame with silver-painted wheels (**23**). It was only sold for three years, so boxed examples are difficult to find. The box we have seen (**22**) was mauve and white with two illustrations, as before, one of the model and one of a realistic farming scene with a tractor and plough at work. Printed on the box was *FARM PLOUGH* and *THIS PLOUGH IS ONE OF THE FARM IMPLEMENTS SUITABLE FOR USE WITH METTOY TRACTORS Nos 3263 & 3264/1.* The plough continued for a further year, until 1958, in set no.3263/2368.

No.3263/73 Tractor and Plough Set (1955–57)

The plough also came out with the new tractor in a Tractor and Plough Set (**24**).

No.3263/2368 Farm Tractor and Implement Set (1955–58)

This was an updated version of no.3262/2468 containing the new no.3263 Tractor and no.3273 Plough, together with the disc harrows, tipping trailer, hay cutter and a plastic driver (**25**).

The final year for the large-scale tinplate tractors and their implements was 1958. Meanwhile Mettoy had also been running a separate series of smaller scale tinplate tractors based on the Fordson Power Major with a more limited range of implements.

Tractors based on the Fordson Power Major

No.6435 Mechanical Farm Tractor or Clockwork Tractor (1950–54)

For Christmas 1950 a tinplate tractor with more rounded bodywork, similar to the Fordson Power Major, was introduced. It was coloured green and red in the catalogue for 1951, with a brake and with a separate tinplate slot-in driver (**27** top right). We have not so far seen the no.6435 in these colours, and it is possible that it actually came out in green and yellow (**28**) and that the 1951 catalogue was misleading. It was shown in green and yellow on the same page in the 1951 catalogue, towing the trailer and the hay cutter. It was also shown in green and yellow in 1952. According to the 1953 and 1954 catalogues, the colours were then changed to red and yellow, but we have never seen these colours on the clockwork version, only on the later friction-drive model (see no.6436 below). The wheels were orange and yellow, and *MADE IN GT BRITAIN* was on the right side of the bonnet.

No.6435/116 Farm Tractor and Trailer Set (1951–53)

The no.6435 Clockwork Tractor in green and yellow was sold in a boxed set with a new two-wheeled *FRUIT & VEGETABLES* trailer (**27** centre and **29**). The Fruit & Vegetables trailer was designed specifically for the small no.6435 tractor and was not sold separately.

No.6435/77 Farm Tractor and Mechanical High Speed Hay Cutter Set (1951)

The no.6435 Clockwork Tractor in green and yellow also came out in a set with a hay cutter. The hay cutter was the same as the no.3278 High Speed Hay Cutter but with tinplate wheels. There were two variations of this hay cutter, with plain yellow wheels (**30**) and with decorated yellow-and-blue wheels (**31**). The illustrated box showed the model on the top, and a farming scene on the sides with a tractor and hay cutter with spoked wheels. The moving parts on the cutter were emphasised on the box with *MECHANICAL ACTION* on the top beside the picture of the model.

No.6438 Tractor with Caterpillar Drive (1951–54)

For four years from 1951 there was also a clockwork crawler, the no.6438 Tractor with Caterpillar-drive (**27** top left), but we do not have a photograph of this model.

No.6430 "Happy Hayseed" and his Bucking Tractor (1951–52)

The early 1950s was clearly the high point in Mettoy's production of tinplate farm toys, with four clockwork tractors and their implements on the market. In 1951 the small-tractor range was extended further with another version of the no.6435, the no.6430 "Happy Hayseed" and his Bucking Tractor, 'an amusing novelty' with a 'powerful clockwork motor' (**32**), apparently inspired by D.C.M.T.'s Harry Hayseed on his Tricky Tractor, which was on the market by 1949 (see the D.C.M.T. and Lone Star chapter in Volume 1). The Mettoy version was not advertised after 1952 and good examples are hard to find. The tractor was red and yellow with a blue steering wheel, bare metal seat and orange-and-yellow wheels. The driver had blue trousers with several patches, white shirt, blue sleeves and a long red nose. *MADE IN GT. BRITAIN* was on the right-hand side of the bonnet. Unlike the other tractors which were wound on the left, the fixed key was underneath.

No.6436 Tractor with friction-drive mechanism (1956–58)

In 1956 the no.6435 Clockwork Tractor was replaced by no.6436, a friction-driven version. It was coloured red and yellow, with orange and pale blue wheels. *MADE IN GT. BRITAIN* was on the right side of the bonnet (**33**).

No.6439 Caterpillar Tractor with friction-drive mechanism (1956)

The no.6438 Tractor with Caterpillar Drive was changed in 1956 to no.6439 friction-driven Caterpillar Tractor (**34**). It was coloured blue and yellow with yellow plastic wheels and rubber tracks. The wheels on the box were red, but we have not seen this colour on a model. On the right-hand side of the bonnet was *MADE IN GT. BRITAIN*. There were two illustrations on the box, one of the model and one of a farming scene with a green crawler which did not look at all like the model.

It is interesting that only shortly before the tinplate tractor range was due to come to an end, new models were still coming on the market. However, the no.6439 Caterpillar Tractor was not repeated in 1957, and the no.6436 Tractor was withdrawn after 1958.

Heavy Cast Mechanical Toys and their Implements

No.860 Mechanical Farm Tractor with Brake (1949–51 and 1953)

The first Mettoy diecast clockwork tractor, issued in 1949, was an orange-and-yellow tractor with bare metal steering wheel and hubs, and rubber tyres. It was based on the Ferguson TE20 (**35** to **37**). Underneath was *METTOY MADE IN GREAT BRITAIN BRITISH AND FOREIGN PATENTS APPLIED FOR* and *MTY 860* was cast as front and rear number plates. The model had a lift-up bonnet (**37**), and the key fitted over a long bar which projected to the left just in front of the rear wheel. While the tractor looked like the Ferguson at the front, the rear had a strange extension to hold the clockwork mechanism, not like the real Ferguson (**36**).

The illustrated box showed two farming scenes of ploughing and pulling a self-binder, with the tractor coloured orange and green. The tractor was sold separately and also in a boxed set with a yellow trailer with red raves, probably numbered 861 (not illustrated). These were late additions to an original group of six cars and vans in a new no.800 series first launched in 1948 and later known as Castoys. The series was not a success and most items were withdrawn by 1951. The tractor was in the catalogue from 1949 to 1951 and, after a gap of one year, again in 1953, when it was described as being 'available again' as the no.860 Heavy Die-cast Clockwork Farm Tractor, 'modelled after the famous Ferguson Tractor'.

No.862 Castoys Tractor (1958–59)

Five years later the tractor was re-introduced as a non-mechanical version, in red and blue with a black steering wheel and red hubs, numbered 862, without the rear extension and with just a rear loop to hold the trailer pin (**38**). Cast underneath this tractor was *METTOY CASTOYS MADE IN GT. BRITAIN*, and *MTY 862* was cast as the front number plate.

No.863 Castoys Tractor, Driver and Trailer Set (1958–59)

No.862 Castoys tractor was also sold in a boxed set with a trailer and a plastic driver as the no.863 Castoys Tractor, Driver and Trailer set (**39**). Underneath the trailer was *METTOY CASTOYS MADE IN GT. BRITAIN*. The Castoys models appeared in the 1958 and 1959 catalogues only (**40**), so they were short-lived and examples, particularly of the tractor and trailer set, are now rare.

Miniature Clockwork Models

No.660 Tractor (1950–53)

Released in 1950 was a much smaller diecast tractor in a new Miniature Clockwork Models Series, the no.660 Tractor, with a green body, tinplate mudguards, plastic hubs and bare metal steering wheel. The same casting was used for no.663 Road Roller (**41** top right). The winding key was inserted into the tractor on the left. The tractor was sold in a window box (**41** top left and bottom), and the plastic hubs varied in style and colour (**42**). To sell items in a window box was an innovation for Mettoy, not used on the larger farm items. The tractor was made in plastic from 1952 when a driver was added (**44**).

No.660/661 Mechanical Model Tractor with Wagon and Hay Brackets (1950–53)

The no.660 Tractor was also sold with a yellow trailer with red raves (**43**). The tractor and trailer were replaced in plastic in 1952, when the tractor was given a plastic driver, but the trailer retained its red metal raves (**44**). The tractors had no identification on them, but the yellow metal trailer had *MADE IN GT. BRITAIN C8693* cast underneath. The production of these miniature items had a short life and they were withdrawn after 1953.

Mettoy's early attempts at clockwork diecast models had not been a great success, but no doubt this was all a part of the company's learning process before Corgi Toys were launched without clockwork in 1956 (see the Corgi chapter in Volume 1).

Model List

Serial No.	Model	Dates
Tinplate Tractors		
3262	Farm Tractor, called the Heavy Tractor in the catalogue (silver and dark green with clockwork motor, long air cleaner and exhaust, re-issued in 1951 in light green with a yellow grille) Plastic driver added in 1952 when the tractor became no.3262/1	1948-54
3263	Mechanical Tractor, called Farm Tractor in the catalogue (orange, yellow and grey with clockwork motor, grey plastic engine, air cleaner and no exhaust)	1955-58
3264	Mechanical Farm Tractor, called Giant Tractor in the catalogue (red and blue with clockwork motor, rubber skin on rear tyres and long air cleaner and exhaust, colours changed to red and green in 1956) Plastic driver added in 1952, when the tractor became no.3264/1	1951-56
6430	"Happy Hayseed" and his Bucking Tractor (red and yellow with clockwork motor, attached driver with a red nose)	1951-52
6435	Mechanical Farm Tractor, called Clockwork Tractor in the catalogue (green and yellow with clockwork motor, brake and slot-in driver)	1950–54
6436	Tractor (red and yellow with friction-drive mechanism for forward and reverse, brake and slot-in driver)	1956-58
6438	Caterpillar Tractor, called Tractor with Caterpillar Drive in the catalogue (blue and yellow with clockwork motor, rubber tracks, a brake and a slot-in driver)	1951-54
6439	Caterpillar Tractor (blue and yellow with friction-drive mechanism for forward and reverse, brake, rubber tracks and slot-in driver)	1956

Serial No.	Model	Dates
Tinplate Implements		
3270	Mowing machine (with rotary blades)	1948-51
3272	Disc Harrows (yellow or green frame with red discs) sold separately	1948-51
	Thereafter only in sets	to 1958
3273	Farm Plough (orange frame with black levers, silver wheels and mouldboards or red frame with black levers, yellow wheels and silver mouldboards)	1955-57
	Thereafter in set no.3263/2368	to 1958
3274	Rake (orange and green or orange and yellow) sold separately	1948-51
	Thereafter only in the Farm Tractor and Implement Set after which it was replaced by the plough	to 1954
3276	Farm Trailer sometimes called Farm Tip Trailer in the catalogue (red with green or black raves, or red with rare yellow end boards and black raves)	1948-57
	Thereafter in sets	to 1958
3278	Mechanical High Speed Hay Cutter (dark green and then light green body with reciprocating blade in a cutter bar with metal and then plastic spoked wheels)	1951
	Thereafter in sets	to 1958
Tinplate Boxed Sets		
3262/76	Mechanical Tractor and Trailer Set, with plastic driver added in 1952	1949-54
3262/0246	Mechanical Farm Tractor Set (with disc harrows, mowing machine with rotary blades, tipping trailer and hay rake)	1949-50
3262/2468	Farm Tractor and Implement Set (with disc harrows, mowing machine with reciprocating blades in cutter bar, tipping trailer and hay rake). Plastic driver added in 1952	1951-54
3263/2368	Farm Tractor and Implement Set (with disc harrows, mowing machine with reciprocating blades in cutter bar, tipping trailer and plough)	1955-58
3263/73	Tractor and Plough Set	1955-57
3263/76	Farm Tractor and Trailer Set	1955-58
6435/77	Farm Tractor and Mechanical High Speed Hay Cutter Set	1951
6435/116	Farm Tractor and Trailer Set (with a *FRUIT & VEGETABLES* trailer)	1951-53
Heavy Cast Mechanical Toys		
860	Mechanical Farm Tractor with Brake (clockwork version in orange and yellow)	1949-51 & 1953
861?	Mechanical Farm Tractor and Trailer Set	1949-51?
Castoys		
862	Non-mechanical version of the Farm Tractor (red and blue)	1958-59
863	Tractor and Trailer Set (non-mechanical tractor in red and blue with driver, yellow trailer with red raves)	1958-59
Miniature Clockwork Models		
660	Tractor (diecast, made in plastic with driver from 1952)	1950-53
660/661	Mechanical Model Tractor with Wagon & Hay Brackets (diecast, made in plastic with driver from 1952)	1950-53

Further Reading

Cooke, David, 2008, *Corgi Toys*. Shire Publications.

Van Cleemput, Marcel R., 1989, *The Great Book of Corgi 1956–1983*. New Cavendish Books.

1. *Mettoy tinplate no.3262 Farm Tractor, or Heavy Tractor, in an illustrated box, silver and dark green steerable tractor with red seat and bare metal steering wheel, long exhaust and air cleaner and clockwork motor with wind-up left-hand rear wheel (195 mm).*

2. *Mettoy tinplate no.3262/1 Farm Tractor, or Heavy Tractor, in an illustrated box, yellow and light green steerable tractor with black seat and bare metal steering wheel, exhaust and air filter, clockwork motor with wind-up left-hand rear wheel and a plastic driver (195 mm).*

3. *Mettoy tinplate no.3270 Tractor-Mower in an illustrated box, orange with green blades and bare metal wheels (210 mm).*

4. *Mettoy tinplate no.3274 Rake, orange with yellow tines and bare metal spoked wheels (210 mm).*

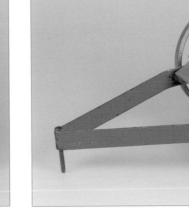

5. *Mettoy tinplate no.3274 Rake, as (4) but with green tines.*

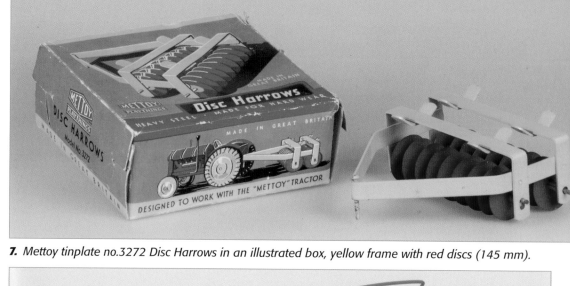

7. *Mettoy tinplate no.3272 Disc Harrows in an illustrated box, yellow frame with red discs (145 mm).*

6. *Mettoy tinplate no.3274 Rake, as (5) but with light brown plastic spoked wheels.*

8. *Mettoy tinplate no.3272 Disc Harrows, as (7) but green frame.*

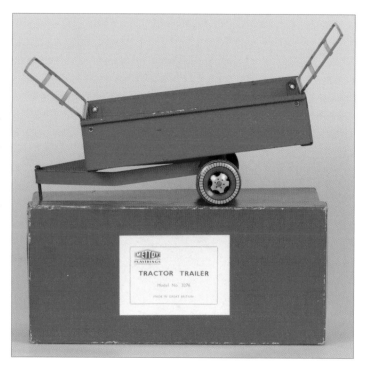

9. *Mettoy tinplate no.3276 Farm Trailer, or Farm Tip Trailer, in an illustrated box. The yellow chassis and raves illustrated on this box and on the boxes for (12) and (13) were not found on models. Red trailer with green raves and yellow, grey and black tinplate wheels (300 mm).*

10. *Mettoy tinplate no.3276 Farm Trailer, or Farm Tip Trailer, in an un-illustrated red box. Trailer as (9)*

11. *Mettoy tinplate no.3276 Farm Trailer or Farm Tip Trailer in box as (10). Trailer as (10) but with yellow end boards and black raves.*

12. *Mettoy tinplate no.3262/76 Mechanical Tractor and Trailer Set in an illustrated box, with tractor as (**2**) and trailer as (**9**).*

13. *Mettoy tinplate no.3262/76 Mechanical Tractor and Trailer Set in an illustrated box as (**12**) but with driver. Note the addition of a driver in the box illustration.*

Tractor Sets

No. 3262/76

MECHANICAL TRACTOR with farm trailer.

TRACTEUR MECANIQUE avec remorque agricole.

TRACTOR MECANICO con remolque agrícola.

16¼ x 5⅝ x 4⅛ inches. 41 ozs.
410 x 140 x 105 mm. 1160 gms.

No. 3262/0246

MECHANICAL FARM TRACTOR SET, comprising tractor with mechanism, farm trailer, mowing machine, hay rake and disc harrows.

GROUPE DE TRACTEUR AGRICOLE MECANIQUE, comprenant tracteur avec son mécanisme, remorque agricole, faucheuse, rateau-faucher et pulvérisateur.

JUEGO DE TRACTOR AGRICOLA MECANICO, consistiendo en un tractor con mecanismo, remolque agrícola, guadañadora mecánica, rastro, y trilladora de discos.

16¼ x 11⅜ x 4⅛ inches. 74 ozs.
410 x 290 x 105 mm. 2100 gms.

HEAVY STEEL TOYS

17

14. *From the Mettoy 1949 catalogue, p.17 showing the no.3262/0246 Mechanical Farm Tractor Set, comprising tractor, mower with rotary blades, hay rake, disc harrows and trailer and also the no.3262/76 Mechanical Tractor and Trailer Set as in (**12**).*

15. *Mettoy tinplate no.3264/1 Mechanical Farm Tractor, or Giant Tractor, in an illustrated box showing a green Giant Tractor, while the tractor in the box was blue and red. (230 mm).*

Tractor Sets

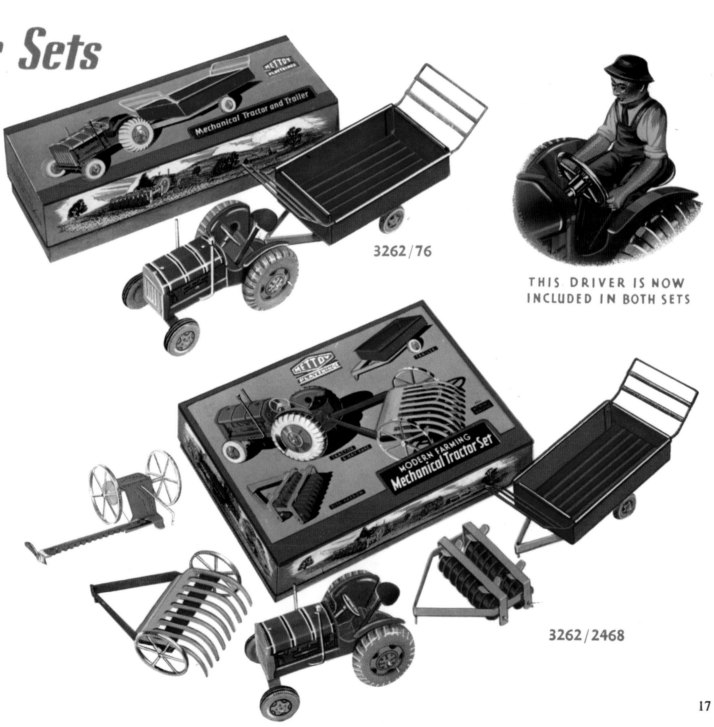

3262/76

Tractor and Trailer, removable driver. Powerful clockwork contained in rear wheel. Detachable heavy tip trailer. Attractive box as illustrated.

Tracteur et remorque.

Tractor y remolque.

16" x 5½" x 4". 41½ ozs.
405 x 140 x 100 mm. 1174 gms.

3262/2468

Farm Tractor and Implement Set. Contains 3262/1 Clockwork Tractor, driver, heavy trailer, hay rake, high speed hay cutter and disc harrow. Packed in an attractive box as illustrated.

Tracteur agricole et outils.

Tractor agricola y juego de implementos.

16½" x 11½" x 4". 69 ozs.
415 x 285 x 100 mm. 1952 gms.

3262/76

THIS DRIVER IS NOW INCLUDED IN BOTH SETS

3262/2468

16. *From the Mettoy 1952 catalogue, p.17 showing the no.3262/2468 Farm Tractor and Implement Set, comprising tractor, high speed hay cutter as (17) to (19), hay rake as (4), disc harrows as (7) and trailer as (9) and also the no.3262/76 Tractor and Trailer Set as (12). Note the mention of the driver which was added to the sets from 1952.*

17. *Mettoy tinplate no.3278 High Speed Hay Cutter in an illustrated box, hay cutter with yellow reciprocating blade in a red cutter bar driven off a single red notched plastic wheel on the axle, light green body and bare metal spoked wheels (125 mm long and 205 mm wide).*

18. *Mettoy tinplate no.3278 High Speed Hay Cutter, hay cutter as (17) but with a double yellow notched plastic wheel on the axle.*

19. *Mettoy tinplate no.3278 High Speed Hay Cutter, hay cutter as (17) but light brown plastic spoked wheels.*

20. *Mettoy tinplate no.3263 Mechanical Tractor, or Farm Tractor, in an illustrated box, orange and yellow steerable tractor with grey plastic engine block, black seat, bare metal steering wheel and air cleaner, and wind-up left-hand rear wheel (195 mm).*

21. *Mettoy tinplate no.3263/76 Mechanical Tractor and Trailer Set in an illustrated box, with tractor as (20) and trailer as (9) but with black raves (450 mm).*

22. *Mettoy diecast no.3273 Farm Plough in an illustrated box, red frame, black levers, yellow wheels and silver-painted mouldboards (270 mm).*

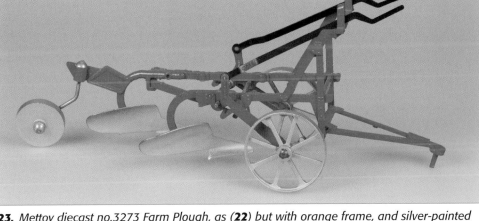

23. *Mettoy diecast no.3273 Farm Plough, as (22) but with orange frame, and silver-painted wheels (270 mm).*

24. *Mettoy tinplate no.3263/73 Mechanical Tractor and Working Model Plough Set in an illustrated box, with tractor as (20) and plough as (22).*

Mechanical Tractor Set

3263/2368 FARM TRACTOR AND IMPLEMENT SET

Contains 3263 Clockwork Tractor and Driver. Heavy Tip Trailer, Plough, Disc Harrow and Mowing Machine. Packed in strong presentation box as illustrated.

16½" x 11¼" x 4"   48 ozs.
410 x 285 x 100 mm.   1,360 gms.

3263/2368

3273 FARM PLOUGH

Heavy metal construction with real working action. Brightly coloured. Designed for use with 3263 or 3264/1 Tractors. Each in decorated box.

8½" x 3¾" x 3¼"   7 ozs.
215 x 100 x 80 mm.   200 gms.

3276 FARM TIP TRAILER

Heavy steel construction with raves and ends to open. Each in decorated box.

12" x 4½" x 5½"   15½ ozs.
305 x 115 x 140 mm.   440 gms.

Tractor and Plough

3263/73 TRACTOR AND PLOUGH

No. 3263 Tractor together with detachable Farm Plough. Packed in attractive box as illustrated.

15" x 5" x 4"   21¼ ozs.
380 x 125 x 100 mm.   615 gms.

3263/73

Page 19

25. *From the Mettoy 1955 catalogue, p.19 showing the no.3263/2368 Farm Tractor and Implement Set, comprising tractor, high speed mower, plough, disc harrows and trailer, and no.3263/73 Tractor and Plough Set.*

26. *Mettoy tinplate no.3264/1 Giant Tractor. A rare example of the large red-and-green tractor with orange grille issued as a successor for one year only in 1956 to the red-and-blue no.3264/1 Giant Tractor as in (15) (195 mm).*

Farm Equipment

6435

Clockwork Tractor with driver. Front wheel steering, brake and towing hook. Packed each in a decorated box.

Tracteur mécanique, avec conducteur. Commande de direction par roue d'avant. Frein et crochet de remorque. Mis en une boîte attrayante.

Tractor mecánico, con conductor. Mando de dirección por rueda delantera. Freno y gancho de remolque. Puesto en una caja atractiva.

5"x 3½"x 4".   6 ozs.
125 x 90 x 100 mm.   170 gms.

6438

Tractor, Caterpillar-drive. Powerful motor, brake and towing hook. Packed each in attractive box.

Tracteur pour terrain accidenté. Moteur puissant, frein et dispositif de remorque. Mis en boîtes decoratives.

Tractor de orugas. Motor poderoso, freno y gancho de remolque. Puestos en cajas decorativas.

5"x 3"x 3½".   5¼ ozs.
125 x 75 x 90 mm.   150 gms.

6435/116

Tractor and detachable Trailer packed in decorated box.

Tracteur et remorque détachable. Mis dans boîtes decoratives.

Tractor y remolque separable. Puestos en cajas decorativas.

11½"x 3½"x 4".   7½ ozs.
290 x 90 x 100 mm.   210 gms.

6435/77

Tractor and mechanical high-speed Hay Cutter. Packed in decorated box.

Tracteur et dispositif mécanique à haute vitesse pour couper le foin. Mis en boîtes attrayantes.

Tractor y dispositivo cortador de heno. Puestos en cajas decorativas.

10"x 8¼"x 4".   9 ozs.
255 x 210 x 100 mm.   255 gms.

6438

6435

METTOY REGD. PLAYTHINGS

FRUIT & VEGETABLES WHOLESALE RETAIL

6435/116

6435/77

19

27. *From the Mettoy 1951 catalogue, p.19 showing the no.6435 Clockwork Tractor, the no.6438 Tractor with Caterpillar Drive, the no.6435/116 Tractor and Trailer Set and the no.6435/77 Tractor and High Speed Hay Cutter Set.*

28. Mettoy tinplate no.6435 Mechanical Farm Tractor, or Clockwork Tractor, in an illustrated box, green-and-yellow clockwork tractor with slot-in driver.

29. Mettoy tinplate no.6435/116 Farm Tractor and Detachable Trailer Set in an illustrated box, green and yellow tractor (with slot-in driver missing in this example) and FRUIT & VEGETABLES trailer. The key is hanging from the drawbar *(photo by Ben Griffin)*.

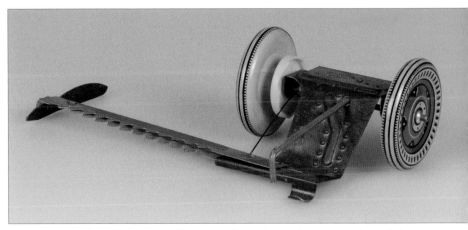

30. *Mettoy tinplate no.6435/77 Farm Tractor and High Speed Hay Cutter, with the no.6435 tractor as (28) and the hay cutter with a light green body, red and yellow cutter bar driven off a single yellow plastic drive wheel and yellow tinplate wheels. The key is hanging from the front axle.*

31. *Mettoy tinplate Mechanical High Speed Hay Cutter from set no.6435/77, cutter as (30) but with a dark green body and blue, black and yellow tinplate wheels (115 mm long and 205 mm wide).*

32. *Mettoy tinplate no.6430 "Happy Hayseed" and his Bucking Tractor, in an illustrated box, clockwork red-and-yellow tractor with attached driver with a red nose (145 mm).*

33. *Mettoy tinplate no.6436 Tractor, red-and-yellow friction-driven tractor with slot-in driver (140 mm).*

34. *Mettoy tinplate no.6439 Caterpillar Tractor in an illustrated box. Blue-and-yellow friction-driven crawler with slot-in driver (140 mm).*

35. *Mettoy diecast no.860 Mechanical Farm Tractor with Brake in an illustrated box, orange-and-yellow clockwork steerable diecast Ferguson tractor with lift-up bonnet, brake lever and rubber tyres (150 mm).*

36. *Mettoy diecast no.860 Mechanical Farm Tractor with Brake in (**35**) showing the rear projection containing the clockwork mechanism.*

37. *Mettoy diecast no.860 Mechanical Farm Tractor with Brake in (**35**) with the bonnet raised to show engine detailing.*

38. *Mettoy Castoys no.862 non-clockwork steerable version of the Ferguson in red and dark blue with a loop at the rear to take the trailer drawbar (140 mm).*

39. *Mettoy Castoys no.863 Tractor and Trailer, red and dark blue non-clockwork Ferguson tractor, rubber tyres and plastic driver, and registration number MTY 862. Yellow trailer with red raves.*

863 Tractor, Driver and Trailer

A fine detailed model of the famous Ferguson Tractor. Die cast construction with lifting bonnet to disclose detailed engine. Steering wheel control. Pivoted front axle beam. Heavy duty type moulded rubber tyres on authentic cast hubs. Two colour enamel finish. Detachable moulded driver. Heavy die cast trailer with removable raves—lever control for stand when NOT attached to tractor—moulded tyres on cast hubs —packed in decorated box.

13¼″×4″×4¼″   24 ozs.
330×100×105 mm.   680 gms

Opening Bonnet

862 Tractor

As contained in No. 863. Driver not included. Each in decorated box.

5⅞″×4″×3″   15 ozs.
150×100×75 mm.   425 gms.

842 Articulated Lorry

Six wheel lorry. Driving cab fitted transparent windows. Heavy die cast construction. Enamelled two tone colours. Moulded tyres—each in decorated box.

7 11/16″×2⅝″×2⅝″   11 ozs.
195×65×65 mm.   310 gms.

A14

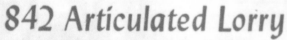

40. *Mettoy 1959 catalogue, p.A14 showing the no.862 Castoys Tractor and the no.863 Castoys Tractor, Driver and Trailer set.*

Miniature Clockwork Models

660
Tractor with clockwork mechanism and adjustable steering. Rubber tyred wheels. Packed each in attractive box.
Tracteur à mouvement d'horlogerie et commande de direction réglable. Roues caoutchouté. Chacun mis en une boîte attrayante
Tractor con mecanismo relojero y cambio de dirección adjustable. Ruedas con neumáticos. Cada uno puesto en una caja atractiva.
$2\frac{5}{8}$" x $1\frac{3}{4}$" x $1\frac{3}{8}$". $1\frac{3}{4}$ ozs.
65 x 40 x 35 mm. 50 gms.

663
Road Roller with clockwork mechanism. Adjustable steering. Each in an attractive box.
Rouleau compresseur à ressort. Commande de direction réglable. Mis en une boîte attrayante.
Cilindro de camino, con movimiento relojero. Cambio de dirección ajustable. Puesto en una caja atractiva.
$2\frac{7}{8}$" x $1\frac{1}{2}$" x $1\frac{1}{2}$". 2 ozs.
75 x 40 x 40 mm. 57 gms.

660/661
Tractor with clockwork mechanism. Detachable Farm Trailer. Adjustable steering. Rubber tyred wheels. Packed each in attractive box.
Tracteur à mécanisme d'horlogerie, avec remorque agricole détachable. Commande de direction réglable. Roues caoutchoutés. Mis dans une boîte attrayante.
Tractor mecánico con remolque agrícola separable. Mando de dirección ajustable. Ruedas con neumáticos. Puestos en cajas atractivas.
$6\frac{1}{4}$" x $1\frac{1}{2}$" x $1\frac{3}{8}$". $4\frac{1}{2}$ ozs.
160 x 40 x 35 mm. 120 gms.

3022BA
Motor cycle and Side-car. Clockwork mechanism. Packed assorted "Clown" and "A.A.". Each in decorated box.
Motocyclette et Side-car. Mécanisme à ressort. Assortiment Clown et A.A. Mis en boîtes individuelles.
Motocicleta con side-car. Movimiento relojero. Surtido Clown y A.A. Puestos en cajas individuales.
$3\frac{3}{8}$" x $1\frac{3}{4}$" x 2". $1\frac{3}{4}$ ozs.
85 x 45 x 50 mm. 50 gms.

2005
Monoplane with clockwork mechanism and brake. Packed each in a decorated box.
Monoplan mécanique, avec frein Mis en une boîte embellie.
Monoplano, mecánico con freno. Puesto en caja decorativa.
5" x 4" x $1\frac{1}{4}$". 1 oz.
125 x 100 x 30 mm. 28 gms.

660

663

METTOY REGD. PLAYTHINGS

3022 B A

660/661

2005

41. *Mettoy 1951 catalogue, p.7, showing the Miniature Clockwork Model no.660 Tractor in its innovative window box.*

42. *Mettoy no.660 Miniature Clockwork Model Tractor, three diecast tractors with variations in colour and style of plastic wheel hubs (65 mm).*

43. *Mettoy no.660/661 Mechanical Model Tractor with Wagon and Hay Brackets in a window box. Green diecast clockwork tractor with red tinplate mudguards, yellow plastic hubs and plastic tyres, yellow diecast trailer with red raves, red plastic hubs and plastic tyres.*

44. *Mettoy no.660/661 Mechanical Model Tractor with Wagon and Hay Brackets, as (**43**) but in plastic with white plastic driver, yellow plastic hubs on the trailer but with red metal raves.*

Micromodels

Company History

Micromodels were originally produced in the early 1940s by a company named Modelcraft Ltd. Formed in 1938 by H.S. Coleman, Modelcraft were based at 77 Grosvenor Road, London SW1. The company's business was supplying plans to model makers, and at the instigation of its chief designer of card models, Geoffrey Heighway, some of Modelcraft's plans were miniaturised and issued as printed card models.

The first product issued was a set of six ships and galleons, ranging from a Roman ship to a paddle steamer. It was followed by seven sets featuring contemporary subjects, such as tanks, fighting ships and aeroplanes, all utilising original Modelcraft plans. The models were a great success; they were affordable at a time of austerity, and were highly portable. The models and tools required to build them could be taken almost anywhere; in fact one of the company's slogans was 'Your Workshop in a Cigar box'. Heighway left Modelcraft in 1947, taking the Micromodels designs with him. Modelcraft seem not to have registered ownership of the Micromodels name, as Heighway used it both for his new company and its trade mark. Micromodels used an office at 6 Racquet Court, Fleet Street, London EC4 for marketing and administration, and all the preparation of artwork and printing of the finished products was subcontracted. The first post-Second World War models were all of locomotives, but new items for 1948 included architectural subjects. In 1949 nineteen sets were issued, including more military and railway-related subjects, London landmarks, novelty moving toys and the Set T.O Threshing Outfit.

Micromodels continued to introduce new models over the next few years, most of which were of either railway or architectural subjects, and included a 10-card set issued for the 1951 Festival of Britain. But in 1956, Heighway sold Micromodels and it became part of Broadway Approvals, a London-based company dealing in postage stamps. The business was in decline; the last new model, of the Mayflower sailing ship, was issued in 1957, and the last advertisement appeared in *Meccano Magazine* in 1959. Micromodels closed in the early 1960s, having produced 173 different models, issued in 82 sets.

The Micromodels stock held at the time of its closure was later sold off by Watford Model Supply Company during the 1970s. Subsequently, all Micromodels designs and production materials were purchased by D. G. Models (Autocraft), of Kinver in Staffordshire, and Micromodels products are still available from that company.

Model

Set T.O Threshing Outfit

The Set T.O Threshing Outfit (**1**) comprised six printed cards, two each for the traction engine (**2**), threshing machine (**3**) and straw elevator (**4**). Each card measured 130 mm x 90 mm. The parts of the models were cut from the card and assembled by folding and gluing together the cut pieces. The colours of the items remained constant throughout the life of the product.

Further Reading

World of Micromodels website: www.worldofmicromodels.com.
Autocraft website: www.autocraft.plus.com.

1. Cover of Micromodels Set T.O Threshing Outfit.

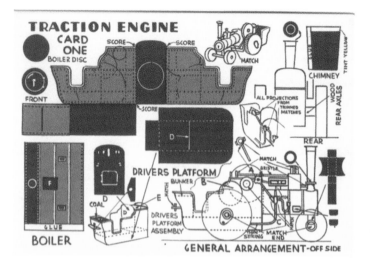

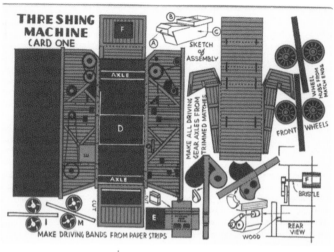

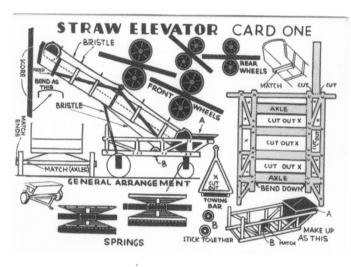

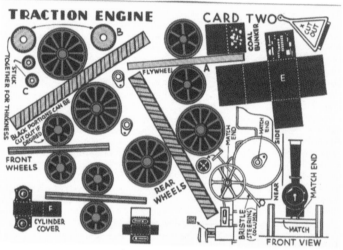

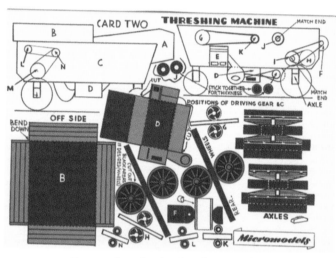

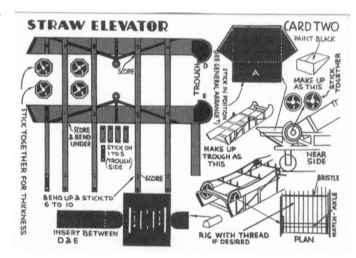

2. *Two-card set for the traction engine.*

3. *Two-card set for the threshing machine.*

4. *Two-card set for the straw elevator.*

5. *The completed Threshing Outfit (approximately 125 mm) (photo courtesy of Rod Moore).*

612

Mills

Company History

Mills Brothers, based in St Mary's Road, Sheffield 2, was founded in 1919 by three sons of a master cutler. From 1939 to 1949 they were listed in telephone directories as Mills Bros, Model Engineers, but from 1950 as Mills Bros (Sheffield) Engineers. They made woodworking tools and model railways, including clockwork tinplate locomotives, marketed under the name Milbro. These models were of very high quality and were often built under commission. Mills Brothers also had a showroom at 2 Victoria Colonnade, Victoria House, Southampton Row, London WC1. They advertised their locomotives as 'MILBRO TRUE-TO-SCALE replicas for reliability'.

In the immediate post-war years Arnold Lewis Hardinge, an engineer, went into business with the brothers. He subsequently bought the name, and set up a factory at 143 Goldsworth Road, Woking, Surrey, although Mills were still in Sheffield in 1959 (and maybe later) according to telephone directory information. He discontinued the railway models and focused on producing prototypes of high-precision components and compression-ignition engines for models, particularly aeroplanes (Harding, 2006). The patent for the engines was applied for in 1946. Although these engines were popular, the company ceased trading in the 1960s.

It was probably because of the precision quality of their models, that the Mills Brothers were asked by Harry Ferguson to make demonstration models of the Ferguson TE20 and its revolutionary three-point linkage system.

Models

The Original Dearborn Model

Perhaps the most famous model tractor ever made was the one which featured in photographs of the historic meeting in 1938 at Dearborn, Michigan, U.S.A., between Harry Ferguson and Henry Ford, when their 'handshake' agreement allowed Ford to incorporate the Ferguson three-point linkage into the Ford 9N (**1** to **9**). The event is described in detail by Fraser (1972, pp.103-7). This clockwork model was used that day by Ferguson to great advantage to promote his three-point linkage. Ferguson also had two model single-furrow ploughs: a red trailed plough and a grey plough for mounting on the three-point linkage. When the tractor pulled the trailed plough along a track the front end of the tractor reared up dangerously when the point of the plough met the obstacle of a wooden block inside the track. But, when pulling the grey plough mounted on the three-point linkage the tractor did not rear up, as described by Stuart Gibbard (2006, p.48). Ford was obviously impressed.

The original demonstration set comprised a wooden carrying case containing the tractor, the trailed and mounted ploughs, a two-wheeled trailer and a cardboard box of hand tools and spare springs, and it all still survives. We do not know who manufactured this kit. The words *CONVERSION SET FOR HYDRAULIC PUMP ASSEMBLY 8507* printed on the cardboard box were not relevant to the model (**2**). The lid and the bottom of the wooden box could be fixed together to produce the track for the demonstration. The tractor was a crude, hand-built, tinplate model with soldered joints using Meccano wheels and black rubber tyres with *DUNLOP CORD*, *MECCANO* and *MADE IN ENGLAND* (**3** to **5**) moulded on the tyres. A drawing of a Meccano clockwork tractor from 1935-36 using wheels identical to those used on the Ferguson model is illustrated by Love and Gamble (1986, p.149). The tractor was driven by a coiled spring mechanism wound up by the crank handle and a series of Meccano gears to drive the rear wheels (**6**) The red single-furrow trailed plough used the same Meccano wheels as on the front of the tractor, but without tyres (**7**). This model set later featured in Ferguson's legal battle with Ford over the patent rights to the three-point linkage and his hydraulic system. It might also have been used to support patent applications in the U.S.A. (Turner, 2007, pp.92-5). It then disappeared from view until a description by Alan Turner was published in *Tractor and Machinery* magazine in November 2007. We are grateful for his permission and the permission of the owner to reproduce some of the pictures here (**1** to **9**).

The Ferguson Demonstration Models

After the Second World War, when the Ferguson TE20 went into production in Coventry, the original model became the pattern for a series of demonstration kits consisting of a tractor and the two ploughs made by Mills Bros. The new model tractor (at 1:12 scale) was a precision model, again powered by a coiled spring wound up by the starting handle (**11** and **12**). This drove the rear axle via a crown-wheel and pinion (**13** and **14**). The sets were issued by Ferguson for use by his salesmen worldwide. The model tractor was similar to the original, but it was a more accurate replica of the real thing with more engine detail and it was on *GOODYEAR* tyres. As before, the kit came in a wooden carrying case (**10**) containing a demonstration track, a trailed plough with solid wheels (**15**) and a mounted plough (**16**). With it was an instruction booklet (**17**) with 'The Ferguson Demonstration Model ... ITS USE IN DEMONSTRATING AND LECTURING ON THE FERGUSON SYSTEM' on the cover. This publication number was FP51, and it was probably issued in about 1949.

A clear video sequence showing how the demonstration model works is included in Thorne (2005, DVD, Extras: Three-point linkage).

Ferguson Toy Tractors

Mills Bros also produced two toy models of the TE20, sold probably to promote the Ferguson tractor, although examples are rare. The one pictured differed from the demonstration model in having headlights and less engine detail (**18**). It had a very simple drive mechanism using just an elastic band (**19**), also wound up by the crank handle. A similar device was used by Chad Valley for its wooden tractors (see the Chad Valley chapter in Volume 1). There was no three-point linkage – just a hole punched in the extended rear plate for implements (**20**). Otherwise it was obviously the same model, and its use to promote the Ferguson tractor is supported by the one example we have seen which has a transfer on the right-hand side showing that the tractor was supplied by Hoggarths Ltd, tractor dealers of Kendal and Ulverston (**21**).

The second was described as friction-driven but we have only found this in an advertisement in *Punch* magazine (believed to be 1949) from the hardware and toy wholesalers, E.W. Wagstaff Ltd of Shrewsbury. The advert described the toy as 'AN IDEAL PRESENT FOR A BOY THE TRACTOR IS A MODEL OF THE FERGUSON ONE TENTH SCALE SELF-PROPELLED FRICTION DRIVE (No clockwork mechanism – which is always likely to break down and fail very quickly)' (**22**). There was a Wagstaff crest on the left-hand side, and we know that this wholesaler frequently applied their crest to models they were distributing (see the Childs & Smith chapter in Volume 1). Readers are asked to look out for an example and to publish a picture in an appropriate magazine when one can be found.

Further Reading

Fraser, Colin, 1972, *Harry Ferguson: inventor & pioneer*. John Murray. Paperback edition 1998, Old Pond Publishing.

Gibbard, Stuart, July 2006, 'Tractor Collectables', *Old Tractor* Issue 34 pp.48-9.

Harding, M., 2006, *The Original Mills History*. www.modelenginenews.org.

Love, Bert and Gamble, Jim, 1986, *The Meccano System and the Special Purpose Meccano Sets*. New Cavendish Books.

Thorne, Mike, 2005, *Massey Ferguson Tractors 1956-1976* (DVD). Old Pond Publishing.

Turner, Alan, November 2007, 'Piece of History', *Tractor and Machinery* Vol.13 pp.92-5.

1. *The well-known picture of Harry Ferguson (left) and Henry Ford at the Ford family's Dearborn Estate in the U.S.A. in October 1938 at the time of the famous 'handshake agreement', with the model which demonstrated the advantages of the Ferguson System (photo courtesy Stuart Gibbard).*

2. *The Dearborn model as it is now, still with its black wooden carrying case, the tractor, the red trailed plough, the grey plough for mounting onto the Ferguson three-point linkage, a trailer and a cardboard box of hand tools.*

4. The Dearborn tractor rear Meccano wheel with a rubber tyre carrying MECCANO MADE IN ENGLAND DUNLOP CORD around the circumference.

3. The Dearborn tinplate tractor with Meccano wheels on its wooden demonstration track.

5. The Dearborn tractor front Meccano wheel, again with a MECCANO DUNLOP tyre.

6. The Dearborn tractor, underside of the tractor showing the drive mechanism of a coiled spring wound up by the crank handle to drive the rear wheels through Meccano gears.

7. The Dearborn tractor with the red trailed plough.

8. The Dearborn tractor with the grey mounted plough.

9. *The Dearborn tinplate tractor with the grey two-wheeled trailer.*

10. *Mills Ferguson Demonstration Model consisting of the tractor and the two ploughs held firmly in their carrying case with wooden lugs.*

12. *Mills Ferguson Demonstration Model tractor rear view showing the three-point linkage operated by a lever to the right of the driver's seat.*

11. *Mills Ferguson Demonstration Model tractor showing the radiator grille, engine detailing, and GOODYEAR rubber tyres (262 mm).*

13. *Mills Ferguson Demonstration Model underside showing the simplified coiled spring drive mechanism wound up by the crank handle.*

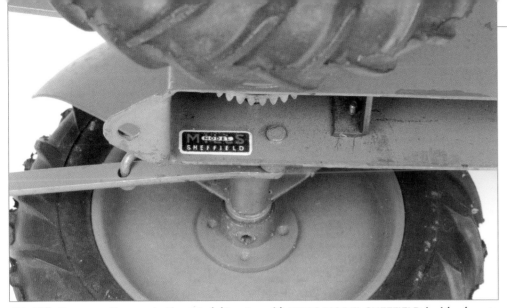

14. *Mills Ferguson Demonstration Model tractor with* MILLS MODEL SHEFFIELD *inside close to the rear wheels.*

15. *Mills Ferguson Demonstration Model tractor with a trailed plough on the demonstration track built into the lid of the black wooden carrying case.*

16. *Mills Ferguson Demonstration Model tractor with a mounted plough on the demonstration track.*

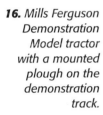

17. *The Ferguson Demonstration Model handbook explaining '...ITS USE IN DEMONSTRATING AND LECTURING ON THE FERGUSON SYSTEM'. This booklet was Ferguson Publication 51 (FP51).*

18. *Mills Ferguson Demonstration Model tractor (left) as (11) and the promotional Ferguson toy tractor (right) with headlights and less engine detail.*

19. *Mills promotional Ferguson toy tractor showing how the coiled spring drive mechanism had been replaced by an elastic band (the band missing in this example).*

20. *Mills promotional Ferguson toy tractor rear view showing the absence of three-point linkage; there was just a hole in an extended rear plate for implement attachment.*

RETAIL PRICE

67/6

INCL. TAX
Post
Free

AN IDEAL PRESENT FOR A BOY

THE TRACTOR IS A MODEL OF THE FERGUSON **ONE TENTH SCALE**

SELF-PROPELLED FRICTION DRIVE
(No clockwork mechanism—which is always likely to break down and fail very quickly).

Overall Size 10″ × 6″ × 5″.   Weight 2½ lbs.

Very strongly constructed, all metal and fitted with real GOODYEAR 1/10 scale rubber tractor tyres. Each tractor boxed in strong carton.

Trade Enquiries invited:

E. W. WAGSTAFF LTD.,
MONKMOOR ROAD,
SHREWSBURY
Telephone 2834

21. *Mills promotional Ferguson toy tractor as (**20**) with a dealer's transfer reading* SUPPLIED BY Hoggarths Ltd KENDAL ULVERSTON.

22. *Advertisement in Punch magazine, believed to be from 1949, by E.W. Wagstaff Ltd of Shrewsbury for 'AN IDEAL PRESENT FOR A BOY. THE TRACTOR IS A MODEL OF THE FERGUSON ONE TENTH SCALE SELF-PROPELLED FRICTION DRIVE'. The Wagstaff crest is shown on the left-hand side of the engine block.*

Moko

Company History

The trade name Moko was first used in the late nineteenth century by the toy wholesaler Moses Kohnstam of Fürth, Germany. His son Julius established a branch of the business in London, and the company J. Kohnstam Ltd was incorporated on 31 December 1923. However it was not till after the Second World War, with Julius's son Richard Kohnstam in charge of the company, that J. Kohnstam Ltd started to market their own lines of diecast toys under the Moko name. Moko was registered as a trade mark in 1949.

In February and May 1948 Kohnstam advertised the 'The Farminit Series of Farm Implements' in the trade magazine *Games & Toys*. This series was produced by Metal Tube Products of Woking (whose trade name was Kayron), and distributed by A.V.H. Productions Ltd (see the Kayron chapter for more details).

One of the earliest Moko diecast toys was the Heavy Tractor, a Blaw-Knox Tractor scaled down from the Dinky Toy design. The Dinky appeared in 1948 and the Moko model followed in 1949, in both tractor and bulldozer versions. Kohnstam was always a distributor rather than a manufacturer, so the actual maker of the tractor is unknown. Kohnstam was based at 393 City Road, London EC1, on the northern outskirts of the City of London, and it seems likely that Moko diecast toys were sourced from some of the small diecasting companies that were becoming established in north and east London. Several Moko items were made by Lesney Products & Co. Ltd, but Kohnstam also used other firms. Some collectors tend to equate the name Moko with early Lesney toys, but just as there were several Moko toys which were not made by Lesney (including all the farm items in this chapter), there were also early Lesney toys which were not sold by Moko. Much of Lesney's history has been well researched, and careful reading of the literature on the subject should avoid confusion.

The next Moko farm toy was a rather nice Ferguson tractor with working steering, for which two different trailers were produced, as well as a hay rake with tines which could be raised and lowered by moving a lever.

Kohnstam struck gold with the miniature Coronation Coach produced by Lesney in 1953, of which over a million were sold, and went on to market Lesney's Matchbox Series (see the Matchbox chapter for relevant farm models). They also marketed the large Lesney Massey-Harris tractor (see the Early Lesney Toys chapter). The success of Matchbox inspired several other ranges of miniature models, including the Moko Farmette Series of horse-drawn items produced around 1954-55. However the Matchbox business quickly dwarfed any other of Kohnstam's lines. J. Kohnstam Ltd was finally bought out by Lesney in 1959, giving Lesney control of their own marketing and distribution, and Richard Kohnstam went on to start his own company distributing plastic kits with the trade name Riko.

Models

Heavy Tractor

This model was first advertised in May 1949 in *Games & Toys*, in both tractor and bulldozer versions, although the latter is outside the scope of this book, being construction plant rather than a farm vehicle. It must have been a popular toy because examples are still easy to find, and probably the number produced was in the hundreds of thousands. A clockwork motor was added to produce a mechanised version, and this was advertised in April 1950 as 'available shortly'. The model was probably discontinued in 1951 due to the ban on the use of zinc for toy production (see the Introduction in Volume 1, page 13).

The model was the subject of an apology letter to Kohnstam from another wholesaler, Morris & Stone, and Kohnstam published the letter in an advertisement in *Games & Toys* November 1949, as follows:

> '19 September 1949
>
> Dear Sirs,
>
> We hereby apologise for the fact that we have sold model tractors and bulldozers not of your manufacture or merchandise in cartons bearing your regd. trade mark MOKO. We undertake that we shall not hereafter sell, offer for sale or advertise such models or other toys not of your manufacture or merchandise under such trade mark and that we shall not in any other way infringe your rights in such mark. You are at liberty to publish this letter if you think fit.
>
> Yours faithfully,
>
> S.Morris
>
> Director, Morris & Stone (London) Ltd.'

This intriguing letter suggests that there were two versions of the Moko tractor, one marketed by Kohnstam and the other by Morris & Stone, but both using the Moko name. Close examination of the models reveals that there were indeed two different dies for the body casting, but both dies used the same wheels, driver and bulldozer blade, so both types must have been made by the same manufacturer. Both dies started out with no identification on the model and then had the *MOKO* name added to the casting. Kohnstam probably paid for the original die in order to have the tractor as an exclusive line, so possibly the only way that Morris & Stone could persuade the manufacturer to supply the tractors was to pay for a second die to be made. From the manufacturer's point of view, this may have seemed fair; Morris & Stone could hardly be criticised for copying a

Kohnstam design which itself was copied from a Dinky! Alternatively the second die may simply have been made to speed up production by having two diecasting machines in operation. However, for Morris & Stone to label their product with a competitor's brand name was clearly grounds for legal action, and may have prompted the registration of the Moko trade mark, as well as resulting in the apology and undertakings given by Sam Morris. Thereafter, Morris & Stone used their own Morestone brand name on their toys, and later went on to produce Budgie Toys (see the Budgie chapter in Volume 1).

After Morris & Stone withdrew from selling the tractors, the second die was modified to accept a clockwork motor, so that mechanical models could be produced alongside the non-mechanical version, which continued to use the original die. It is puzzling that the modification to the second die also included removing the *MOKO* name and replacing it with a blank panel on each side of the casting, even though the mechanical model was sold in good quantities by Kohnstam, to the extent that there are four different *MOKO* box variations for the clockwork model. Perhaps this is further evidence that the second die was not owned by Kohnstam, hence the manufacturer felt that it should not carry the *MOKO* name. Models from the two different dies are easily distinguished – the first die had the fuel filler cap in a forward position on the fuel tank, whereas on the second die the filler cap was closer to the rear of the fuel tank (**1**). Another easily visible difference affected the panel in front of the driver, which had a plinth at the bottom. On the first die, the plinth was carried round to the sides of the model, whereas on the second die the plinth stopped at the edge of the chequerboard markings on the floor. There were many other differences too, but the filler cap position and the plinth in front of the driver are memorable and consistent means of identification.

The models can be found painted orange (shades vary) or red, and we believe that the red versions were intended to be bulldozers and had a blade fitted, whereas the tractors were orange. The bulldozer blade can easily be unclipped so that red tractors and orange bulldozers (with red blades) can be created. However the last version of the tractor (from the first die) was intentionally produced in red (a trade box has been seen) as well as orange and blue. This last version had the *MOKO* name removed from the die, as for the mechanical model, but in this case it probably indicates that the manufacturer continued producing the toy in an anonymous guise after Kohnstam had discontinued their branded line with its individual box.

The separate driver was always painted dark brown with a light green shirt and pink face and hands. The rubber tracks were usually green, less commonly in natural off-white rubber. Reproduction tracks have been produced, which tend to be thicker than the originals and a lighter shade of green. They are made of plastic rather than rubber, and often have a mark where the plastic moulding was attached to the sprue, which would never be seen on original rubber tracks.

The sequence of variations on the models is set out below. Remember that both dies were used concurrently and red models should have bulldozer blades (apart from number (v)). There were also several minor casting variations which have not been mentioned.

First die (with forward fuel filler cap)

(i)   Orange body and wheels, green rubber tracks, no lugs for blade attachment (i.e. produced as a tractor only, all other variants had the lugs). No lettering cast on the model (**2**).

(ii)   Orange or red body with wheels in body colour (**3**), or orange with black wheels (**4**), green rubber tracks, with lugs for blade, no lettering cast.

(iii)   As (ii) but *MOKO* cast on the left side and *MADE IN ENGLAND* cast underneath (not illustrated).

(iv)   As (ii) but *MOKO* cast on both sides and *MADE IN ENGLAND* cast underneath (**5**).

(v)   Orange, blue or red body with unpainted wheels, or orange body with black wheels, off-white rubber tracks, blank panel in place of *MOKO* lettering on both sides, *MADE IN ENGLAND* cast underneath but the lettering can be faint due to die wear (**6** to **8**).

Second die (with rearward fuel filler cap)

(i)   Orange or red body, wheels in body colour, green rubber tracks, no lettering cast (**9**).

(ii)   As (i) but *MOKO* cast on both sides and *MADE IN ENGLAND* cast underneath (not illustrated).

(iii)   Body casting heavily modified and clockwork motor fitted, orange or red body, black wheels, green rubber tracks, blank panel in place of *MOKO* lettering on both sides, *MADE IN ENGLAND* removed from the die due to the casting of the bonnet being thickened (**10** to **15**).

The models were individually boxed, and there were several box variations which are best understood by studying the picture captions. In addition, a trade box of six has been seen of the final version of the first die, containing red tractors with unpainted wheels and off-white tracks.

Tractor

This was clearly modelled on a Ferguson, although not attributed on the box. Unlike the previous model, no contemporary references to this tractor have been found, so the dates of production are uncertain. One clue is the box style, with *Moko* in script, a style of lettering which was first used on the box for the Lesney miniature Coronation Coach. The Ferguson tractor and its three trailers were therefore probably sold by Moko at some time during the period 1952 to 1955, between the ending of the zinc ban and the huge expansion of the Matchbox range. The tractor was not a new model, having been produced in the late 1940s, and our best guess is that the manufacturer was Pure Rubber Products of Walthamstow, who used the name Benbros from 1951. It is assumed that Benbros produced the Moko tractor for Kohnstam. The earlier version of the tractor had unpainted metal wheels and is described in the Benbros chapter in Volume 1.

The Moko tractor had several differences from the earlier model. The wheels were a new design, always with a painted finish, and the rear wheels were driven tightly onto the rear axle, which was a thicker gauge of wire than the crimped nail used previously. The two halves of the body casting simply pushed together on the earlier model, having three locating pins cast on one side and three corresponding holes on the other. The Moko model had an additional locating pin cast with the left side casting, which passed through a hole in the right side casting and was riveted over. The riveted end can be seen on the right side of the model just below the bonnet and forward of the driver's position.

The model was painted either light blue, with red or orange wheels (**16** and **17**), or orange with red or grey wheels (**18** and **19**). Silver trim was applied to the engine details and radiator grille. The land-girl driver was painted brown or grey or an in-between shade of grey-brown, and no paint trim was applied (**20**). See the Benbros chapter for a greater discussion of the variations of this driver.

The tractor was boxed either singly (**16**) or with the farm trailer (**21**). No identification was cast on either model.

Tractor and Farm Trailer

See above for details of the tractor. As far as we know the trailer was only sold in a boxed set with the tractor. It was painted a matching orange with grey wheels (**19**) or light blue with red wheels and raves (**21**). The raves are missing from the orange trailer illustrated, so we do not know whether they were grey, to match the wheels, or red, as on the light blue trailer.

Hay Rake

Like the tractor, this model was previously issued by Pure Rubber Products with different wheels (see the Benbros chapter for details). It had a separate land-girl operator and the tines could be raised and lowered by moving a lever. The Moko model was painted red or orange with light blue wheels and tines, and the operator was grey, grey-brown or pink without trim. There was no identification on the model, but it came in an attractive Moko box (**22** and **23**).

Halesowen Trailer

It is strange that another trailer was produced in this small series, but perhaps the original trailer suffered die problems or was thought to be too boxy and unrealistic. The new trailer was loosely copied from the Dinky Toy no.27b Halesowen Trailer, but was somewhat shorter and lacked the detail of the Dinky. We do not know whether raves were provided with the model. It was painted green, with an orange chassis, and wheels in orange or red. The wheels were the same pattern as the other items in the range, and as no boxed examples have been seen, this is the only way we can attribute this model to Moko (**24** to **26**).

Farmette Series

This series comprised six miniature models of horse-drawn farm vehicles. The first three numbers in the series (**1** to **3**) each had two tandem horses, and the models were quite heavy lead castings. The other three models, each with a single horse, were numbered 10, 11 and 12, and while the horse was still cast in lead, the wagons were zinc diecastings and quite often suffer from metal fatigue. The single-horse models were packed in *TREASURE CHEST* boxes and were introduced in 1955. The models with tandem horses probably appeared a year or two earlier. The feather-legged horses were always painted brown with white feather, and were in cart harness. In reality the leading horse of a tandem pair would be in trace harness, but as so often occurred, this detail was not replicated in the model in order to save expense and complexity by having just one die for the horse.

No.1 Log Wagon and Two Horses

The leading horse was connected to the shafts by unpainted wire traces. The rest of the model was painted green, with red wheels. The log was a real piece of wood dowel, with the surface painted dark brown to represent tree bark. No identification was cast on the model (**27**).

No.2 Farm Wagon and Two Horses

This model used the same horses, shafts and front axle chassis as no.1, but had larger rear wheels on the wagon. Most examples were painted dark blue with red wheels and yellow raves (**28**). A rare variant was issued in light blue, with red wheels, shafts, chassis and raves (**29**). No identification was cast on the model.

No.3 Bull Cart with Bull and Two Horses

The bull wagon is much harder to find than the first two Farmette models. Again the horses, shafts and chassis were the same as no.1, but the model used the larger wheels on both axles. There was a separate figure of a bull (painted brown) and a removable tailgate on the cart. There were two colour variations: (i) a green wagon with yellow shafts, chassis and wheels (**30**), and (ii) a green wagon, chassis and shafts with red wheels (**31**). No identification was cast on the model.

No.10 Farm Cart

The single-horse carts shared a common casting for the shafts and chassis, with *ENGLAND* cast underneath. The farm cart was painted orange with green wheels and raves and *MOKO* was cast under the body (**32**).

No.11 Water Wagon

This model was painted green with red wheels and red trim on the tank lid. *MOKO* and *ENGLAND* were cast underneath (**33**).

No.12 Miller's Cart

The miller's cart was painted light blue with orange wheels and was supplied with a load of four white-painted metal sacks, two with *OATS* and two with *CORN* cast on them. *MOKO* and *ENGLAND* were cast underneath (**34**).

Model List

Serial No.	Model	Dates
	Heavy Tractor	1949–51
	Heavy Tractor (clockwork version)	1950–51
	Tractor	1952?
	Tractor and Farm Trailer	1952?
	Hay Rake	1952?
	Halesowen Trailer	1952?
Farmette Series		
1	Log Wagon and Two Horses	1954?
2	Farm Wagon and Two Horses	1954?
3	Bull Cart with Bull and Two Horses	1954?
10	Farm Cart	1955
11	Water Wagon	1955
12	Miller's Cart	1955

Further Reading

McGimpsey, Kevin and Orr, Stewart, 1989, *Collecting Matchbox Diecast Toys The First Forty Years*. Major Productions Ltd.

1. Moko Heavy Tractors showing the first die on the left, with the fuel filler cap towards the front of the fuel tank, and the second die on the right, with the filler cap closer to the rear of the tank. The model on the right is a clockwork version, with the body extended at the rear to accommodate the motor (non-mechanical version 76 mm, clockwork version 80 mm).

2. Moko Heavy Tractor, orange with orange wheels and green tracks, first die, early version without lugs for the bulldozer blade, early plain box with dark blue printing (76 mm).

4. *Moko Heavy Tractor, as (3) but black wheels.*

3. *Moko Heavy Tractor, orange with orange wheels and green tracks, first die, with lugs for the bulldozer blade, illustrated box (76 mm).*

5. *Moko Heavy Tractor, as (3) but with MOKO cast on each side.*

6. Moko Heavy Tractor, orange with unpainted wheels and off-white tracks, first die, blank panel in place of MOKO lettering (76 mm).

9. Moko Heavy Tractor, orange with orange wheels, second die, no lettering cast. Note the filler cap is closer to the rear of the fuel tank, and the plinth in front of the driver does not wrap round to the sides of the model (75 mm).

7. Moko Heavy Tractor, as (6) but painted red.

8. Moko Heavy Tractor, as (7) but painted blue (driver repainted and non-original tracks) (photo by Hans and Marco Ludwig).

10. Moko Heavy Tractor with clockwork motor, orange with black wheels and green tracks, second die, blank panel in place of MOKO lettering, early plain box with dark blue printing (80 mm).

11. Moko Heavy Tractor with clockwork motor, as (**10**) but box with Moko *in script on the bottom corner of the box face* (photo by Hans and Marco Ludwig).

12. *Detail of the box in (**11**)*
(photo by Hans and Marco Ludwig).

13. Moko Heavy Tractor with clockwork motor, as (**10**) but illustrated box with MECHANISED *on a paper label – compare with the box in (**3**). This model has unusual double tracks with an inner band of plain black rubber, but we are not sure whether this is original.*

14. Moko Heavy Tractor with clockwork motor, as (**10**) but illustrated box of a new design, WITH CLOCKWORK MOTOR *on the end flaps. The box pictured is thought to be a black-and-white photocopy reproduction, but we have not seen an original.*

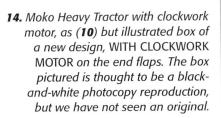

15. *Reverse of the box in (**14**).*

16. Moko Tractor, light blue with red wheels and brown driver, individual illustrated box (86 mm).

17. Moko Tractor, light blue with orange wheels (86 mm).

18. Moko Tractor, orange with red wheels (86 mm).

19. Moko Tractor and Farm Trailer (missing the raves), orange with grey wheels and brown driver (tractor 86 mm, trailer 90 mm).

20. Colour variations of the Moko land-girl tractor driver, (left to right) grey, grey-brown and brown.

21. Moko Tractor and Farm Trailer, light blue with red wheels and raves, brown driver, illustrated box (tractor 86 mm, trailer 90 mm).

22. Moko Tractor and Hay Rake, light blue tractor with red wheels and brown driver, red hay rake with light blue tines and wheels, grey rake operator.

23. Moko Hay Rake, orange with light blue tines and wheels (lever broken off), pink operator, illustrated box (90 mm).

24. Moko Halesowen Trailer, green with orange chassis and wheels (102 mm).

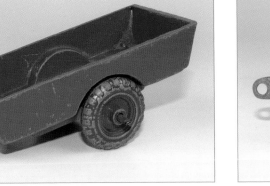

25. Moko Halesowen Trailer, as **(24)** but red wheels.

26. Moko Halesowen Trailer, underneath view of **(25)**.

27. *Moko Farmette Series no.1 Log Wagon and Two Horses, green with red wheels and brown horses, with illustrated box (110 mm).*

28. *Moko Farmette Series no.2 Farm Wagon and Two Horses, dark blue with yellow raves, red wheels and brown horses, with illustrated box (110 mm).*

29. *Moko Farmette Series no.2 Farm Wagon and Two Horses, as (28) but light blue with red raves, wheels, shafts and front axle chassis.*

30. *Moko Farmette Series no.3 Bull Cart with Bull and Two Horses, green wagon with yellow wheels, shafts and chassis, brown horses and bull, with illustrated box (110 mm).*

31. Moko Farmette Series no.3 Bull Cart with Bull and Two Horses, as (**30**) but green wagon, shafts and chassis, red wheels.

32. Moko Farmette Series no.10 Farm Cart, orange cart, green raves and wheels, brown horse, TREASURE CHEST SERIES box (55 mm).

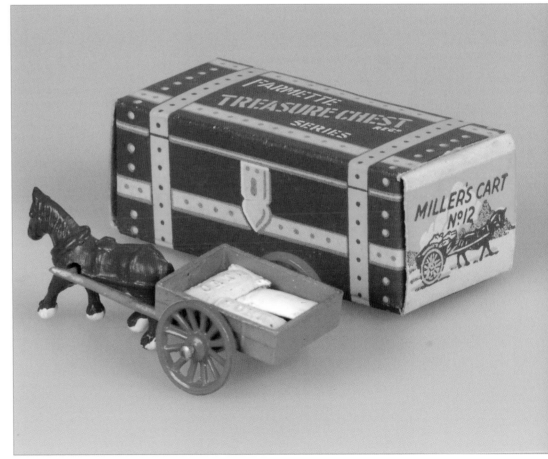

33. Moko Farmette Series no.11 Water Wagon, green cart with red trim on the tank lid, red wheels, brown horse, TREASURE CHEST SERIES box (55 mm).

34. Moko Farmette Series no.12 Miller's Cart, light blue cart with orange wheels, white sacks, brown horse, TREASURE CHEST SERIES box (50 mm).

Nicoltoys

Company History

Nicoltoys, a maker of wooden toys based in Robertsbridge in East Sussex, was originally part of the firm of L.J. Nicolls which had been making cricket bats in Robertsbridge since 1876. Between the First and Second World Wars, L.J. Nicolls started to make wooden toys which were sold under the Nicoltoys brand, and in 1941 L.J. Nicolls and Nicoltoys were acquired by H.J. Gray & Sons of Cambridge, another famous maker of cricket bats. Under war-time exigencies, H.J. Gray had been appointed a 'nucleus' firm, whereby small companies operating in related industries would have their production concentrated into a small number of larger units in order to enable more efficient production and free up much-needed wartime production capacity. Under this scheme, H.J. Gray acquired the sports products manufacturers Hazells (which made hockey and tennis products), Fulcrum, Odd & Sons and S.M. Wainwright, in addition to Nicolls.

The firm of L.J. Nicolls had previously been owned by Len Newbery and Don Bridger, who was related to the founder, L.J. Nicolls. In 1949 the firm of Gray-Nicolls sold their shares in Nicoltoys back to the original owners, Messrs Newbery and Bridger, who continued to make wooden toys under the Nicoltoys name. However, the heyday of wooden toys had passed, and in 1979 Nicoltoys was closed down. The company was re-acquired by Gray-Nicolls, which moved its snooker-related production into the former Nicoltoys factory in Robertsbridge. Finally, in 1982, in order to keep up with increased demand for its cricket bats and cricket-related products, Gray-Nicolls moved into the Robertsbridge factory from which it still operates today.

The centrepiece of the Nicoltoys wooden toy range was their Multi-Builder constructional toy, which allowed a broad range of vehicles, including a tractor, to be built from a small number of pre-formed wooden parts, wooden dowel and metal split pins. The Multi-Builder range consisted of sets M.B.1 and M.B.2 and accessory set M.B.1.A, which converted set 1 into set 2. Other toys known to have been made by the company were wooden horses and carts, and wheeled cars and trains. The wooden toys we illustrate would have been made during the period 1945 to 1979, and it is likely there are other Nicoltoys products to be found. The company's logo was derived from the coat of arms of the ancient county of Sussex, which consisted of six martlets (heraldic swallows) on a shield. On the examples we have seen, Nicoltoys used a gold shield with *ROBERTSBRIDGE SUSSEX* printed across the top of the shield, *NICOLTOYS* added in red in a diagonal band across the shield, and *ENGLAND* printed at the bottom (**2**). The larger toys made by the company displayed the logo.

Models

Wooden Horse and Cart

The cart was cut out of board 22 mm thick, with solid wooden wheels and with dowels for shafts which were held onto the horse with metal clips. The horse was made from thicker board and was set on wooden rollers. A piece of string attached via a hole though the horse's mouth allowed the horse and cart to be pulled along the ground (**1**). We have also seen the same horse pulling a different style of wooden cart.

Tractor made from Multi-Builder set

Unfortunately we have been unable to find set 2, which included the correct contents to make the tractor, so we have merely shown the illustration from the Multi-Builder set instructions (**3** to **5**). The very rudimentary tractor was number 25 in the series of over 30 models that could be made with the Multi-Builder set.

Further Reading

The Gray-Nicolls company website: www.gray-nicolls.co.uk.

1. *Nicoltoys wooden horse and cart. White horse with blue dappling, red mane and tail and red details to the fetlocks, on red rollers, pulling a yellow cart on red wheels (405 mm).*

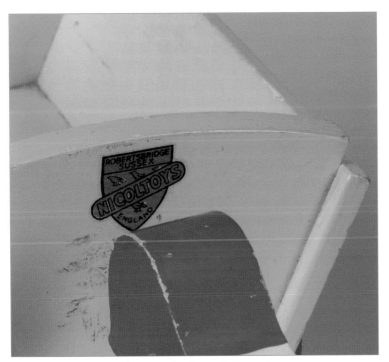

2. *Nicoltoys company logo on the front of the cart.*

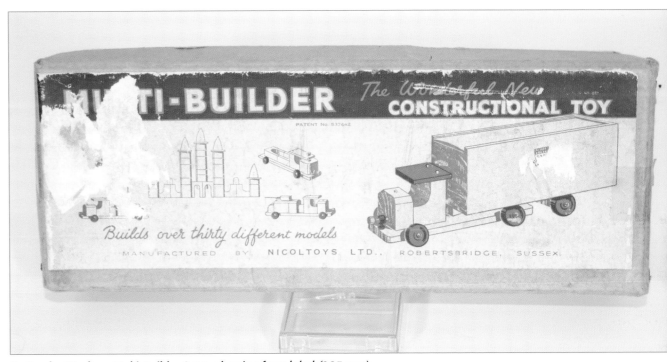

3. *Box for Nicoltoys Multi-Builder Set 1, showing front label (285 mm).*

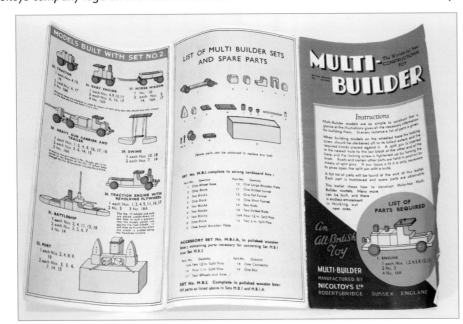

4. *Instructions for Nicoltoys Multi-Builder set.*

5. *Detail of Nicoltoys Multi-Builder set instructions, showing assembled tractor.*

Paramount

Company History

Paramount Plastic Products Ltd appeared in telephone directories at Hawthorn Grove, London SE20, between 1948 and 1956 and at addresses in Anerley Station Road, London SE20, between 1957 and 1983. Later years are not included in the telephone directory archives, but Companies House records show that Paramount continued to file accounts till 1997 and the company was dissolved in 2000. The company was started by William Lucas, who had previously been an employee of Toy Importers Ltd (Timpo), and had started Sacul Playthings Ltd in 1951. Until it closed in 1954, Sacul made a small range of lead figures similar to the products of Timpo. Although the Paramount company name does imply that it sold only plastic products, originally metal figures were produced that were very similar to the Sacul range, especially figures of medieval knights, and Sacul boxes have been found containing figures marked Paramount. The early years of the Paramount business would seem to have overlapped the period when Sacul was still trading, so probably Paramount had been registered as a business before Sacul closed, as the company quickly changed over to the production of plastic toys.

Among the plastic toys was a range of kits in two series – horse-drawn vehicles and horse-drawn coaches – and a series of simple plastic toys such as carpet trains. They also produced a range of 54 mm cowboys and Indians which were often copies of other makers' products, but with minor variations to the originals. The figures were marked under the base with the Paramount name above a mountain range.

A range of farm vehicles was produced, called the Little Farmer Series (**1**), which consisted of a small tractor with implements for it to tow, and these smaller items were very similar to models produced by T.N. Thomas (see the T.N. Thomas chapter). The company of T.N. Thomas was in business between 1958 and 1974 and was linked to a company named Thomas & Jones which was itself a sister company to Poplar Plastics (see the Poplar Plastics chapter). In the 1940s and 1950s both Thomas & Jones and Poplar Plastics were heavily involved in swapping moulds with other U.K. plastics manufacturers. Poplar Plastics was a major U.K. producer of plastic products, so it is probable that the moulds for the Little Farmer Series were actually originally made by Thomas & Jones or Poplar, and some kind of commercial arrangement between the companies allowed Paramount to make its own versions. Alternatively, of course, it is quite possible that Paramount created the range and, having achieved limited success, sold the moulds to T.N. Thomas, but without more evidence we cannot be certain exactly what the situation was. When found, Paramount packaging is always marked *Made in England*, whereas T.N. Thomas packaging is marked *Made in Great Britain*.

A number of the smaller toys have been found in cellophane packs with a card stapled to the top labelled *LITTLE FARMER SERIES Made in England* (**1**). It is reasonable to suppose that they were all sold in such packaging, as a small kit of parts to be fitted together by the buyer. This was certainly the case with the prize made by Paramount for 'Oxydol' laundry detergent (**3**), which is too big to fit in its box when assembled (**4**). Many of the models, though not all, had *PARAMOUNT PR MADE IN ENGLAND* moulded into the plastic of the model, whereas the Thomas versions of the same vehicles are always found unmarked. Consequently, unmarked implements such as the tipping trailer – unassembled in photo (**1**), assembled in photo (**20**) – are impossible to attribute definitively to either company when found loose. It is not known why the Thomas range differs slightly from the Paramount range, but Thomas seem to have made more implements, and some of them, like the triple-gang mower, are more complicated than items made by Paramount.

In addition to the series of small farm vehicles, Paramount made one large tractor in plastic for which we have found no implements.

With the information we have, none of the Paramount plastic toys can be positively dated to later than the late 1950s, so it would seem that the move into plastic toys was not a commercial success for Paramount. It is likely that the company had ceased toy production by the early 1960s, but continued producing other plastic products.

Models

Plastic Construction Kit No.1

Of the horse-drawn vehicle series, the only one relevant to this book was the *BROAD WHEELED SUSSEX WAGGON as used in the counties SUSSEX & KENT Southern England. 1700 A.D. SCALE MODEL 1/30 HORSE DRAWN VEHICLE SERIES KIT No. 1.* It was a plastic kit of white parts to make up the four-wheeled waggon, two horses and carter, and black parts on a sprue to make up the harness (**5**). It came in a flimsy card box with a colourful illustration of the waggon pulled by the two horses in tandem on the lid.

Small Plastic Tractor

This was normally sold with a farm implement, packaged together in a cellophane bag with a card header (**1** and **2**). The tractor was a simple toy made from eight components: the tractor body, a driver, steering wheel, two separate front wheels which pushed onto stub axles, and two separate rear wheels fixed to the ends of a metal axle (**3**). The Paramount tractor was usually found with a roundel on the underside of the bonnet, which read *PARAMOUNT MADE IN ENGLAND* and there was a small *PR* in the centre of the roundel. Although the tractor was very similar to the tractor by T.N. Thomas there were some minor variations to aid identification when an unmarked example was found. The characteristics of the Paramount examples were that they had a larger radiator cap than the Thomas version; Paramount tractors had three lines down the grille whereas Thomas tractors had five, and the spokes on the Paramount steering wheel were slightly thinner than on the Thomas version. Lastly, although both types had a small rectangle with a raised lip on the left mudguard, possibly meant to represent a seat, the Paramount examples were plain with no detail whereas the Thomas examples had a grid of raised lines moulded in the plastic.

In addition to being sold as a toy under the Paramount brand, the tractor was given away as a prize with Oxydol, a laundry detergent made by the Procter & Gamble company. When given as a prize, it was packaged in a flimsy illustrated box carrying the *FREE TOY TRACTOR from Oxydol* message. The tractor was supplied in kit form (**3**). The assembled tractor was too big to go in the box (**4**).

The Paramount tractor body, driver and steering wheel were moulded in red, blue, green or yellow plastic, resulting in numerous possible colour combinations, as the parts could be swapped around (**6** to **12**). The wheels were always black plastic.

Hay Rake

The hay rake came in four parts: the triangular drawbar and axle, the two wheels and the tines. Similar to the tractor, the hay rake can be found in a range of colours, including yellow drawbar and red tines with blue wheels (**13**); yellow drawbar, blue tines with red wheels (**14**); and red drawbar, yellow tines with blue wheels (**15**). *PARAMOUNT PR MADE IN ENGLAND* was moulded into the top of the wide bar connecting the tines.

Spike-toothed Harrow

This red three-piece harrow is a relatively rare find, and we have seen it in yellow (**16**), green (**17**) and blue, all with similar red wheels with wide rims. *PARAMOUNT PR MADE IN ENGLAND* was moulded onto the upper surfaces of the inner pair of horizontal cross bars of the frame.

Tipping Trailer

The tipping trailer consisted of four pieces: the trailer body and the chassis, which also formed the drawbar, and two small black wheels which were the same wheels used on the front of the tractor and on the four-wheeled trailer. We have seen the trailer body in red, green and dark blue, with green, yellow and red chassis (**18** to **21**). Although this trailer is always found unmarked, we have been able to attribute some examples to Paramount because they have been found in original packaging.

Four-wheeled Trailer

The four-wheeled trailer was made from seven pieces which consisted of the front axle and drawbar, the rear axle, four small black wheels and the trailer body. We have seen this trailer in green with a green drawbar and axles (**22**), dark blue with yellow drawbar and axles (**23**) and red with blue drawbar and axles (**24**), all with the standard small black wheels as used on the front of the tractor and on the tipping trailer. The front axle and drawbar pivoted on a central pin which protruded below the trailer body. The Paramount roundel, as on the tractor, was on the underside of the trailer body.

It is highly likely that other colour variations of all the above tractors and implements were produced.

Large Plastic Tractor

This was a larger, more sophisticated model, with greater detail and many more parts.   No packaging for this model has been found to date, consequently it is not known if this model was sold complete or unassembled. It was pale blue with yellow wheels, belt pulley and headlights, dark blue steering wheel and red seat and driver. We are also aware of a version with red wheels, belt pulley and headlights, and it is likely that other colour combinations were made. It had a towing eye but no implements for it to tow have been found to date (**25** and **26**).

Further Reading

Joplin, Norman, 1993, *The Great Book of Hollow-Cast Figures*. New Cavendish Books.

Plastic Warrior, Dec 2003, *The Plastic Warrior Guide to UK Makers of Plastic Toy Figures*. Plastic Warrior Publications.

1. Paramount Little Farmer Series cellophane pack with a Paramount blue tractor with yellow steering wheel and a green tipping trailer with red chassis.

2. Paramount Little Farmer Series Small Tractors with header card from the plastic bag in which they were sold. Yellow, blue and red tractors all with yellow steering wheels and either yellow or blue drivers (75 mm).

4. Paramount Small Tractor as (**3**) but assembled and with green steering wheel.

3. Paramount Small Tractor in red with yellow steering wheel and blue driver, in kit form, packaged in Oxydol box. Tractor as illustrated on the box.

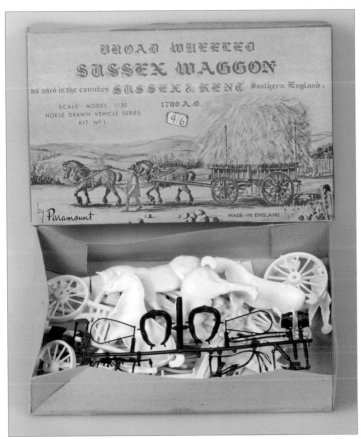

5. *Paramount Kit No.1 Broad Wheeled Sussex Waggon. White plastic components to make up the wagon, horses and carter, and black plastic components on a sprue to make up the harness. A colourfully illustrated box with a lift-off lid showed the wagon in use.*

6. *Paramount Little Farmer Series Small Tractor in red, with yellow steering wheel and yellow driver.*

7. *Paramount Little Farmer Series Small Tractor, as (6) but with blue steering wheel and red driver.*

8. *Paramount Little Farmer Series Small Tractor, as (6) but with green body and driver (missing the steering wheel).*

9. *Paramount Little Farmer Series Small Tractor, as (6) but with blue body and yellow steering wheel and driver.*

10. *Paramount Little Farmer Series Small Tractor, as (6) but with yellow body and red steering wheel (missing the driver).*

11. *Paramount Little Farmer Series Small Tractor, as (10) but with green steering wheel and driver.*

12. *Paramount Little Farmer Series Small Tractor, as (10) but with yellow steering wheel and blue driver.*

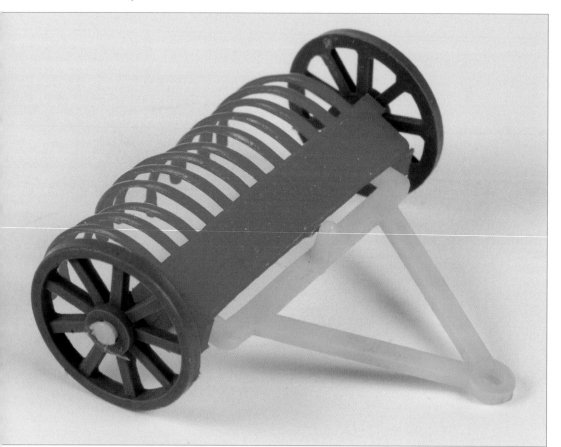

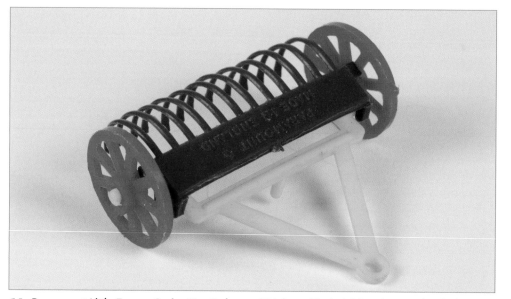

14. *Paramount Little Farmer Series Hay Rake, as (13) but with dark blue tines and red wheels.*

13. *Paramount Little Farmer Series Hay Rake with yellow drawbar and axle, red tines and dark blue wheels (56 mm).*

15. *Paramount Little Farmer Series Hay Rake, as (13) but with red drawbar and axle, yellow tines and dark blue wheels.*

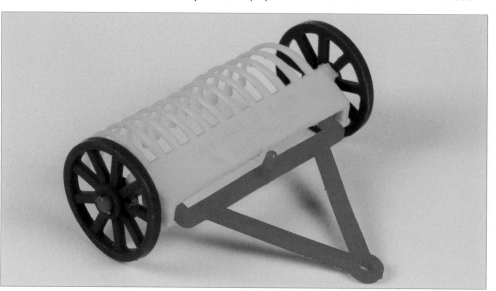

16. *Paramount Little Farmer Series Spike-toothed Harrow, yellow with red wheels behind a green tractor (harrow 64 mm).*

17. *Paramount Little Farmer Series Spike-toothed Harrow, as (16) but green with red wheels.*

19. *Paramount Little Farmer Series Tipping Trailer, as (18) but with red body and yellow drawbar and chassis.*

18. *Paramount Little Farmer Series Tipping Trailer with red body and green drawbar and chassis (90 mm).*

20. *Paramount Little Farmer Series Tipping Trailer, as (18) but with green body and red drawbar and chassis.*

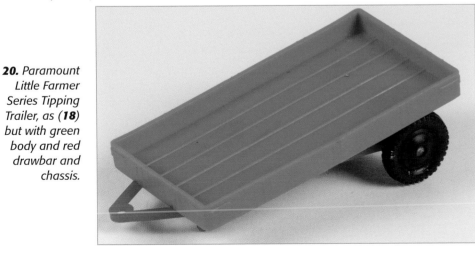

21. *Paramount Little Farmer Series Tipping Trailer, as (18) but with dark blue body and red drawbar and chassis.*

22. *Paramount Little Farmer Series Four-wheeled Trailer with green body and green axles and drawbar behind a red tractor (trailer 93 mm).*

23. *Paramount Little Farmer Series Four-wheeled Trailer, as (**22**) but with dark blue body and yellow axles and drawbar.*

24. *Paramount Little Farmer Series Four-wheeled Trailer, as (**22**) but with red body and dark blue axles and drawbar.*

25. *Paramount Large Tractor. Light blue tractor with dark blue steering wheel, red seat and driver, and yellow wheels, headlamps and belt pulley. Paramount logo stamped into the plastic on the inside of the vehicle (140 mm).*

26. *Paramount Large Tractor. Rear view of (**25**) showing towing eye.*

Passall

Company History

The Passall trade mark was registered on 14 April 1949 by Freeman & Brown Ltd, of Eastbourne, Sussex. The company was started soon after the Second World War by Mr Freeman and Mr William Brown to manufacture a range of toy hot air engines. The toy engines were originally manufactured in very small quantities by Freeman & Brown, but were subsequently made for them by the firm of Hotchkiss & Son, based at 41 Ashford Road, Eastbourne. Production by Hotchkiss commenced during 1947.

Hotchkiss & Son had been in business since 1885, originally starting as general maintenance and jobbing engineers involved with oil, gas, hydraulic and steam engines. The company was listed in trade directories just after the Second World War as Boilermakers and Engineers, and as a jobbing engineer it was involved in automotive and farm machinery repair, as well as the manufacture of concrete makers and pulverisers. It also briefly undertook subcontract manufacture of metal toys including lead soldiers, and this was the connection with Freeman & Brown.

In the late 1940s Hotchkiss & Son, in financial difficulties, was bought by the Ohly family, from 1948 appearing in trade directories as Hotchkiss (Eastbourne) Ltd, Engineers, at the address in Ashford Road, Eastbourne.

The Hotchkiss Group is still privately owned and now has extensive worldwide interests in air conditioning and ventilation ductwork contracting, and the manufacture and distribution of specialist ventilation equipment.

The Passall range was very small, as far as we are aware consisting only of the tractor shown and a stationary hot air engine. The business was not a success, and during 1950 Hotchkiss took over ownership of the Passall range from Freeman & Brown. In an effort to expand the market for the stationary engine, an example was mounted in a boat and is known to have been exhibited at model shows in the Brighton area in 1950. However, this did not lead to the hoped-for expansion in sales and the Passall range was discontinued by Hotchkiss during 1951.

Model

The tractor was made from sheet metal and had rubber tyres set on sheet metal hubs (**1**). The body, fuel tray and steering wheel were painted green, and the seat, front axle, wheel hubs and rear bulwark of the engine compartment were red. The tractor's engine compartment contained the hot air engine with an unpainted exhaust set almost exactly in the middle. At the front on both sides of the engine was the Passall trade mark, which consisted of the *PASSALL* name in red letters, *REGD. TRADE MARK*

beneath, and two yellow-and-red wings above, all set against a dark green background within an eccentrically shaped panel with a yellow border. A heavy metal heat sink was used for the rear bulwark of the engine compartment, and this, combined with unpainted heat sinks in the engine compartment itself, stopped the rear of the tractor from becoming too hot for safety (**2**). The tractor had a very simple bent metal seat with a steering wheel offset to the right and a stop/go switch on the left of the seating area. Push rods from the engine drove the rear wheels of the tractor via a series of toothed cogs and gear wheels (**3**).

It would have been an expensive toy when new, not just because of the number and complexity of the components used but also because of the materials involved and the time that would have been required to assemble the finished item. At 265 mm long it was a substantial and heavy toy and, as one of the disadvantages of hot air engines *vis-a-vis* steam engines is that they produce limited torque, the tractor was probably not a particularly successful design.

Further Reading

Hotchkiss Group website: www.hotchkissgroup.co.uk.

1. *Passall Hot Air Tractor side view (265 mm).*

2. *Passall Hot Air Tractor, overhead view of (**1**), showing heat sinks, seating area, push rods and cog wheels etc.*

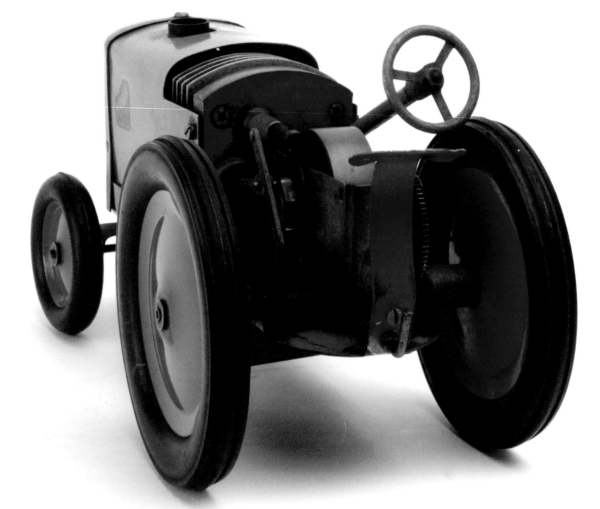

3. *Passall Hot Air Tractor, rear view of (**1**), showing seat detail, simple tow hook and detail of drive mechanism.*

Peter Ward

Company History

We have found a tractor, lorry and lorry trailer each branded *P. WARD*, and we are not clear if these were scratch-built home-made items or were from limited production runs. Some of the cast items such as the grille on the tractor appear to be quite sophisticated castings, so we include the tractor here in case others can be found. The history of these items is largely limited to hearsay; a previous owner believed they may have come from the Newcastle area, but that cannot be validated.

Model

Tractor

This large, heavy, steerable model of a Fordson tractor was made with wood and metal (**1** to **4**). The engine block and bonnet were wood; the grille, exhaust, steering wheel and rear axle were brass, and the rest of the bodywork was sheet metal. The wheels were solid cast metal. The paintwork was in various shades of blue, except for the seat, steering column and dashboard which were red. Grey levers on either side served no obvious purpose. The hubs were blue outlined with a gold circle. Brass plates on each side of the engine block were inscribed *P. WARD*, and brass registration plates at each end carried the number *FT 107*.

The lorry (not illustrated), also made of wood and sheet metal with the same wheels, had a brass plate with *PETER 1948* over the cab and carried the registration number FT 108. The lorry trailer had a brass plate with *P. WARD. HAULAGE CONTRACTOR* (**5**) and the registration number *FT 106*. We have not so far found a record of a haulage contractor called Peter Ward active in 1948.

1. Peter Ward tractor with light blue engine block and hubs, dark blue bonnet and mudguards, red seat, steering column and dashboard and grey levers. The blue hubs were outlined with a circle in gold (268 mm).

2. *Peter Ward tractor, rear view of (**1**).*

3. *Peter Ward tractor, brass plate on engine block of (**1**).*

4. *Peter Ward tractor, underside showing steering mechanism of (**1**).*

5. *Peter Ward trailer, brass plate on side of trailer.*

Poplar Plastics

Company History

Kenneth D. Brown, in his seminal work, *The British Toy Business, A History since 1700* (1996), states that 'Poplar Plastics began in 1945 as a one-man operation' using the demobilisation grant of Major Eric Jones. A company letterhead for Poplar Plastics, dated June 1948, had the company address as York Rd, Trading Estate, Bridgend, Glamorgan, exactly the same address as for Thomas & Jones Injection Moulders (see the T.N. Thomas chapter), and with the same phone number. On the letterhead the proprietor of Poplar Plastics was Major E.W. Jones, who was also one of the founding directors of Thomas & Jones, with his cousin, Islyn Thomas. Thomas already owned a plastics manufacturing company, Thomas Manufacturing Corporation, in Newark, U.S.A.

A later letterhead dated October 1953 had an address of New Street, Trading Estate, Bridgend, with the same phone number as previously and, although the proprietor of Poplar Plastics on the letterhead had changed to Poplar Playthings Ltd, Eric Jones was still in charge. It is probable that Poplar Playthings was another joint venture between Jones and Thomas, but while Jones undoubtedly utilised his cousin's expertise in injection moulding, we have been unable to find material directly linking Thomas with ownership of either Poplar Plastics or Poplar Playthings. Islyn Thomas would have been busy with Thomas Manufacturing in the United States until it was sold in 1960, and from 1958 the company of T.N. Thomas was also selling plastic toys from the same Bridgend, Glamorgan area as Poplar. So it is possible that his involvement with those two entities precluded his taking a legal role within Poplar.

Poplar also had a subsidiary company in Europe, Poplar Plastics Continental p.v.b.a. of Lokeren in Belgium. Although we do not know when the Belgian company started trading, it is known to have been in business during 1965. However, it is thought to have been just a sales office as we are not aware of any toys or packaging for Poplar marked Made in Belgium.

The company name of Poplar was a clever play on words; 'Poplar' really meant 'Popular', and one of the strap lines used to promote Poplar was 'Popular Plastic Toys'. In the years immediately after the Second World War a significant proportion of the Poplar range was based on toys originally made by Thomas Manufacturing in the United States from where the original moulds were brought to the U.K. for reproduction. The business was very successful, as by the late 1940s it had 110 workers (Brown, 1996).

Increasingly, Poplar developed its own products and made toys specific to the U.K. market, although it is known that some toys designed by Poplar in the U.K. were actually manufactured first at the Thomas Manufacturing plant in Newark, U.S.A., before being made in the U.K. by Poplar. The late 1940s and the 1950s were periods of extensive plagiarism of products but also pre-arranged mould swapping, both within the U.K. and also between U.K. plastic toy companies and their American and Continental counterparts.

Poplar regularly marketed a sample range of American-made products to U.K. retailers. If initial interest was good, the moulds would be borrowed and the item would be produced for a month or so, enabling a year's demand to be met, before the moulds were sent back to the American owner. Until its sale in 1960, Thomas Manufacturing similarly reciprocated with Poplar products. Poplar had a substantial business in the U.K. with F.W. Woolworth, the high street chain, but Poplar's products and prices were regularly challenged by other U.K. plastic manufacturers offering similar items at a lower cost, even though the alternative manufacturer had produced the items by merely copying the Poplar or Thomas Manufacturing originals.

The range of plastic toys produced by Poplar in the period between the late 1950s and early 1970s was very broad. It ranged from plastic figures and vehicles in 1:32 scale, produced in cellulose acetate by the injection-moulding process, through to large polythene-based ride-on children's toys such as pedal trikes and grand prix cars, produced by the blow-moulding process introduced in the mid-1960s. Additionally the company made garden toys such as swing seats and spades, etc.

It is known that at some point Eric Jones's son Robert took over the Poplar business, but although Poplar Playthings was registered as a business in Glamorgan up to 1984, and Poplar Plastics up to 2001, it is not known how long toy vehicles were in production.

Models

Donkey Cart

This was originally based on a very popular toy made in the United States by Thomas Manufacturing Corporation, and is therefore likely to have been one of the first items made in the U.K. by Poplar (**1**). An advertisement in *The Toy Trader & Exporter* of March, 1950 (**2**) described the Donkey Cart as 'DONKEY AND CART. White Donkey. Carts in bright and assorted colours. Each is packed in a three colour box.'

The box was marked *DONKEY CART BRITISH MADE* and *A POPLAR*

PLASTIC PRODUCT on one end flap (**1**), and *A COLOURFUL AND UNIQUE TOY WITH MANY HOURS OF CHILD PLAY VALUE* on the other. The box was always found illustrated with a yellow cart with white wheels being pulled by the ubiquitous white donkey. However, *pace* the advert, the only other colour combinations we have found are green with white wheels (**3**), and a lighter green with yellow wheels (**4**). Most carts had open-ended shafts as per the box illustration, but on some the ends of the shafts were joined together, and looped round the front of the donkey (**4**). Moulded underneath the cart was MADE IN ENGLAND on a roundel, but the donkey was never marked. The cart can be found made of a glossy but inflexible cellulose acetate-type plastic as in (**3**), and also of a duller, more flexible polystyrene-based plastic as in (**4**), proving its longevity and evolution within the Poplar range.

The donkey was used in another Poplar product, when it was ridden by a Mexican with a large sombrero (**2**). The Mexican was made from rubber and, minus his hat, was also used in another very popular Poplar product, as a speed cop riding a motorbike.

No.551 Tractor and Snowplough

A Poplar Plastics catalogue illustration for 1967 showed the headed polythene bag and its contents of a tractor complete with snowplough (**5**). The catalogue stated 'Polythene replica of Tractor in contrasting colours, fitted with detachable Snowplough front.' Unfortunately, from the illustration it is not obvious how the plough was attached. To date we have found only one example of this tractor complete with its header card (**6**). This was a brightly coloured plastic tractor with a red chassis and mudguards, and a blue bonnet. All the accessories such as the radiator cap, headlights, exhaust, steering wheel, gear lever, seat and belt pulley were in yellow. The wheels were black and made from plastic with metal axles. The illustrated header in (**6**) stated that the contents were from the *JUNIOR FARMER SERIES*, but it is not known what other items were sold as part of the range. Also on the header were A *POPLAR PLASTIC PRODUCT MADE IN ENGLAND* and the Poplar Plastic logo of *PP* in a circle with spikes. The statement 'Made in England' was not strictly correct. In the years after the war it was very difficult to obtain any raw materials in Britain unless the end product was being exported to help pay Britain's huge debt. 'Made in England' was probably seen as a better marketing ploy than 'Made in Wales' would have been, since the latter was less well-known around the world.

No.554 Tractor and Hay Wagon

This set used the tractor from set no. 551 but without a snowplough, and added a hay wagon with raves (**7**). The tractor was identical to the one used in set no. 551, except that we have found examples with both smooth (**7** to **9**) and treaded (**10**) rear wheels. Both tractor and wagon were made from the

same brightly coloured plastic in a variety of colours, though both items are always found with black wheels. The tractor has been found with red chassis, yellow bonnet and blue or yellow accessories, pulling a yellow wagon with green or yellow raves (**7** to **9**), and also with the tractor body colours reversed, pulling a green wagon with yellow raves (**10**).

The illustrated box was as brightly coloured as the tractor, reflecting the new company's need to compete for shop shelf space for its plastic products. The box had *TRACTOR AND HAY WAGON No. 554 flexible hygienic polythene A POPLAR PLASTIC PRODUCT* and *MADE IN ENGLAND* on two sides and also carried the *PP* logo. On the other two sides was *TRACTOR and HAY-WAGON* to the left of an illustration of the tractor pulling a hay wagon that was fully loaded. In the illustration the tractor had *NO.554* on the front number plate (**7**), was coloured completely red and had a driver, as had the tractor illustrated in set no. 551 (**6**). However, we do not believe that Poplar made a tractor driver for either set. On the end flaps of the box was *TRACTOR and HAY WAGON No.554* and *A POPLAR PLASTIC PRODUCT.* Although there was no maker's mark on either item, *MADE IN ENGLAND* was moulded underneath the tractor.

No.1000 Giant Farm Set

The Poplar Plastics 1967 catalogue featured this set, actually described on its box as *GIANT FARM SERIES* (**11**). According to the catalogue, the set contained a 'polythene Land Rover in contrasting colours, Tractor and Haywagon, Snowplough, Chicken Pen, containing three chickens and three geese; horse and foal; cow and calf; two sheep and a lamb.' The tractor, hay wagon and snowplough were the items from sets nos. 551 and 554. The box lid had an illustration of the tractor with snowplough attached and with a Jeep in the background. Unfortunately we have not found an example of this set to illustrate in more detail.

No.1017 Giant Tractor

To date we have found only one example of this tractor, and it was possibly made from one of the moulds that originated in the U.S.A. (**12**). At over 332 mm long, it was nearly twice the size of the previous tractors and appeared in the 1967 Poplar Plastics catalogue where it was described as having a 'large blown polythene body in red, showing full details of an ordinary tractor. Fitted with steering wheel and rotating front axle suspension in contrasting colours. Large plastic wheels on rear of tractor and two similar plastic wheels on front axle.'

Our example is red with black wheels, the rotating front axle suspension is blue and the small steering wheel and headlights are white. Both front and rear wheels are attached to the tractor by means of metal axles. The box had *GIANT TRACTOR* and the *PP* logo on two sides, against a large background illustration of the toy. The other two sides had a smaller illustration depicting the tractor in a ploughed field, with *HYGIENIC*

POLYTHENE, again extolling the virtues of blow-moulded polythene, and *GIANT TRACTOR, A POPLAR PLASTIC PRODUCT* to one side of the illustration. On the ends was a line drawing of the tractor, *GIANT TRACTOR ITEM No.1017 A POPLAR PLASTIC PRODUCT MADE IN ENGLAND* and the *PP* logo.

No.1114 Goliath Tractor and no.1217 Goliath Tractor and Trailer

Unfortunately we have only copy pages from the 1973 Poplar Plastics catalogue with which to illustrate these toys (**14** and **15**). At first it might seem that the toys' names were chosen because Poplar had run out of superlatives, but we know from the dimensions shown in the catalogue that these were substantial toys, as the tractor was 390 mm long and the overall length of tractor and trailer was 790 mm (**13**).

The catalogue described the Goliath tractor as, 'A strong polythene model of a Farm Tractor. The body is blow moulded in blaze orange with contrasting black polythene blow moulded wheels. Each wheel is held firmly on to the metal axle by the polythene hub cap. The front wheels can be turned by a movement specially attached to the steering wheel. All accessories are moulded in polythene in a contrasting lime green.'

The trailer is separately described as, 'moulded in a chrome-yellow

polypropylene with two black polythene wheels held in place with a metal safety clip.'

When sold individually the tractor was packaged in a polythene bag; when the two items were sold together as set no.1217 they were 'Individually boxed in printed carton.' From their size and garish colours we can only assume these products were aimed at pre-school children.

Further Reading

Brown, Kenneth D., 1996, *The British Toy Business, A History since 1700*. The Hambledon Press.

O'Brien, Richard, 1992, *Collecting Toy Soldiers*. Books Americana Inc.

Plastic Warrior, 2003, *The Plastic Warrior Guide to UK Makers of Plastic Toy Figures*. Plastic Warrior Publications.

Plastic Warrior, 2007, *Poplar Plastics Special*. Plastic Warrior Publications.

Young, S. Mark et al., 2001, *BLAST OFF! Rockets, Robots, Ray Guns and Rarities from the Golden Age of Space Toys!* Dark Horse Comics, Inc.

U.S. Dimestore toys website: www.usdimestore.com. Copyright 2005 by Bill Hanlon Enterprises.

1. Poplar Plastics Donkey Cart. White donkey pulling a yellow cart with white wheels, showing the box (cart 155 mm).

3. *Poplar Plastics Donkey Cart as (1) but green cart.*

2. *Poplar Plastics advertisement from* The Toy Trader & Exporter, *March 1950 showing Donkey and Cart and similar donkey used as part of Pancho and Donkey.*

4. *Poplar Plastics Donkey Cart as (3) but lighter green with yellow wheels, and shafts joined in front of the donkey.*

551. TRACTOR AND SNOWPLOUGH

Polythene replica of Tractor in contrasting colours, fitted with detachable Snowplough front. Metal Axles fitted with polythene wheels.

Overall length : 7″ : Width : 4″ : Height : 4″

5. *Part page from Poplar Plastics 1967 catalogue showing contents of set no.551*

6. *Poplar Plastics* JUNIOR FARMER SERIES *Tractor showing illustrated header card for set no.551. Tractor with red chassis, blue bonnet and yellow accessories on black wheels (probably missing the snowplough) (170 mm).*

7. *Poplar Plastics boxed set no.554 Tractor and Hay Wagon. Tractor with red chassis, yellow bonnet and blue or yellow accessories. Wagon in yellow with green raves. Both tractor and wagon with black plastic wheels. Note that on this example the left-side headlight is actually a right-hand light that has been incorrectly assembled on the left side so that it faces backwards (Tractor 170 mm, hay wagon 195 mm).*

8. *Poplar Plastics no.554 Tractor and Hay Wagon as (7) but showing box end and belt pulley on left side of tractor.*

9. *Poplar Plastics no.554 Tractor and Hay Wagon as (7) but tractor with blue radiator cap and correct headlights, and trailer with yellow raves.*

10. *Poplar Plastics no.554 Tractor and Hay Wagon as (9) but tractor with body and bonnet colours reversed and trailer in green with yellow raves. Tractor is missing steering wheel and seat.*

11. *Part page from Poplar Plastics 1967 catalogue, showing set no.1000* GIANT FARM SERIES.

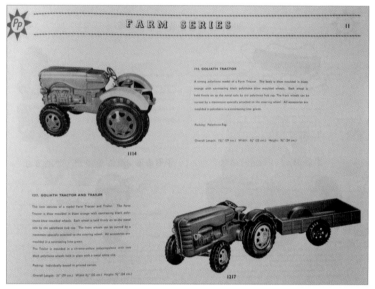

13. *Page from Poplar Plastics 1973 catalogue, showing no.1114 Goliath Tractor, and no.1217 Goliath Tractor and Trailer.*

12. *Poplar Plastics boxed no.1017 Giant Tractor. Tractor in red with blue front suspension, white headlights and steering wheel and black wheels in a red-and-blue illustrated box (332 mm).*

14. *Detail of (13) showing no.1114 Goliath Tractor.*

15. *Detail of (13) showing no.1217 Goliath Tractor and Trailer.*

Company History

Primus Engineering was a constructional toy similar to Meccano but using wooden parts as well as metal. An editorial item in *The Toy and Fancy Goods Trader*, October 1914, (reproduced in Brown, 2007, p.45) described the toy as follows.

'PRIMUS ENGINEERING OUTFITS

Amongst the large number of constructional toys at present on the market, we think the most complete set is that comprised by the "Primus Engineering Outfit". These outfits, which are in attractive boxes, contain metal strips with holes to include screws, and also wheels, angle pieces, and a varied assortment of parts not found in any other set of constructional toys. In addition to the metal parts, these outfits also include wooden sections, and these enable composite models to be constructed.

The models which can be made with Primus Engineering Outfits are unique, in so much as it is impossible to compare them with other constructional toys. Not only can model trucks be built, but also complete railway stations and railway carriages, in fact the Primus Constructional Outfits might almost be described as the most advanced construction which a boy can obtain from this interesting type of toy.

These outfits are supplied in five sizes, ranging from 6s. to £2 5s., are attractively boxed, are provided with full instructions for building the various models, and are a line which every toy dealer would be well advised to include in his stock.'

The success of Meccano (introduced around 1902) inspired many competitors, evidenced by the article's mention of the 'large number' of such toys. We have not found any reference to Primus earlier than 1914, but the manufacturers W. Butcher & Sons Ltd were listed in telephone directories from 1902 as Photographic Apparatus Manufacturers. Their address was Camera House, Farringdon Avenue, London EC4, where they remained till 1926, so it seems reasonable to assume that the Primus sets were produced throughout the period 1914 to 1926.

The building of road vehicles from Primus sets was helped greatly by the Motor Chassis Outfit (**1**), available by about 1920. This set built up into a chassis as shown in the photograph, with wheels, mudguards, seat, bonnet and steering assembly. Different bodies could then be added by using other Primus sets or home-made parts, and some possibilities were shown on the box lid. The instruction booklet included what Sue Richardson called a 'weird looking tractor' (Richardson, 1984, p.8), presumably the model we have described below (or an earlier version of it), but unfortunately we have not seen an example of the instruction booklet to check this point.

Model

Primus Tractor

According to Sue Richardson, a late introduction to the Primus Engineering range were Boys' Own Ready Made Models – a series of models supplied ready-built but easily dismantled. We think that this tractor must have been one of those ready-built models, although Sue did not list it. The model used the bonnet and radiator grille from the Motor Chassis Outfit, together with other parts, some of which were specific to the tractor. The grille was a fine wire mesh with a small hole in the middle, for no obvious reason. The front wheel steering gear could be operated from the steering wheel via a metal rod. The front wheels were Bakelite, while the rear wheels (like the rest of the model) were tinplate. There was no motor. *PRIMUS* was stamped on the radiator surround above the grille, and the model was painted brown. It is a rare item, and almost certainly was the first ever British-made toy tractor (**2** to **5**).

Further Reading

Brown, Kenneth D., 2007, *Factory of Dreams – A History of Meccano Ltd.* Crucible Books.

Richardson, Sue, 1984, 'Primus Engineering', *Modellers' World* Vol.13 no.2.

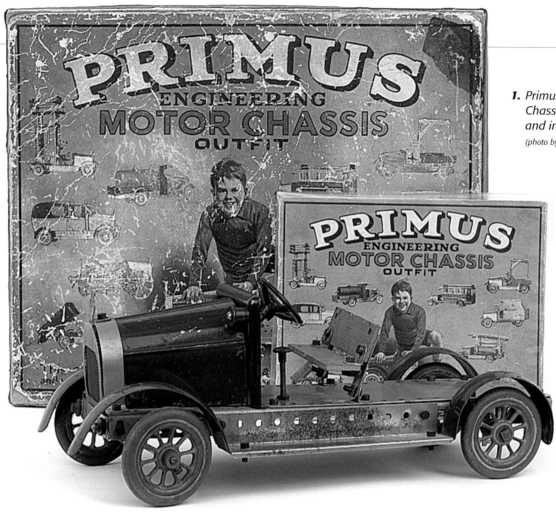

1. *Primus Engineering Motor Chassis Outfit with box and instruction booklet*
   (photo by Vectis Auctions).

2. *Primus Engineering Tractor, brown (195 mm).*

3. *Primus Engineering Tractor, as (2), showing the seat and towing hook.*

*4. Primus Engineering Tractor, as (**2**), showing the mesh grille with PRIMUS stamped above.*

*5. Primus Engineering Tractor, as (**2**), underneath view showing the steering mechanism.*

Promotional Model Tractors

We have found two plastic model tractors without any indication of make which we believe were promotional models for the full-sized machines. It is possible that they were made by the same manufacturer, but without any indication on either model it is difficult to be certain. It is interesting that both are rare which does rather confirm that they were never distributed through toy shops.

Models

David Brown Cropmaster

The first is a boxed example of a David Brown Cropmaster 6 (**1** and **2**); the Cropmaster range was in production from 1947 to 1953. The box gives little away about its origins, with *SCALE MODEL OF THE DAVID BROWN Cropmaster 6 BRITISH MADE* on all faces, and a black-and-white illustration in a white circle on the top. The model was moulded in red plastic and had rubber tyres, with *DUNLOP FARM TRACTOR 11 25-28* moulded on the rear tyres and *DUNLOP FARM TRACTOR 6 00-19* on the front. There was a *DAVID BROWN* transfer down the grille and *Cropmaster* on both sides of the bonnet, and it had only one headlight, on the right-hand side. The leaflet in the box described it as a scale model but gave no scale and the leaflet was undated. The leaflet had some details about the tractor and a cut-away labelled diagram of the tractor 'Courtesy of BRITISH FARM MECHANISATION'. On the back the same information was given in Spanish and French. The leaflet asked 'HAVE YOU GOT YOUR DAVID BROWN BADGE?', and it is possible that the model was produced for the David Brown Drivers' Club.

International Harvester Farmall M

The other model was an International Harvester Farmall M, unboxed and also in red plastic. This tractor was first made at Doncaster from 1949 and was an enduring feature of British farms in the 1950s (**3** and **4**). The model had rubber tyres and a black steering wheel with *INTERNATIONAL HARVESTER FARMALL M DONCASTER ENGLAND* on transfers on both sides of the bonnet. The rubber tyres had no markings on them.

1. *Promotional red plastic model by an unknown manufacturer of a David Brown Cropmaster 6 tractor in a box. Red body with a black seat, steering wheel, tool box, single headlight, exhaust and belt drive at the rear (207 mm).*

2. *Promotional model of a David Brown, rear view of (1) showing the absence of a drawbar.*

3. *Promotional red plastic model by an unknown manufacturer of a Farmall M tractor with a black steering wheel and* INTERNATIONAL HARVESTER FARMALL M *on the bonnet. The part of the transfer reading* DONCASTER ENGLAND *is missing on this example (195 mm).*

4. *Promotional model of a Farmall M, rear view of (3).*

Shackleton

Company History

The well-known Shackleton clockwork David Brown Trackmaster 30 model is one of the most sought-after models for toy tractor collectors. It is a superb piece of engineering, although rather heavy and hardly suitable to be played with as a toy. Not many examples of this fine model come on the market. It was manufactured primarily as a promotional model for the David Brown Trackmaster 30 which was sold to farmers for three years from 1950. The background information for this chapter comes from two articles by John Ormandy in *Modellers' World* Vol. 12, No.4 and Vol.13, No.1, who obtained the information from Maurice Shackleton who ran the company.

James Shackleton & Son was a small company formed by Maurice Shackleton before the Second World War to make wooden toy lorries and doll's houses and other wooden toys in premises behind the Wheatsheaf Hotel in Sandbach, Cheshire. In 1948 the company moved to a larger factory at Malkins Bank, Sandbach with two diecasting machines to start making Foden FG 6-wheel lorries for the toy trade. It was recorded in telephone directories at these addresses from 1948 to 1952. All parts of the lorries were made on the premises. The prototype was exhibited at the 1949 *Daily Mail* Ideal Home Exhibition, and to start with models were sold through freelance salesmen. Later they were marketed through Abbey Corinthian Games Co., which placed adverts for the Foden lorries in the *Meccano Magazine*.

By 1951 the government restrictions on the supply of zinc were causing real problems for Shackleton. Nevertheless, it was persuaded by the David Brown tractor company to design and manufacture a clockwork model of their Trackmaster 30 crawler for promotional purposes. John Ormandy says that by October 1952 about 150 had been made, of which 100 had been bought by the David Brown company before Shackleton collapsed. The other 50 were disposed of through general distribution at a retail price of 63s 0d. All the Shackleton dies and machines were sold to the Chad Valley company. However, this is probably not the whole story, because the boxed model illustrated here has the standard small circular Chad Valley sticker found on their tractors on the Shackleton box (see the Chad Valley chapter in Volume 1), and the large key for the clockwork engine is a Chad Valley key. The slip of paper in the box reads:

> 'Inspection note
>
> If there should be any complaint regarding this article, please state the nature of the complaint, and return this slip to:- THE CHAD VALLEY CO., LTD., Harborne, Birmingham, 17.
>
> Inspected by No. 169' (the number in pencil).

So, did Chad Valley continue production or just sell off their purchased Shackleton stock in Shackleton boxes? Probably the latter, otherwise Chad Valley would have produced their own boxes. However, enough of these models have turned up in auction rooms in recent years to make one wonder if the Shackleton production of the David Brown was quite as low as 150 before the company closed its doors.

Other toy manufacturers survived the 1950s, and one wonders why Shackleton could not have survived as well, particularly if they had had an export drive. The answer may well be that their models were just too good, too well engineered and too expensive to compete on the world market.

Model

The David Brown Trackmaster 30 crawler was made partly of tinplate and partly of diecast components. It was usually all-red except for the black metal exhaust, a small black plastic battery box to the right of the seat, two black plastic push-in headlights and black rubber tracks (**1**). However, we have seen several auction photographs of examples with the two control levers painted black (**2**). There was a white-on-black *DAVID BROWN* transfer on both sides of the bonnet, but nothing else was written on the model. The rear wheels were chain driven from a clockwork motor under the bonnet wound by a key on the left-hand side. The model must have contained many components and would have been time-consuming to assemble.

Inside the box was a large bare metal key with *"0" GAUGE CHAD VALLEY HARBORNE ENGLAND* and also a hollow brass screwdriver with a removable top inside which was a smaller screwdriver and inside that a yet smaller one.

On the yellow, red and white box was *DAVID BROWN Trackmaster 30 A CONSTRUCTIONAL MODEL MECHANICALLY DRIVEN A SHACKLETON MODEL* Also on the box was a small circular sticker with *CHAD VALLEY MADE IN ENGLAND*.

Further Reading

Ormandy, John, 1983, 'Shackleton', *Modellers' World* Vol.12 no.4 and Vol.13 no.1.

2. *Shackleton David Brown Trackmaster 30 crawler, as (1) but with black control levers* (photo by Vectis Auctions).

1. *Shackleton David Brown Trackmaster 30 crawler, a boxed example made of red tinplate and diecast components, with black metal exhaust, a small plastic black battery box to the right of the seat, removable black plastic headlights and black rubber tracks (220 mm).*

Sontaw

Company History

Sontaw toys were made by the firm of W.H. Watson, of Brighton in West Sussex, the brand name being derived from the owner's name. Little is known about this manufacturer other than what can be found in local trade directories. The first mention of Watson & Co. was in the *Towners Brighton Directory* for 1908-09. The company was based at 15 North Road, Brighton and was described as manufacturers of shop fittings and the Sontaw Bust. An accompanying advertisement in the directory stated 'You should have a Sontaw Bust to make your blouses on'.

The listing for the *Towners Brighton Directory* for 1914 showed that the company then occupied 13–15 North Road, Brighton and was the Sontaw Bust Metal Co. It was still supplying shop fittings and busts, and between 1917 and 1935 it was variously listed in the trade directories of the time as suppliers of screws, tools and metal.

Finally, in the 1935 to 1939 editions of *Kelly's Directory*, William Henry Watson was listed as a Metal Dealer, while under the Toy & Fancy Repositories section of *Pikes Brighton Directory (The Blue Book)* for 1935-36 appeared the business of W.H. Watson Toys of 13 North Road, Brighton. When looking at the materials used for the Sontaw products illustrated, which included the use of an old meat tin as a cart, it is easy to see the connection between a metal dealer and a small manufacturer of metal and wooden toys.

The company seems not to have survived the Second World War as there are no records of it in any of the post-war directories.

The items shown are the only examples of farm vehicles that have been found bearing the Sontaw name. However, W.H. Watson did make other toys, and these included wooden and tin wheel-less 'armoured cars', again using meat tins, and various small farm buildings such as pigsties and hen coops, which also utilised scrap tin and odd pieces of wooden board. The fact that the Sontaw products we have found all use scrap materials implies that Sontaw was a very small, opportunistic company, with a limited product range which was almost custom-made and dependent on the availability of the right scrap materials.

Models

Horse and Cart

The cart was made from a meat tin, with holes punched through the front to allow attachment of the wire shafts (**1**). The horse used was a standard Britains pre-Second World War cart horse, with holes in its sides. It is impossible to know whether Sontaw bought the horse in especially for the purpose of supplying it with the cart or indeed whether it was originally supplied by Sontaw. The shafts hooked over a wire pushed through the holes in the horse. The wheels for the cart are similar but not identical to those made by Britains, Charbens and Johillco, so were possibly specially made.

Milk Cart and Horse

The milk cart was crudely made from wood, with a tinplate baseplate to allow attachment of the wire shafts (**2** and **3**). The horse used was again a standard Britains pre-Second World War cart horse, with holes in its sides. The shafts hooked over a wire pushed through the holes in the horse. The wheels for the milk cart were the same as those used on the other cart.

The milk cart contained three solid lead milk churns, probably made by the company and quite crude in manufacture. Underneath was a label with *W.H. WATSON, 13 NORTH RD., BRIGHTON. SONTAW FARM & RAIL MODELS.* printed on it (**3**).

Further Reading

Towners Brighton Directory, various editions between 1908 and 1935.

Kelly's Directory for Brighton, editions from 1935 to 1939.

Pikes Brighton Directory (The Blue Book), 1935–36 edition.

1. *Sontaw Horse and Cart. Blue-painted re-cycled meat tin on red wheels with a white Britains horse (175 mm).*

3. *Sontaw Milk Cart in (2) – underside view showing baseplate to attach shafts, and nailed-on manufacturer's label.*

2. *Sontaw Milk Cart and Horse. Wooden cart painted sand colour on the exterior and green on the rim, interior and underneath, with metal shafts, on red wheels with three solid bare-metal churns with blue-painted tops and a brown Britains horse (155 mm).*

Speedwell

Company History

The name of the company is said to derive from the 'Speedwell' London telephone exchange, although Speedwell telephone numbers related to the Golders Green area and not to the company's addresses. Speedwell is assumed to have been one of the small plastics manufacturers that sprang up in the 1950s as raw materials became more available in post-war Britain. However, Speedwell Manufacturing Co., Plastic Toys, 34 Settles Street, London E1, only appeared in the telephone directory between 1960 and 1963 and Speedwell Manufacturing, Nelson Place, London N1, from 1964 to 1969.

The company is known to have made a range of plastic soldiers, Arabs and North American Cowboys and Indians, including 'swoppet' type figures, similar to those made by other U.K. makers such as Timpo. Some plastic animals were also made and a boxed set containing several animals and the farm cart is known.

Models

Plastic Horse and Cart

The only known farm vehicle produced by this company was a cart with silver chassis and shafts, a blue body and unrealistically small black wheels. It was 'hitched' to an accurate model of a carthorse, but there was no harness on the horse and no obvious method of attaching the cart to the horse. The cart and the horse were unmarked. The cart has been found in a small open-fronted boxed set that also included some farm animals. It is not known if the cart was ever sold separately.

Further Reading

Plastic Warrior, 2003, *The Plastic Warrior Guide to UK Makers of Plastic Toy Figures.* Plastic Warrior Publications.

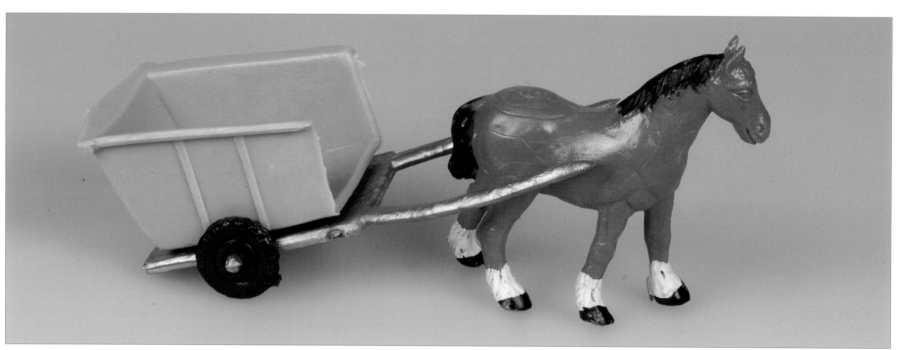

1. *Speedwell Manufacturing Horse and Cart. Crude plastic model of a blue cart with a silver-coloured chassis, short shafts and small black wheels. On this example the cart has been mounted the wrong way round on the chassis (120 mm).*

Spot-On

Company History

In 1959 Tri-ang, which until then had been producing tinplate and plastic vehicles (see the Tri-ang chapter), introduced their first diecast models under the new Spot-On brand. At 1:42 the models were slightly bigger than their Corgi and Dinky equivalents. It was a clever name which gave the impression of very accurate representation of the real thing. The Spot-On range was made in a new extension to the Lines Bros factory at Castlereagh Road in Belfast, and they were outstanding models, produced to a high standard of realism.

However, in 1964 Tri-ang made the mistake of buying Meccano, which was by then a failing company (see the Dinky chapter), and tried unsuccessfully to run the Spot-On and Dinky ranges in parallel. The last new Spot-On model was introduced in 1967, but production in the U.K. came to a close soon after. In 1968, diecasting machines and moulds for part of the range were sent to Lines Bros (New Zealand) Ltd, where Spot-On models were made for a further three years before the series finally came to an end.

A comprehensive history and detailed review of all the Spot-On models is provided by *The Ultimate Book of Spot-On Models Ltd* by Brian Salter (2013).

Model

No.137 Massey Harris Ferguson 65 Tractor (1963)

Only one farm item came out in the Spot-On range, and that was a rather fine, and now quite rare, Massey Harris Ferguson 65 Tractor (**1**). It was only produced for one year, so this tractor is difficult to find and one of the biggest prizes for any serious farm model collector. The Spot-On catalogue stated 'This Spot-On model is designed to pull a range of Spot-On implements', but none was ever produced.

The box carried the distinctive Spot-On pair of dividers on one face and a grid of white lines representing graph paper on another, all giving the impression of precision engineering. There was a red spot in the first *'O'* of *SPOT-ON*, and below was *models by Tri-ang*. On the end flap there was a white square with *MASSEY HARRIS FERGUSON 65 TRACTOR No.137*.

The model was a very detailed miniature of the real thing. The grey engine block, gearbox and rear axle were cast as a single piece split longitudinally and held together with two bare metal rivets, one through the engine block and one through the gearbox. The three-point linkage was well replicated, although the lift arms were fixed, and there was also a tow hook. The rear mudguards were also grey, and the diecast hubs, which had rubber tyres, were a distinctive bright yellow. The bonnet was the correct Massey Ferguson red, and the silver-painted grille was perfectly formed. *MASSEY FERGUSON 65* was applied to both sides of the bonnet with transfers. The steering wheel was unpainted metal and the seat was black. Nothing was written underneath the tractor, but on the inside of both mudguards was *Tri-ang MODELS by SPOT-ON*. On the outside of the mudguards, partly hidden by the tyres, was *MADE IN THE UK*. The model was steerable through an external connecting rod which ran between the front axle and the steering column.

There was also a rare colour variation with red mudguards and a black steering wheel (**2**).

Spot-On also produced collector's cards with a picture of the full-sized machine, in this case one of the publicity photos used by Massey Ferguson in their 1958 sales literature (**3**). On the back was: 'Massey-Harris Tractor A familiar sight on the fields of Britain, the Massey-Harris '65' has proved itself capable of round the clock working under the most arduous climatic conditions. A fine example of British engineering skill.' This was followed by the technical data on the tractor covering engine bore, stroke, compression and capacity. The cards were rolled up and placed in the boxes with the models. Each card had punched holes on the left side to allow them to be inserted into a Spot-On collector's album which could be purchased separately.

Further Reading

Salter, Brian, 2013, *The Ultimate Book of Spot-On Models Ltd*. 'In House' Publications.

Thompson, Graham, 1983, *Spot-On Diecast Models by Tri-ang; A Catalogue and Collectors'* Guide. Haynes Publishing Group.

2. *Spot-On No.137 Massey Harris Ferguson 65 Tractor, as (1) but with red mudguards and a black steering wheel.*

1. *Spot-On No.137 Massey Harris Ferguson 65 Tractor with grey engine block, gearbox and mudguards, yellow hubs, red bonnet, black seat and bare metal steering wheel (79 mm).*

3. *Spot-On collector's card for the Massey Harris Ferguson 65 Tractor, with the picture captioned 'MASSEY-HARRIS TRACTOR'.*

S.T.

Company History

S.T. (Wembley) Ltd were making large wood and tinplate toys in Greenford in Middlesex in the late 1940s and then at Herne Bay in Kent in the early 1950s. The range of their models was depicted on a pair of labels stuck to the sides of most boxes (**1** and **2**) and it appears that they only made farm toys. The company was incorporated in late 1947 or early 1948. We have not found them in telephone directories for Greenford, but they were listed as Toy Manufacturers at Gilmore House, Greenhill Road, Herne Bay in Kent from 1951 to 1953. The company presumably closed down after 1953. Their models usually had paper stickers underneath showing either their Greenford (**3**) or Herne Bay (**4**) addresses. Their toys are rare, but do not attract much interest because, by today's tastes, they are rather over-large and inelegant.

The company's logo was an inter-twined *S* and *T* on the boxes and on the paper stickers, and everything was *An ST Product* (**3**). They made great play of the combination of these two letters to describe their toys which were labelled on the boxes as *"Super" Toys for mechanically minded boys!* (**1** and **2**), *A STEERING TOY* (**5**), *A STRONG TRAILER* (**10**), *A SPECIAL TOY* (**11**), *A Satisfactory Toy* (**14**) or *A SENSATIONAL TOY* (**18**).

Models

We have found examples of the full range of the company's farm models as illustrated on the box labels in (**1**) and (**2**) with one exception. Their main model was the tractor with a strange upright steering column (**5** to **8**). To be pulled by this tractor we have found so far a flatbed hay trailer called a Farm Wagon (**10**), a Reaper (**11** to **13**), a low-level elevator powered off the rear axle called a Hay Loader (**14** to **17**), a Hay Rake (**18**), a two-wheeled Tipping Trailer (**19**) and a pair of Disc Harrows (**20**). The one item we cannot trace is a high-level adjustable elevator illustrated second from the right in (**2**). All were supplied in plain card boxes with illustrated red labels. Most of the models had numbers stamped onto their axles, and these may have been model numbers, although there was never a number printed on a box. The numbers on the axles have been used in the headings below, although their significance is not entirely clear. No catalogues for S.T. have so far been found.

No.20 Tractor

The tractor was described as *A STEERING TOY*, and the label on the end of the box described it as a *TRACTOR* followed by a stamped colour description, such as *GREEN* in (**5**). The only other colour we have seen was red (**6** and **7**). Both had yellow hubs and grilles (**5** and **6**). The illustration on the lid showed a smartly dressed boy with a tie and short trousers with a blue tractor with red hubs and grille, but this is a colour combination we have not seen on models.

The main body of the tractor was made of wood with a yellow tinplate grille, bare aluminium mudguards and yellow tinplate seat. The rear hubs were wood with brass centres and the front hubs were cast metal, both yellow. The wheels were all held in position with small split pins. The axles were flat-sectioned lengths of brass, and the number *20* was stamped into the rear axle of (**5**) and (**6**). The tall steering column was turned wood, and the steering wheel was an all-rubber wheel, presumably for a different model, with the treads painted yellow around the rim. The tyres were rubber and there was nothing written on them. The treads on the front wheels ran around the rims of the tyres. The starting handle was represented by a piece of bent wire, and there was a wire tow hook at the rear screwed onto the wooden base.

The steering system had a metal rod which ran down through the wooden steering column to a right-angle below the tractor. This was linked to the front axle by an external connecting wire. Brass screws represented a radiator filler cap and air cleaner. The crudeness of the toy contrasted with the highly detailed dashboard with its four dials, which were difficult to see behind the steering column (**7**). Stuck underneath on (**5**) to (**7**) was a paper *GREENFORD* label (**3**) - the same address as on the box in (**5**).

No.30 Tractor

This red tractor (**8** and **9**) had a number a of significant differences from the first two. Stamped on the rear axle was the number *30*. The grille was bare metal, and the radiator cap was a small slotted screw. The steering wheel was larger and bare metal, and the air cleaner was represented by a steel, not a brass, screw. The most interesting difference was on the rear tyres which had *S.T. TRACTOR 10-50-20 SPANDIT* moulded around both of them inside and out. As this model did not have a box or paper label underneath, it is not clear if this was a product of the Greenford or Herne Bay factories.

No.23 Farm Wagon

The flatbed trailer with *23* stamped on the rear brass axle was clearly from the Herne Bay factory because the *GREENFORD* label on the box had a *HERNE BAY* sticker over it, and there was a *HERNE BAY* label (**4**) stuck underneath the wagon. On the end of the box in pencil was written *FARM*

WAGON and *1548* near the bottom. We have not been able to identify the significance of that number. The trailer base was made of wood with red metal wire raves slotted in at either end (**10**). The front axle could swivel and there was a red wire drawbar. The wheels had red wooden hubs with brass centres, and the tyres were rubber with treads without writing on them. The wheels were fixed onto the axles with split pins. The illustration on the box showed a blue trailer with red axles and red hubs as on the model, but with smooth tyres.

No.16 or 43 Reaper

We have seen two versions of this quite good representation of a reaper binder. One had a flat leading edge to each sail and *16* stamped on the axle (**11** and **12**) and the other had the leading edge made just of wire, like the rest of the reel, with *43* stamped on the rear axle (**13**). The tyres on (**12**) were smooth, and on (**13**) they had treads. On the end of the box in (**11**) the label was stamped *REAPER* and *SPARE DRIVING BAND*. The rubber driving band survives on both models. It ran between an extended front wooden wheel hub to another wooden wheel which drove the reel. The pillar separating the two was of wood. The wheels had wooden hubs with brass centres held on with split pins. Most of the visible parts of the model were coloured tinplate, the seat was the same tinplate as seen on the tractor, and the axles were flat-sectioned brass. The drawbar was made of wire, as on the trailer, painted red in (**13**) but not in (**12**). A yellow wooden cylinder on the back of both represented the container to hold the string for binding the sheaves. The red body with yellow sails and hubs was as illustrated on the box in (**11**), but the lid of the box for (**13**) does not survive. The model in (**13**) had a *HERNE BAY* sticker underneath, but the other had no sticker.

No.3 Hay Loader

One unboxed example of the elevator (**15** and **16**) had a *3* stamped on the flat-sectioned front axle. It had a continuous brown rubber belt with thin u-section metal strips glued at 55 mm intervals to provide support for the items being lifted up the belt. The belt was on a yellow tinplate frame with a thick yellow wooden roller at the bottom and a thinner one at the top. The flat-sectioned brass front axle, with an unpainted wire drawbar, swivelled under the red frame which formed a chassis. The rear wheels had red wooden hubs and rubber tyres which were the same size as the rear tyres on the tractor and on the hay rake, but there was nothing written on them. The front wheels had smooth tyres with red wooden hubs. The wheels had brass centres and were held on with split pins. The belt was driven by the turn of the rear axle which ran through the bottom roller. The elevator belt could only be turned by pulling along the whole machine. A *GREENFORD* sticker was located under the tinplate conveyor frame.

The second example (**17**) had a silver-coloured rubber belt with flat, light blue plastic strips glued onto it. The hubs were all-wood, painted red, and

there was no sign of a number stamped onto the front axle, but on the large rear tyres was *S.T. TRACTOR 10-50-20 SPANDIT* as on the tractor tyres in (**8**) and (**9**). The smaller front tyres had treads running around the rim. A *HERNE BAY* label was stuck under the conveyor frame.

No.114 Hay Rake

This box had the *HERNE BAY* label stuck over the earlier *GREENFORD* label on the lid, and there was nothing on the end except for the price in pencil of *24/6*. The illustration showing the hay rake on the label on the lid was in black and white. Inside the box was a corrugated card liner, presumably to prevent the tines of the hay rake perforating the box. The model was a blue adjustable hay rake with a red seat and hubs, bare metal tines and large rubber tyres, and there was *114* stamped under the axle (**18**). There were twenty bare metal tines pushed through holes in a bar attached to the axle. A hooked wire joined this bar to a control lever with two positions, for up and down. The lever mechanism was riveted to a wide horizontal plate onto which the usual tinplate seat was part-riveted and part-screwed. The *HERNE BAY* sticker was attached to the underside of this plate, and the heavy tyres, identical to those on the no.30 tractor, had *S.T. TRACTOR 10-50-20 SPANDIT* moulded on both sides. The hubs were again wood with brass centres, and the wheels were held on with split pins. The painted wire drawbar pivoted underneath the horizontal plate to which the lever mechanism and seat were attached, and this was held in place by the same fixing screw which attached the seat.

No. 36 Tipping Trailer

The blue box-shaped trailer was made of wood with a flat tinplate tailgate held in position by resting in a groove in the trailer floor and by a pair of swivelling metal red clips at the top. The trailer could tip when a red clip fixed under the trailer floor was turned thus releasing the red wire drawbar from the underside of the trailer. The plain rubber tyres ran on red wooden hubs held on with split pins (**19**).

Pair of Disc Harrows

This pair of disc harrows was linked so that one was only slightly offset when pulled behind the other. Neither had a number stamped on the brass flat-sectioned top bar to the frame, which was held rigid to the drawbar by two triangular plates. The individual heavy metal discs, 14 in each set, were held in place on a thick wire. This was actually quite a heavy toy (**20**).

Combine Harvester

We have located one item with a *HERNE BAY* sticker which was not illustrated on the box labels in (**1**) and (**2**). This was a combine harvester, based on the Massey-Harris 726 with an attachment for bagging the grain, manufactured in the U.K. between 1949 and 1953. We believe this model was probably produced shortly before the company closed, and in very limited numbers. It was made almost entirely of wood, painted red and yellow, with a wire reel which was driven by an elastic band from the front left-hand hub. The steering wheel, as used on the tractors in (**5**) to (**6**), controlled the steering on the rear axle (**21**).

Model List

Possible No.	Model
20	Tractor (with rubber steering wheel)
30	Tractor (with metal steering wheel)
23	Farm Wagon
16 or 43	Reaper
3	Hay Loader (driven off the rear axle, not height adjustable)
?	Elevator (manufacturer's description not known; adjusted by a rear crank handle)
114	Hay Rake
36	Tipping Trailer (manufacturer's description not known)
?	Pair of Disc Harrows (manufacturer's description not known)
?	Combine Harvester

1. S.T. label on the side of a card box showing the Farm Wagon, Tipping Trailer, Tractor and Reaper.

3. An S.T. Product *paper label stuck under the tractor in (**6**) showing that it was made in the company's factory at Greenford, Middlesex. The words* PATENT APPLIED FOR FOR *on the bottom line were blanked out.*

2. S.T. label on the side of a card box showing the Hay Rake, Hay Loader, Tractor, adjustable Elevator and Disc Harrows.

4. An S.T. Product *paper label stuck under the Farm Wagon in (**10**) showing that it was made in their later factory at Herne Bay, Kent.*

5. *S.T. no.20 green Tractor with yellow grille, seat and hubs and rubber steering wheel (260 mm).*

6. *S.T. no.20 Tractor as (5) but with red wooden bodywork.*

9. *S.T. no. 30 Tractor rear view of (8) with steering wheel removed to show the four very detailed dials on the dashboard.*

7. *S.T. Tractor, rear view of (6) showing the dials on the dashboard.*

8. *S.T. no.30 Tractor with red bodywork, bare metal grille and large bare metal steering wheel (260 mm).*

11. *S.T. no.16 Reaper with red bodywork, yellow sails and hubs and bare metal drawbar (250 mm).*

10. *S.T. no.23 Farm Wagon with blue wooden trailer bed, red wire raves and drawbar and red hubs (340 mm).*

12. *S.T. no.16 Reaper, rear view of (11) without its box.*

13. *S.T. no.43 Reaper as (12) but with wire used to form the leading edge of the sails.*

14. *S.T. Hay Loader label on the side of the box for (17).*

15. *S.T. no.3 Hay Loader with brown rubber belt, yellow elevator frame and yellow wooden rollers top and bottom, red hubs and bare metal drawbar (304 mm).*

16. *S.T. no.3 Hay Loader rear view of (15).*

17. *S.T. no.3 Hay Loader, as (15) but with silver belt, light blue plastic strips and yellow hubs.*

19. *S.T. no.36 Tipping Trailer, blue with red clips and drawbar (179 mm).*

18. *S.T. no.114 Hay Rake with bare metal tines, blue frame, red seat and hubs and a blue drawbar (225 mm).*

20. *S.T. Pair of Disc Harrows with sets of red discs in blue frames (130 mm each).*

21. *S.T. Combine Harvester, based on the Massey-Harris 726, made largely of wood, painted red and yellow (some yellow components repainted in beige) (325 mm).*

Taylor & Barrett

Company History

Taylor & Barrett were prolific manufacturers of hollow-cast lead figures and vehicles in the inter-war years. According to Norman Joplin (Joplin, 1993, p.217) the firm was founded in 1920, and two of the partners had previously worked for Britains Ltd. Their first mention in the London street directory was in 1927, at Scholefield Buildings, Scholefield Road, Upper Holloway, London N19, and in 1930 they moved to larger premises at Park Works, 125 High Road, East Finchley, London N2.

Unlike Britains, toy soldiers were a less important part of Taylor & Barrett's range, and the civilian farm, zoo, dolls' house and street scene models took prominence. Surprisingly, there is only one farm vehicle to mention, a four-wheeled wagon drawn by two horses hitched to a central pole. It utilised the same casting and horses as the T. & B. Wild West Covered Wagon, but with raves instead of the cloth canopy. The farm

wagon was loaded with four costermonger's baskets and a selection of oversized vegetables, the same items as supplied with the T. & B. donkey-drawn costermonger's cart. The wagon was introduced some time in the period 1935 to 1939 and was probably produced until the Second World War curtailed all toy production. We know the dates because there is a pre-war T. & B. catalogue (which has been reproduced), consisting of a main catalogue, dated positively to 1934, and a supplement of 'Latest New Lines' which is from 1938 or 1939. The 'Wagon and Load' is in the supplement but not in the main catalogue, indicating that it appeared in 1935 or later. It was numbered 196 and the retail price was 2s. 6d.

After the war, the Taylor & Barrett partnership was split up into F. G. Taylor & Sons and A. Barrett & Sons (see the F.G. Taylor & Sons chapter), but the farm wagon was not re-issued.

1. Taylor & Barrett Wagon with Load, dark blue with red raves, wheels and pole, brown horses (182 mm).

Model

No.196 Wagon with Load

The wagon was diecast in lead and painted dark blue with red raves, wheels and pole. The wagon floor was dark blue painted cardboard. There was a black-painted brake lever which operated on the left side rear wheel. *T&B ENGLAND* was cast underneath the wagon and *T&B* (twice) *ENGLAND COPYRIGHT* was cast under the turntable. On the raves, *T&B* was cast twice on one side and *T&B ENGLAND* on the other. The hollow-cast lead horses were brown, with orange and white trim and a piece of string as reins. *COPYRIGHT T&B ENGLAND* was cast underneath each horse. The figure of a walking man had no lettering cast on him, and was painted with a dark blue shirt, brown trousers and cap and a green base. The four costermonger's

baskets were painted cream. According to the catalogue, Taylor & Barrett produced nine kinds of vegetables, but the wagon pictured has six kinds – cauliflowers, marrows, carrots, potatoes, turnips and a cucumber. We have also seen a bunch of bananas. The box was plain card with a lift-off lid and a label on the end as shown in the photograph (**1** and **2**).

Further Reading

Joplin, Norman, 1993, *The Great Book of Hollow-Cast Figures.* New Cavendish Books.

Taylor & Barrett pre-war catalogue, reproduction. Flataus Figures, London.

2. *Taylor & Barrett Wagon with Load, box end with label* (photo by Philip Dean).

F.G. Taylor & Sons

Company History

The factory of Taylor & Barrett at East Finchley was destroyed by wartime bombing and the business closed for the duration (see the Taylor & Barrett chapter). In 1945 the Taylor and Barrett families decided not to re-start their partnership, and instead two new firms were created, F.G. Taylor & Sons and A. Barrett & Sons, each of which inherited some of the old T. & B. moulds. A. Barrett & Sons were at 9 Sonderburg Road, Holloway, London N7. They made very few vehicles and none for the farm.

F.G. Taylor & Sons also set up close by, at 22 Hampden Road, Upper Holloway, London N19. They produced a few road vehicles in the late 1940s, some based on pre-war T. & B. vehicles, but from new dies. The models were slush-cast in a zinc alloy rather than lead, which gave a smoother appearance to the inside of the casting, in contrast to the rough interior of a lead slush casting. One of these items was a caravan, set up as a mobile laboratory for the Animal Health Trust. We may be stretching a point to include it as a farm vehicle, but it does say on the box *You need this to complete your Farm Set!*

Two other F.G. Taylor farm vehicles started out as conventional lead diecastings, but the dies were later converted to produce plastic mouldings. These were a two-wheeled Farm Cart and a Hansom Trap, both drawn by the same pony. The Hansom Trap is not necessarily a farm vehicle, but it is a very pretty and rare model, and since we have included the Britains Farmer's Gig it seems appropriate to include the Hansom Trap as well. The plastic version is slightly less pretty (with its pink pony) but even more rare.

We have also included the four-wheeled Pony Carriage, another pretty plastic model, because it is rare and as far as we know has not been pictured anywhere before.

Dates of production of all these items are vague. The metal items probably date from the late 1940s or 1950s, and the plastic models from the late 1950s or 1960s. We know that the plastic Farm Cart continued into the 1970s, because a trade box of six has been found with decimal prices (which were introduced in 1971) and mention of V.A.T. (introduced in 1973). The box of six was marked *ADDINGTON MODELS*. A. Barrett & Sons had moved to Vulcan Way, New Addington, Surrey in 1971, but there is no obvious connection between F.G. Taylor and New Addington. F.G. Taylor & Sons Ltd had become a limited company in 1958, and moved to 13 Long Street, Hoxton, London E2 in 1969. We can only speculate that Addington Models may have been a brand name used by both companies to market some of their products jointly. The last phone directory entries for the firms were in 1980 for F.G. Taylor and 1983 for Barrett.

Models

No.168 Animal Health Trust Mobile Laboratory (including Vet and Sick Animal)

This set was probably introduced in the late 1940s. The caravan had *MOBILE LABORATORY* cast on each side, moveable support legs front and rear, a towing eye at the front and diecast hubs with black rubber tyres. It was painted light green with yellow trim around the edge of the roof. Paper labels were applied, reading *ANIMAL HEALTH TRUST* on one side and *MOBILE LABORATORY* on the other, and can be found fitted either way round. The vet and sick calf were hollow-cast lead. The lift-off lid box was covered in yellow paper with an illustrated label on the front and *MOBILE LABORATORY No.168* on one end, and a descriptive leaflet was enclosed with the model (**1** to **3**).

Hansom Trap

FG TAYLOR & SONS ENGLAND was cast on the trap and *FGT & SONS ENGLAND* under the pony. The all-lead version was painted red with grey trim, cream wheels, brown driver and a dark brown pony with string for reins. The lift-off lid box was covered in yellow paper with an illustrated label (**4** and **5**). The lead model was succeeded by a plastic version with lead wheels. The plastic trap was yellow with black trim and red wheels, and had a pink plastic pony and dark brown plastic driver (**6**).

No.546 Farm Cart

FG TAYLOR & SONS LONDON ENGLAND was cast underneath the cart and *FGT & SONS ENGLAND* under the pony. The all-lead version was painted dark blue with cream chassis and shafts, red wheels and a brown pony (**7**). This was followed by a plastic model with lead wheels, which had either a yellow body with brown chassis and shafts, or a green or blue body with yellow chassis and shafts. Wheels were light brown or red, and the pony was white plastic (**8** and **9**). The final version was all-plastic, including the wheels, and came in a trade box of six with *ADDINGTON MODELS* on a paper label (see the Company History). The cart was green or light brown, with yellow chassis and shafts, red wheels and a white pony (**10** and **11**).

No.370 Pony and Carriage Ride

This set consisted of a pair of ponies pulling a four-wheeled carriage with four seated figures. The model was moulded in polythene, hence the word *Unbreakable* on the box label, except for the wheels which were cream-painted metal. The ponies were white plastic with light brown trim on their manes and tails. They were attached to a central pole which was integral with the turntable carrying the front axle, and the turntable was riveted to the carriage body. The body, turntable and pole were moulded in brown plastic. On each side of the model were black plastic mudguards which were a plug-in fit to the body. There was no identification on the model. The four seated figures were as follows: firstly, an adult male in a cap, moulded in grey plastic. He was similar to other F.G. Taylor zoo-keeper figures, reflecting the fact that the model was intended to be a zoo ride rather than a farm vehicle. There were also two seated girls, one in green and one yellow plastic, and finally a small schoolboy in shorts and a cap, moulded in dark brown. All the figures had paint trim added. The lift-off lid box had the lid covered in yellow paper, which was typical of F.G. Taylor boxes. The model details were shown on a label on one end (**12**).

Model List

Serial No.	Model	Dates
168	Animal Health Trust Mobile Laboratory (including Vet and Sick Animal)	1940s?–50s?
	Hansom Trap	1940s?–60s?
370	Pony and Carriage Ride	1950s?–60s?
546	Farm Cart	1940s?–70s

Further Reading

Joplin, Norman, 1993, *The Great Book of Hollow-Cast Figures*. New Cavendish Books.

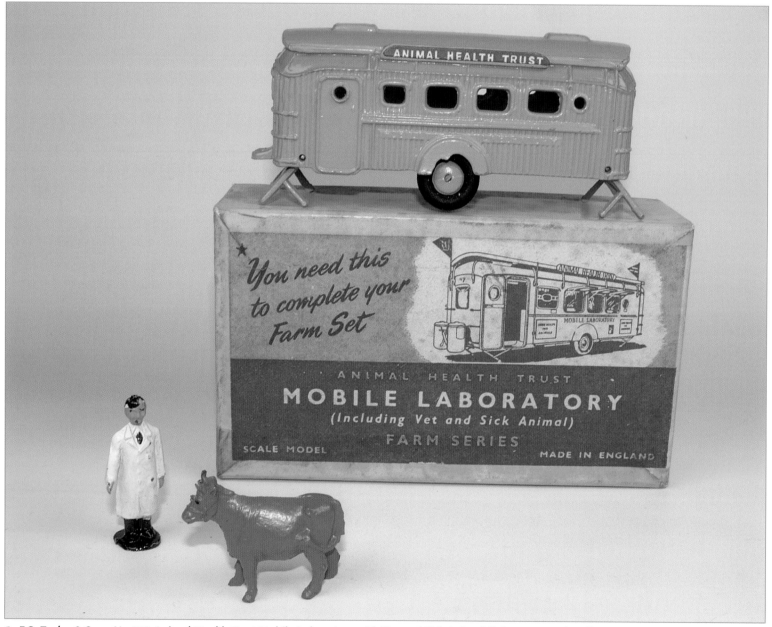

1. *F.G. Taylor & Sons No.168 Animal Health Trust Mobile Laboratory with Vet and Sick Animal, light green with yellow trim, lift-off lid box (106 mm).*

enemies are ranged on the opposing side, and what weapons (a vaccine, perhaps, or an anti-serum ; penicillin or one of the " M & B " drugs) you or your veterinary surgeon are going to use against them.

Inside the caravan laboratory you can imagine microscope and laboratory instruments and glassware. Remember that this is a model of the real thing, and is used by the Animal Health Trust to help fight disease on the farm, and so save food and money and prevent suffering. If you would like to help the Trust, please write to them at Abbey House, Victoria Street, London, S.W.1, enclosing a 2½d. stamp and particulars will be sent to you.

THE ANIMAL HEALTH TRUST'S MOBILE LABORATORY

No model railway is really complete without its mobile breakdown crane and engineers' van, and no model of a battlefield should be without a R.A.M.C. doctor, a stretcher-bearer party and—tucked away safely " behind the lines "—a military hospital flying its Red Cross flag, with an ambulance or two to visit the front and bring back casualties.

Perhaps you are beginning to wonder what all this has to do with a model farm, with its horses and cows grazing peacefully in the fields ? The answer is that every farm is a great deal less " peaceful " than it looks. Of course, there is no din going on—only the lowing of cows and perhaps the sounds of tractor engines and other farm machinery. There are no explo-

2. Leaflet enclosed with the Mobile Laboratory.

sions, no guns, no bombs, no bullets. Yet a battle is going on, all the same. A battle, not between armed men, but between the farm animals and the parasites which attack them. Some of these parasites are germs—either bacteria which can be seen under the microscope, or viruses which are too small even for that—and can only be photographed under the electron microscope. Some parasites are 6-legged insects like the flea ; others are 8-legged mites, microscopic in size. Others, again, are minute forms of plant which you can see in colonies—fungi. Yet others are tiny animals which live inside the animals and make them ill.

All the time the battle is going on. The enemies, mostly invisible, are sure to be there—some of them, anyway. Most of the time the farm animals are winning, but if they do not get enough good food to eat, the battle may turn against them, and the parasites start winning. You can tell when this is happening because the cow is often less bright about the eye, her coat less

smooth and shining, and she begins to get thinner. Sometimes the farm may be invaded by powerful enemies from outside—germs which accompany a newly bought animal, or are brought in on somebody's boots, or arrive invisibly in the water of the stream. Then there may be a very big battle indeed. You need two or three loose-boxes, which you can easily make, to put the sick animals in. And you ought to have a model to represent your veterinary surgeon. He will arrive by car not only when a big disease battle is on, but also occasionally during the more " peaceful " times —just to see that all is well and to inoculate a few animals, which is really like arming them against the invading germs.

As the latter are invisible, you cannot show them on your model farm. This is where the model of the MOBILE LABORATORY comes in so handy. With that around, it makes it obvious that some disease battle is in progress, and that efforts are being made to discover just what invisible

3. Reverse side of the leaflet in (2).

4. *F.G. Taylor & Sons Hansom Trap, lead version, red with grey trim and cream wheels, brown driver and dark brown pony (115 mm).*

5. *F.G. Taylor & Sons Hansom Trap, as (4) with lift-off lid box* (photo by Philip Dean).

6. *F.G. Taylor & Sons Hansom Trap, plastic with lead wheels, yellow with red wheels, pink pony and dark brown driver (115 mm).*

7. *F.G. Taylor & Sons No.546 Farm Cart, lead version, dark blue with cream chassis and shafts, red wheels, brown pony (108 mm).*

8. *F.G. Taylor & Sons No.546 Farm Cart, plastic with lead wheels, yellow with brown chassis and shafts, red wheels, white pony (108 mm).*

9. *F.G. Taylor & Sons No.546 Farm Cart, as (8) but green with yellow chassis and shafts, light brown wheels.*

10. *F.G. Taylor & Sons No.546 Farm Cart, all-plastic version, trade box of six with a paper label reading* ADDINGTON MODELS. *Note the decimal price of 17p (each). Curiously, the price of £2.04 on the red label is the price for a dozen rather than a half dozen. This label mentions V.A.T., so the box must be from 1973 or later. Green cart with yellow chassis and shafts, red wheels, white pony (108 mm).*

11. *F.G. Taylor & Sons No.546 Farm Cart, as (10) but light brown body.*

12. *F.G. Taylor & Sons No.370 Pony and Carriage Ride, brown plastic carriage with black mudguards, white ponies, cream metal wheels, four seated plastic figures, yellow paper-covered lift-off lid box (135 mm).*

Taylor Toys

Company History

Taylors are recorded in telephone directories in 1928 as L.C. Taylor Ltd, sellers of cane furniture at 90 Highcross Street, Leicester and in 1936 as sellers of handicrafts at 112 Wellington Street, Leicester. From 1939 to 1955 they were sellers of handicrafts at Knighton Lane, Grace Road, Leicester. But from 1957 to 1981 they described themselves as toy manufacturers at Knighton Lane, Leicester. So, it seems that they were the sellers, and probably manufacturers, of general handicrafts until the 1950s when they specialised in manufacturing wooden toys. They appear to have closed in 1981.

Model

The Wellington Tractor

We know of one farm tractor that Taylors produced because of a box found in one collection and a tractor in another. From the dimensions of the two it appears that the tractor does fit the box. The box (**1**) was labelled *The "WELLINGTON" TRACTOR*, named after the company's address in the 1930s. The green wooden tractor with yellow wheels (**2**) had *This is a Taylor Toy. MADE IN LEICESTER, ENGLAND* on transfers on the engine block. There was no tow hook, so there was probably no trailer to go with it, and the front axle swivelled.

1. Taylor Toys green box for The "WELLINGTON" TRACTOR *(140 x 90 mm).*

2. Taylor Toys green wooden tractor with yellow wheels, steering wheel made with a large-headed nail and a transfer on the side of the bonnet stating This is a Taylor Toy. MADE IN LEICESTER, ENGLAND *(130 mm).*

T.N. Thomas

Company History

The firm of T.N. Thomas Ltd, Plastics Manufacturers, is listed in telephone directories between 1958 and 1974 at various addresses on industrial estates in the Bridgend area of Glamorgan, Wales. However, the company does seem to have been inextricably linked with the firms of Thomas & Jones and Poplar Plastics which both existed from the 1940s (see the Poplar Plastics chapter). The firm of Thomas & Jones (Plastic Moulders) Ltd was registered on 30 June 1948, with a Capital of £1,000 in £1 shares. The company's 'objects' were to 'carry on the business of manufacturers of and dealers in plastics of all kinds, machinery, moulds, mouldings, &c.' Interestingly, the company's letterhead was for Thomas & Jones (Injection Moulders) Ltd, and the address was York Road, Trading Estate, Bridgend, Glamorgan. The directors were Islyn Thomas of New Jersey, U.S.A., and Eric W. Jones of 33 Wyndham Crescent, Bridgend, who was Thomas's cousin. In fact, the address and phone number for Thomas & Jones were exactly the same as for Poplar Plastics, and correspondence between Eric Jones and Islyn Thomas dated 23 June 1948 questions which company, Thomas & Jones or Poplar, would own a recently ordered new moulding machine.

Islyn Thomas

Islyn Thomas, born in 1912 in Maesteg, South Wales, emigrated with his parents and older sister Sarah to Scranton, Pennsylvania, U.S.A. when he was eleven. After high school he studied at the Johnson School of Technology where he learned the art of tool and die making. After graduating as president of his class in 1930 he joined Consolidated Molded Products Corp., of Scranton, as a tool engineer. At that time, Consolidated was the largest moulder of plastics in the world, and while working for Consolidated, Thomas continued his education at New York University, Columbia University and the University of Scranton, attaining a doctorate in plastics engineering.

By 1938 Thomas had become chief engineer of Consolidated, and among other tasks was in charge of the continual development and manufacture of the plastic parts used in the Rolls Royce Merlin engine, which had been in production from 1936. He left Consolidated in 1942 to become general manager of the Ideal Novelty and Toy Company's new plastics division. In 1944 he left Ideal and started his own company, the Thomas Manufacturing Corporation, in Newark, New Jersey. In addition to work undertaken on behalf of the U.S. Armed Forces, all three companies, Consolidated, Ideal and Thomas made general household plastic products, as well as full ranges of toys, including vehicles, figures and buildings.

After the Second World War

In the immediate aftermath of the Second World War, Islyn Thomas travelled to Europe and was a key figure in the development of the European plastics industry. He helped set up a number of companies both in the U.K. and in continental Europe, and had direct involvement with many British plastic toy companies, including Poplar Plastics and Poplar Playthings. He was a personal friend of Nicholas Kove of Airfix and was instrumental in his early post-Second World War success, having, reputedly, arranged for Airfix to acquire its first injection-moulding machine on credit from R.H. Windsor, the manufacturer, for whom Thomas was a consultant. He was also a friend of Norman Rosedale, the owner of Rosedale Associated Manufacturers and the Tudor Rose brand, and arranged for mould swaps between the two companies of Thomas Manufacturing and Rosedale (see the Tudor Rose chapter). He also supplied toy moulds to Paramount Plastics (see the Paramount chapter) and Injection Moulders Ltd of London NW9, among others.

In 1947, Thomas wrote *Injection Molding of Plastics*, the first book ever published about injection moulding, and in addition to sharing his expertise in post-Second World War Europe, Thomas arranged for the European companies to borrow moulds and co-operate technically with some of his previous employers. Consequently there are similarities between a huge range of products produced by Ideal and Thomas of the United States, by the above U.K. companies, and by some of their continental counterparts. The original factory of manufacture is quite often unknown with European moulds being used by the American companies, as well as original U.S. moulds being used by European companies.

In the period 1954 to 1960 Islyn Thomas was President of the Newark Die Co., which made injection moulds for many U.S. plastic toy companies, including Louis Marx and Revell. At its peak Thomas Manufacturing had 350 employees. The company was sold to Banner Plastics in 1960, with some Thomas Manufacturing products subsequently being sold under the Alden brand.

From around 1960 Thomas worked as an international freelance consultant, and in 1975 received an O.B.E. for his work as a consultant to the British plastics industry. He died in May 2002.

The decade after the end of the Second World War was one of great change in the use of plastics. Prior to the war the most popular plastics used for toy manufacture were cellulose acetate and polystyrene. Because acetate was relatively brittle, polystyrene had become very popular for use in military equipment during the war, and consequently was in short supply throughout the world. Polythene had only been commercially developed in 1939, and in the early 1950s was beginning to become available in large enough quantities to be used for mass production. At the same time, due to the end of the War, polystyrene was no longer under restricted supply. Each type of plastic had its own individual properties and the users would try to benefit from these to gain some commercial advantage.

T.N. Thomas and Thomas & Jones

T.N. Thomas Ltd was incorporated on 15 April 1958. To date, we have no direct evidence linking T.N. Thomas Ltd to Thomas & Jones, or indeed an explanation as to why the Thomas name was prefixed by the initials T.N. However, given Thomas's ownership of the Thomas Toy brand in the U.S. and the use of that brand by T.N. Thomas in the U.K., in conjunction with evidence of his both having entered into and been subject to litigation to protect his various brand names, it is inconceivable that two unrelated companies would exist in the same business in the same industrial area of Bridgend, selling very similar products under similar trade names. It is possible that initially Thomas & Jones was the legal vehicle used by Islyn Thomas to transfer and charge for his technical help and mould swaps in the U.K. while using Poplar Plastics as his manufacturing division. Later, in anticipation of selling Thomas Manufacturing Corporation in the U.S., it was convenient to have an additional company in the U.K. to keep the Thomas Toy brand alive. One boxed set of military figures and vehicles is known which uses the Taffy Toys name, and the box contains some figures known to have been designed by Thomas Manufacturing in the United States. Both the Taffy Toys box and all the T.N. Thomas items found in original packaging are described as being 'Made in Great Britain', so perhaps Taffy Toys was a name that Thomas also considered using more extensively in the U.K.

These seemingly complicated name and cross-ownership issues would have been familiar to Islyn Thomas. His own company, Thomas Manufacturing Corporation, sold toys under a number of names – Tommy Toy, Tiny Tom, Acme, Baby Gene and so on, whilst as General Manager of Ideal Plastics Corporation, one of his biggest customers was Ideal Plastics' owner, the Ideal Novelty and Toy Company.

In the United States both Ideal Toy Co. and Thomas Manufacturing were suppliers of plastic toys to the F.W. Woolworth chain of stores, and T.N. Thomas supplied the U.K. arm of that retailer with plastic toys such as dolls and helicopters made from moulds originating in the U.S. Although it was in the telephone directories up until 1974, it is not known when T.N. Thomas finished production of plastic toys.

Models

Considering Thomas's central role in the supply of plastic toys by many companies over a period of thirty years, there seem to be a very limited number of toys that can be positively attributed to the firm of T.N. Thomas. In addition to toys sold in packaging marked T.N. Thomas, the company is known to have sold toys in packaging marked Taffy Toys. The T.N. Thomas trade mark was *TNT* in a roundel, but although models are often marked *MADE IN ENGLAND*, very few have been found with the trade mark, and when found the mark is indistinct, so that models are difficult to identify when separated from their boxes.

T.N. Thomas made a range of farm vehicles consisting of a small tractor with a selection of trailers and implements for it to tow, but the models are very similar to a range made by Paramount (see the Paramount chapter). The identification of the Thomas versions depends on one boxed set which represents an open-sided cart shed called *MANOR FARM*, in addition to some differences between the Paramount and Thomas tractors and the fact that no Thomas products are marked, whereas the Paramount items normally are. It is not known why the Thomas range differs slightly from the Paramount range, but T.N. Thomas seem to have made more implements than Paramount and some of them, like the triple-gang mower, are more complicated than any item made by Paramount.

Large Plastic Tractor

This was a model of an International Harvester Farmall C rowcrop tractor, made in the United States after the Second World War. The model had possibly been made previously by both Ideal Toy & Novelty Corporation and Thomas Manufacturing Corporation in the U.S. and a similar model was made by Jouef in France. A version was also made by Tudor Rose in the U.K. (see the Tudor Rose chapter for a boxed example), and both the Tudor Rose and T.N. Thomas models had the letter C at the front on each side, for Farmall C tractor. The T.N. Thomas model had no maker's mark, and was supplied in various colour combinations using red, blue and yellow plastics (**1** to **3**). It differed from the Tudor Rose example by having plastic axles at the rear and by using a separate moulded driver's seat which attached to the tractor by two holes in the seat pushing over raised pips on the seat support. To date, no examples of farm implements for the tractor to tow, which can be positively attributed to T.N. Thomas, have been found.

Small Plastic Tractor

The small tractor was made from eight components: the tractor body, a driver, steering wheel, two separate front wheels which pushed onto stub axles, and two separate rear wheels fixed to the ends of a metal axle. In common with all the following carts and implements made by T.N. Thomas, the tractors are unmarked. We are able to attribute the tractors to Thomas because of small moulding differences between them and Paramount examples.

The characteristics of the Thomas examples are: a smaller radiator cap than the Paramount version; five lines down the grille whereas the

Paramount examples have three lines, and slightly thicker spokes on the Thomas steering wheel than on Paramount's. Lastly, although both types have a small rectangle with a raised lip on the left mudguard – possibly meant to represent a seat – the Thomas examples have a grid of raised lines in the rectangle (**4** and **5**), whereas the Paramount examples are plain. And, of course, the Thomas examples are unmarked.

We have seen three colour variations of the T.N. Thomas small tractor. All have red bodies, but the steering wheels and drivers vary between yellow and blue plastic (**4** to **6**).

The attribution of all the following farm implements to T.N. Thomas has been possible because of the existence of a boxed Manor Farm set (**6** and **7**). The vehicles and implements were attached with small rubber bands to the front flap of the box, which folded down to form a farmyard on which the items sat in front of the 'sheds'. None of the following items is marked, and it is assumed that where an unmarked example of a vehicle made by both companies is found, it can be attributed to T.N. Thomas.

Tipping Trailer

The tipping trailer consisted of four pieces: the trailer body and the chassis – which also formed the drawbar – and two small black wheels which were the same as used on the front of the small tractor, on the four-wheeled trailer, the high-sided trailer and the horse box. We have seen the trailer body in red, green or dark blue, with green, yellow or red chassis (**8** to **11**).

Disc Harrow

The disc harrow was very similar to the design of the Dinky disc harrow (see the Dinky chapter), and was possibly based on it. The Thomas version consisted of a one-piece frame with a set of discs on either side and, as with the Dinky disc harrow, a tow hook allowed another implement to be towed behind. The only example we have seen was blue with yellow discs in the Manor Farm boxed set (**12**).

Three-furrow Plough

As the disc harrow was similar to the Dinky version, so the plough seems to have been based on the Britains three-furrow plough (see the Britains chapter in Volume 1). The plough was made from five pieces. The top frame and drawbar was one piece; the three coulters and mouldboards were pressed into this from below, and on the left-hand side was a large depth wheel. This item had a very wide range of colour combinations; the mouldboards and coulters could all be different colours (**13**) or the same (**14** to **16**). We have seen the plough frame in green, red or blue, and the depth wheel in yellow, blue or red (**13** to **16**).

Forage Harvester

T.N. Thomas was the only toy manufacturer, other than Britains to make a forage harvester. It was assembled from five pieces. The base and drawbar was one, with a pair of small push-on rear wheels, and the chute above was formed from two pieces, either in the same or different colours. We have seen examples in green and red (**17**) green, yellow and red (**18**) and red, blue and lemon (**19**).

High-sided Trailer

This trailer with high sides and a sloping rear section of the trailer floor was presumably intended to represent a livestock transport trailer with the rear tailboard forming a ramp. It was made in three pieces plus the two standard black wheels: the trailer body and drawbar were moulded as one, with a separate tailboard, and an axle which push-fitted into holes in the base of the trailer body. We have seen the trailer with a red body and blue tailboard (**20**) or a blue body and red tailboard (**21**), both versions fitted with a red axle.

Seed Drill

This rather clever device was made from thirteen separate pieces and is consequently quite rare. It consisted of an upper rectangular frame with towing eye attached. Pushed into this from below were four separate seed drill units, each with a small hole in the bottom able to take fine particles, such as salt. On each end of the drills were small wheels which in the real version provided support for the mechanism, and in this toy they do enhance the impression of the implement being a drill. The only example we have seen was in the Manor Farm boxed set, and that had a green frame and wheels and four red drill units (**22**).

Triple-Gang Mower

The rare triple-gang mower was possibly modelled on its Dinky equivalent and, like the disc harrow, it had a rear hook to enable a further implement to be attached. It was made from ten separate parts. In the one example we have seen, the upper frame and drawbar, which also formed the cage around the two rear sets of rotating blades, was blue, the cage for the front set of blades was green, the blades were blue, and there were six small yellow wheels, two to support each set of blades (**23**).

Horse Box

Equally rare was the horse box. In the one example we have seen the trailer body and drawbar were red, the doors were yellow and the roof and mudguards were green. It was on the standard black wheels (**24**). Strictly speaking, horse boxes are outside the scope of this book, but we have made an exception here to publish as complete a record as possible of the T.N. Thomas farm-related toys.

1. *T.N. Thomas International Farmall Rowcrop Tractor. Yellow body with blue wheels, steering wheel and seat (215 mm).*

4. *T.N. Thomas Small Tractor in red with yellow steering wheel and blue driver, showing five lines on grille and grid detail on mudguard (77 mm).*

2. *T.N. Thomas International Farmall Rowcrop Tractor, as (**1**) but red body and yellow wheels, steering wheel and seat.*

3. *T.N. Thomas International Farmall Rowcrop Tractor, as (**1**) but blue body and red wheels, steering wheel and seat.*

5. *T.N. Thomas Small Tractor, as (**4**), but with blue steering wheel (driver missing).*

6. *T.N. Thomas Manor Farm box front view showing drop-down 'farm yard' with a tractor and six implements.*

Further Reading

Hanlon, Bill, 1993, *Plastic Toys – Dimestore Dreams of the '40s & '50s*. Schiffer Publishing Ltd.

O'Brien, Richard, 1992, *Collecting Toy Soldiers*. Books Americana Inc.

Plastic Warrior, 2003, *The Plastic Warrior Guide to UK Makers of Plastic Toy Figures*. Plastic Warrior Publications.

Plastic Warrior 2007, *Poplar Plastics Special*. Plastic Warrior Publications.

Young, S. Mark et al., 2001, *BLAST OFF! Rockets, Robots, Ray Guns and Rarities from the Golden Age of Space Toys!* Dark Horse Comics, Inc.

U.S. Dimestore toys website: www.usdimestore.com. Copyright 2005 by Bill Hanlon Enterprises.

7. *T.N. Thomas Manor Farm box rear view.*

8. *T.N. Thomas Tipping Trailer with red body and green drawbar and chassis (90 mm).*

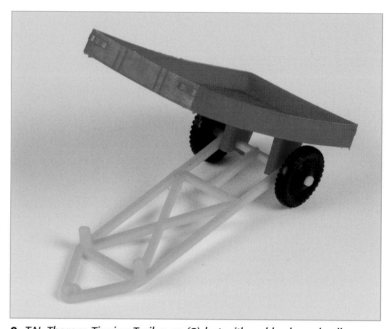

9. *T.N. Thomas Tipping Trailer, as (**8**) but with red body and yellow drawbar and chassis.*

10. *T.N. Thomas Tipping Trailer, as (**8**) but with green body and red drawbar and chassis.*

11. *T.N. Thomas Tipping Trailer, as (**8**) but with dark blue body and red drawbar and chassis.*

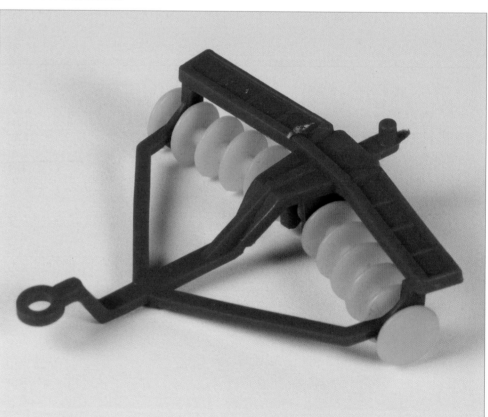

12. *T.N. Thomas Disc Harrow in dark blue with yellow discs (60 mm).*

13. *T.N. Thomas Three-furrow Plough with green frame, red, dark blue and yellow coulters and mouldboards and yellow depth wheel (76 mm).*

14. *T.N. Thomas Three-furrow Plough, as (13) but with red frame and blue coulters, mouldboards and depth wheel.*

15. *T.N. Thomas Three-furrow Plough, as (13) but with red frame, green coulters and mouldboards and dark blue depth wheel.*

16. *T.N. Thomas Three-furrow Plough, as (13) but with dark blue frame and red coulters, mouldboards and depth wheel.*

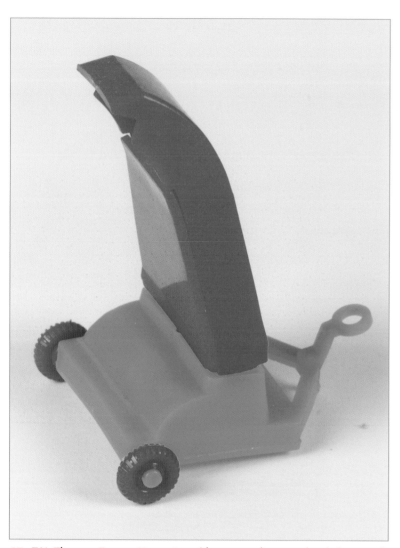

18. *T.N. Thomas Forage Harvester, as (**17**) but with a green base, yellow and red chute and blue wheels.*

19. *T.N. Thomas Forage Harvester, as (**17**) but with a red base, dark blue and lemon chute and yellow wheels.*

17. *T.N. Thomas Forage Harvester with a green base and red chute and wheels (50 mm).*

21. *T.N. Thomas High-sided Trailer, as (**20**) but with blue body and red tailboard.*

20. *T.N. Thomas High-sided Trailer with red body, blue tailboard and red axle (78 mm).*

22. *T.N. Thomas Seed Drill with green frame and red drill units (39 mm).*

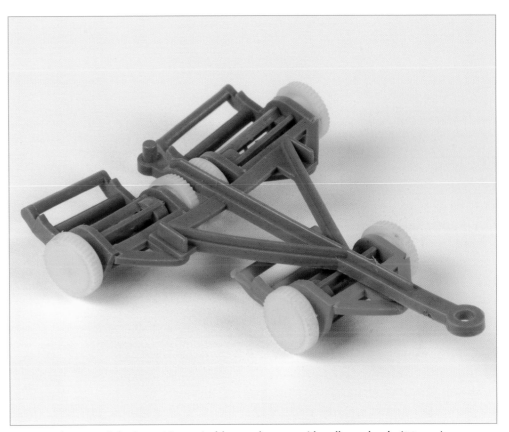

23. *T.N. Thomas Triple-Gang Mower in blue and green with yellow wheels (80 mm).*

24. *T.N. Thomas Horse Box with red trailer body, green roof and mudguards and yellow doors (150 mm). The example in the photograph is being towed by a Paramount tractor.*

Timpo

Company History

The name Timpo comes from Toy Importers Ltd. The business was started in 1938 by Sally (short for Salomon) Gawrylowicz who had left Nazi Germany for France in 1933 and came to London in 1937. His junior partner and co-director was Ernest Hainsfurth. Toy Importers Ltd became a limited company in December 1938, with its head office at 26 Westbourne Grove, Paddington, London W2. They mainly sold toys imported from Holland, but on the outbreak of war importing became increasingly difficult, so Timpo started manufacturing their own lines. Wooden toys, dolls, soft toys, lead figures and vehicles were all produced in four small factories around west London. Timpo advertised regularly to the trade in *Games & Toys* magazine, and the advertisements are very helpful at dating the sequence of events in the company's history. Their first advert was in November 1940. The triangular Timpo Toys logo was first shown in December 1941, and it was registered as a trade mark around the same time.

Mr Sally Gawrylowicz, founder of Timpo. He changed his name to Sally Gee in 1949 and later called himself Ally Gee. Pictured in Games & Toys*, March 1942.*

From 1 January 1942 there was a prohibition of supply of toys containing more than 10% of metal (by weight). This ban remained in place till 1 June 1945. To replace their lead figures, Timpo experimented with what they called 'plastic' figures, first shown at the Manchester toy fair early in 1942. These were more properly called composition figures, made from a mixture of pumice powder and glue. This material made mouldings which were solid and quite durable, but with poor surface detail when compared with metal. Timpo called the material 'Timpolin' or 'Compolin', and by the end of 1942 were offering composition submarines, soldiers, aeroplanes and animals.

On 15 April 1943 they incorporated a new company Timpo (Manufacturers) Ltd to take over the manufacturing side of the business. In 1944 a series of twelve composition vehicles was introduced – these were large models (around six inches long) with wooden wheels, and were very much toys for younger children. They are extremely rare, but fortunately for the farm toy collector there were no farm vehicles included! The composition aeroplanes were much better models and were described as 'scale model planes' in adverts in *Games & Toys* in October and November 1944. This series ran to ten different items; they are not as rare as the vehicles and are quite collectable.

As the war came to an end, Timpo were quick to see the opportunity to introduce zinc diecast vehicles, and these were first advertised to the trade in April 1946. Sally Gawrylowicz realised that as metal once again became available, composition toys would not be able to compete, so he sold the formula for Timpolin to Myer Zang of 48–54 Old Kent Road, London SE1. Zang continued to produce the composition aircraft, and also produced some composition items for Timpo, such as a model of a Crusader tank (with a tinplate base and wooden rollers), and a composition oil cabinet, chauffeur and mechanics, to be included in boxed sets with the new diecast cars. Zang is best known for introducing the Herald series of plastic figures in 1953. Herald was eventually taken over by Britains Ltd.

Back in 1946, Timpo opened a new factory at Mill Street, Slough, but the head office remained at Westbourne Grove. Also about this time Timpo obtained some moulds for hollow-cast lead figures from Kew's Ltd of Endwell Works, Endwell Road, Brockley, London SE4. Kew's had been a prominent pre-war manufacturer of hollow-cast figures. Timpo also acquired the moulds of Stoddarts Ltd, a hollow-cast figures manufacturer going back to 1916, who from 1923 had been at 16 Park Road North, Acton, London W3. Stoddarts had not re-started their business after the war, and the proprietor came to work for Timpo. From about 1947, Stoddarts' premises at Acton were occupied by Hely Trading Co. Ltd, a plastic moulding company which became closely associated with Timpo. Hely made some friction-powered vehicles with plastic bodies, introduced as Timpo Toys in 1951 and later re-branded as 'Elmont' (which was a trade mark registered by Hely in 1949) but still marketed by Timpo. During the 1950s, many of the moulds for Timpo's hollow-cast figures were converted from lead to plastic and then put into production at Hely's premises, and as Timpo's requirement for plastic mouldings continued to increase, they eventually acquired a controlling interest in Hely Trading Co. Ltd.

Also in 1946, Timpo set up an operation in Dublin. The Irish company was called Timpo Toys Ltd, and was listed as 'Toy Makers' in the 1949 street directory, at 1 Mount Brown, Kilmainham, Dublin. Boxed sets of the diecast cars exist with *MADE IN EIRE* on the box, but the cars were marked *MADE IN ENGLAND* as normal, so it seems likely that much of the activity of the Irish company was simply to re-package the English products. However there was at least one item produced locally, a horse-drawn Irish Jaunting Car, and we have decided to stretch the definition of farm vehicles to include it here.

In 1949, Sally Gawrylowicz changed his name to Sally Gee. He had always been known as 'Mr G' at his own request, no doubt because Gawrylowicz was a bit of a mouthful for his British colleagues. In later years he adopted (informally) the first name Ally rather than Sally, possibly because Sally is thought of as a female name in English, and also because his works manager and subsequent business partner was called Sally Lander.

Timpo's diecast cars were at first very successful, although they were rather crude models, but towards the end of the 1940s sales started to suffer competition, as better quality toys from Dinky and others became more widely available again. Timpo decided to expand their range of hollow-cast figures, and many new moulds were produced for them by Wilmor & Sons Ltd of 54 Rosebery Avenue, London EC1, a small firm started by former employees of M. Zang. (Note that Wilmor is the correct spelling, not Willmore, confirmed by contemporary directory entries). One of the model makers at Wilmor was Roy Selwyn-Smith, a skilful sculptor who made many of the Timpo moulds. He later re-joined Zang where he pioneered the Herald series and rose to become joint managing director of Britains Ltd.

Timpo's hollow-cast range at its zenith is shown in a catalogue of which a high-quality reproduction was produced in 1978. The catalogue itself is undated but it is almost certainly from 1952, because the company's name and address for the head office, factory and warehouses are given as Toy Importers Ltd, Devonshire Works, Duke's Avenue, [Chiswick], London W4. Timpo moved their factory from Slough to Chiswick in 1948, but the head office remained at Westbourne Grove till 1952, according to telephone and street directories. If we allow that the directories may be out of date by up to a year, that would accord with Westbourne Grove being vacated in 1951 and the catalogue showing the head office at Chiswick in 1952. The catalogue is unlikely to be 1953 because no Coronation items are shown, and could not be 1954 because by then the company name was Model Toys Ltd. It is unfortunate that there is nothing on the catalogue to say it is a reproduction, so that collectors sometimes think they have found a rare original Timpo catalogue.

Timpo's range of diecast vehicles did not initially include any farm subjects, and it was only when the hollow-cast range was expanded that it was thought necessary to produce some horse-drawn farm models to complement the farm figures. There were five different models, all shown in the 1952 catalogue, but no other advertisements or references to them have been found (**1** and **2**). The models were diecast in zinc with hollow-cast lead horses, and they are all rare because from March 1951 to May 1952 there was a ban on the use of zinc for toy production, due to the Korean war. It is therefore quite unlikely that the models were introduced in 1952 – no new zinc diecast toys could have been launched at the start of that year – so perhaps they first appeared in 1951, and the interruption in zinc supplies accounts for their scarcity. In fact the horse-drawn roller and disc harrow are so rare that we have not been able to locate any examples, and possibly they were never produced as Timpo Toys. The dies for these two items were certainly made because the models exist produced by Benbros, with a different horse and driver, and a blank panel cast underneath where *TIMPO TOYS* would have appeared on the original versions. Benbros acquired the dies for several Timpo diecast farm vehicles and lorries and re-issued the models. See the Benbros chapter in Volume 1 for details.

In 1953 the company structure was altered by winding up Toy Importers Ltd and changing the name of Timpo (Manufacturers) Ltd to Model Toys Ltd (effective 27 February 1953). Sally Gee had been joined as an equal shareholder by his works manager, Sally Lander, and Ernest Hainsfurth was no longer involved with the company. The change of name coincided with a move from Chiswick to 11 Golborne Road, Ladbroke Grove, London W10. During the 1950s, plastic figures gradually replaced hollow-cast, and there was a further move to 275 Kensal Road, London W10 in 1958. Timpo were quick to imitate Britains' 'Swoppet' figures with interchangeable parts (first introduced by Britains in 1958), and Timpo was the first manufacturer to develop over-moulding (in 1962), a technique whereby a single component could be moulded in several different colours of plastic. By a combination of over-moulding and the use of multiple parts for each figure, the need to paint the figures was eliminated.

In 1964, with the assistance of the Board of Trade, Model Toys Ltd moved to a new factory at Torbothie Road, Shotts, Lanarkshire, Scotland, allowing the production which had been carried on at Hely Trading and at Model Toys to be brought under one roof. Then in 1966 Model Toys Ltd merged with Berwick's Toy Co.Ltd to create Berwick Timpo Ltd.

It was not till 1971 that Timpo introduced swoppet-type over-moulded farm figures, together with some plastic farm buildings. These were joined in 1972 by a plastic Ford Tractor and various implements – a rather half-hearted response to Britains' growing and successful series of farm machinery. The Timpo tractor continued until 1979 in the 'Big Farm Set'.

1976 was the last year in which Model Toys Ltd made a profit, so the company was in a poor position to withstand the recession which followed the election of the Conservative government in 1979. There was a sudden decline in sales in the second half of 1979, brought on by a combination of high interest rates which squeezed home demand, and a high exchange rate which hit exports. Model Toys Ltd ceased to trade at the end of 1980, and the holding company Berwick Timpo plc went into receivership in February 1983.

Models

Cast Metal Models

Farmcart with Horse

This was a rather basic farm cart, similar to the lead models by Britains and F.G. Taylor & Sons. The Timpo model was zinc diecast with a hollow-cast lead horse. The first version had no identification cast underneath (**3**), but subsequently *TIMPO TOYS MADE IN ENGLAND* was added, with a single large capital letter *T* serving for both the words *TIMPO* and *TOYS* (**4**). The horse was painted brown (shades vary from dark brown to red-brown) with silver and black trim, some also with dark grey or black trim on the feathered hooves. The horse can be found without lettering, or with *TIMPO TOYS* cast on the left side and *MADE IN ENGLAND* on the right side, or a reverse variation of *TIMPO TOYS* on the right side and *MADE IN ENGLAND* on the left side. The cart was painted red, dark blue or grey, always with the chassis and shafts in green (shades vary), and the wheels were yellow or cream (**5** to **8**). The model was individually boxed in an off-white box with *TIMPO VEHICLES* printed in dark blue. This was a standard box used for several earlier Timpo diecast vehicles such as the removals van and petrol tanker. Sometimes the contents were indicated by a rubber-stamped name in the panel on the front of the box below the triangular logo.

Watercart with Horse

The water cart used the same chassis, shafts, wheels and horse as the farm cart, but all examples we have seen had *TIMPO TOYS MADE IN ENGLAND* cast underneath – we have not seen one with the early chassis without lettering. The water tank was cast in two pieces and riveted to the top of the chassis. Details of the horse and the box are the same as for the farm cart. The water cart was painted red, dark blue, light blue or grey, with a green chassis and shafts (shades vary) and yellow, cream or olive-green wheels (**9** to **12**).

Logwagon with Horses

This model used a pair of the Timpo hollow-cast lead farm horses with separate metal collars. Each collar was attached to the turntable and pole by metal chains, and a further long length of chain was provided to secure the log to the wagon. The horses looked much lighter than the heavy cart-horse used with the other models, and not much like work horses at all. They were clean-legged and had no harness other than the collars. They can be found with no identification cast on them or with *TIMPO ENGLAND* underneath. The wagon had a square-section chassis, cast as one piece with the front bolster. The chassis had *TIMPO TOYS MADE IN ENGLAND* cast on the top surface, and the position of the rear bolster could be adjusted depending on the length of the log, which was a real piece of wood. The rear bolster was secured in position by a diecast peg, which plugged into the top of the bolster and engaged with one of the holes in the chassis. Five colour variations are known: the chassis and rear bolster were either red (**13**), light blue (**14**) or grey (**15**), or the chassis was grey with a red rear bolster (**16**), or the chassis was green with a grey rear bolster (not illustrated). All versions had a green pole and turntable and cream wheels. The horses were brown with dark brown and black trim and black collars. The model came in a plain card lift-off lid box with a small label on the end including a Union Jack in the design. *LOG WAGON* was hand-written on the label.

Farm Roller with Horse and Farmer

Farm Harrow with Horse and Farmer

These two models are so rare that we do not have examples to describe in detail, and can only show them in the catalogue picture (**1**). Possibly they never went into production as Timpo Toys, but the dies were certainly made and later went to Benbros. We also know with certainty that the hollow-cast lead driver was produced (**17**). The two models used the same casting of frame and shafts, which would have had *TIMPO TOYS* cast underneath; the Benbros versions had a blank panel where this lettering was removed from the die. *MADE IN ENGLAND* was also cast underneath. The single axle carried a two-piece metal roller or individually cast discs on the harrow. The horse was the same as for the farm cart and water cart.

Irish Jaunting Car

A jaunting car is a two-wheeled single-horse carriage, usually with back-to-back seating and footboards over the wheels. This rare model was produced by the Irish company Timpo Toys Ltd in Dublin. We have seen it either with the same horse as the British-made horse-drawn farm cart and water cart, or with a different clean-legged hollow-cast horse unique to the Irish model. The Irish horse was painted dark brown with black trim, and had projections on each side of the collar which engaged with holes in the shafts to keep the horse in position. When the British-type horse was used, it was held in place by a length of wire passing through the shafts and the body of the horse. The jaunting car was diecast in lead, making it quite heavy. It had a black-painted chassis, which included the shafts and axle supports, to which was riveted the superstructure, painted brown with red or green trim on the upholstery of the sideways-facing seats. The wheels were painted red or dark brown, and there was a driver (unique to this model) with his left hand in his lap and wearing a bowler hat. The driver had grey trousers, a white shirt, green waistcoat and black hat. No identification was cast on the model, but examples have been seen with half-circle transfers on each side stating *FROM DUBLIN* or *FROM LIMERICK*, or *SOUVENIR FROM IRELAND*, showing that it was sold as a tourist souvenir. The plain card lift-off lid box had a label on the end stating *ONE IRISH JAUNTING CAR* and *MADE IN EIRE TIMPO TOYS LTD* (**18** to **20**).

Plastic Models

No.170 Tractor with Driver

This plastic model of a Ford 5000 tractor was introduced in 1972. It was available individually as no.170, sold from a counter pack of one dozen (**21**), and also in various sets with farm implements (see below for details). It had *SUPER 5000* moulded on each side, but no other identification. Strangely, the headlights were positioned incorrectly at the top of the radiator grille. Earlier models had steering which could be operated from the steering wheel by a metal rod; later this feature was deleted and the steering wheel was attached to a plastic post. The colour combinations of the tractor's various component parts are set out in the table below, and the first column indicates whether the model could be steered from the steering wheel. The main components of the model were glued together, so that while other colour combinations are possible, it is not straightforward for collectors to create new variations, and we have verified all those listed. Shades of the light blue colour could vary.

Steerable	Bonnet	Mudguards	Body	Seat	Front hubs	Rear hubs	Air cleaner and exhaust	Steering wheel, engine, axles	Photo
Yes	Red	Red	Lt blue	Lt blue	Red	Red	Black	Black	(**22**)
Yes	Red	Red	Lt blue	Red	Red	Red	Silver	Black	–
No	Red	Red	Lt blue	Lt blue	Lt blue	Lt blue	Silver	Silver	(**21**)
No	Red	Red	Lt blue	Lt blue	Red	Red	Silver	Silver	(**23**)
No	Red	Red	Red	Lt blue	Lt blue	Lt blue	Black	Black	(**24**)
No	Red	Lt blue	Lt blue	Red	Red	Red	Silver	Silver	(**25**)
No	Lt blue	Lt blue	Red	Red	Red	Red	Silver	Silver	(**26**)
No	Lt blue	Lt blue	Red	Red	Red	Lt blue	Silver	Silver	(**27**)
No	Lt blue	Lt blue	Red	Red	Lt blue	Lt blue	Silver	Silver	(**28**)

All models had black tyres. The driver was moulded in separate single-colour parts, except for the hands which were over-moulded in flesh colour as part of the upper body. It is pointless to list colour combinations, since the parts can easily be swapped around. His trousers can be found in grey, dark blue, yellow, red or brown; his shirt in pale blue, grey, dark blue, yellow or red; his scarf in green, pale blue or yellow; and hat in grey, brown or black. The head was always flesh colour, of course.

No.171 Tractor and Trailer with Hay Bales

This set was introduced in 1972, packed in a green card window box. Details of the tractor are given above under no.170. The low-sided trailer was moulded in grey or brown plastic with a brown chassis and drop-down tailboard, red hubs and black tyres. It came with four yellow hay bales. There was no identification on the model (**28** and **29**).

No.172 Tractor, Plough and Harrow

This set was also introduced in 1972 and used the same box as no.171, identified by a different label on one end flap (**30**). Details of the tractor are given above under no.170. The plough was remarkably similar to the Britains plough, but made in plastic rather than metal. It was black with silver mouldboards, disc coulters and rear wheel. Two moveable levers allowed the height to be adjusted, and no identification was moulded on the model (**31** and **32**).

The harrow had its plastic components moulded in yellow or brown. The colour combinations we have noted are:

(i) brown frame, yellow supports for the discs and yellow discs (**30**);

(ii) yellow frame, yellow supports for the discs and brown discs (**33**);

(iii) yellow frame, brown supports for the discs and yellow discs (**34**).

No identification was moulded on the model.

No.173 Farm Buildings Set 1

In 1973 two Farm Buildings Sets were introduced (nos.173 and 174), the first of which included a tractor and trailer. The catalogues showed either the low-sided trailer (as in no.171) or a new high-sided livestock trailer in the set, together with two buildings, trees, figures and animals, all in a green card window box (**35**). The livestock trailer is described below under no.160.

No.174 Farm Buildings Set 2

This set contained the tractor with a plough and disc harrow (as in no.172), together with two buildings, trees, figures and animals (**36**).

No.160 Big Farm Set

This set included the tractor, livestock trailer, four buildings, two trees, a duckpond, playmat, animals and figures (**37**). A short run was produced in which the duckpond was substituted by the plough and harrow, and the set contained a printed note explaining the substitution. From 1977 the set was given a new style of box and called *MEADOW FARM* (**38** and **39**). The livestock trailer was obviously inspired by the Corgi Toy Beast Carrier and was sometimes also included in set no.173, as an alternative to the low-sided trailer. Colour combinations of the livestock trailer were as follows:

(i)   brown body, chassis and tailgate, light blue hubs (**40**);

(ii) grey body, brown chassis and tailgate, light blue hubs;

(iii) grey body, chassis and tailgate, red hubs (**39**).

(iv) grey body and tailgate, brown chassis, red hubs (**41**).

All models had a light brown plastic net over the top of the trailer, and there was no identification on the model.

Model List

Serial No.	Model	Dates
Cast Metal Models		
	Farmcart with Horse	1951–52
	Watercart with Horse	1951–52
	Logwagon with Horses	1951–52
	Farm Roller with Horse and Farmer	1951–52
	Farm Harrow with Horse and Farmer	1951–52
	Irish Jaunting Car	1951?
Plastic Models		
160	Big Farm Set containing Tractor and Livestock Trailer plus four buildings, duckpond (sometimes replaced by the plough and harrow), trees, figures, animals and playmat	1973–79
170	Tractor with Driver	1972–74
171	Tractor and [Low-sided] Trailer with Hay Bales	1972–77
172	Tractor, Plough and Harrow	1972–74
173	Farm Buildings Set 1 containing Tractor and Low-sided or Livestock Trailer, plus two buildings, trees, figures and animals	1973–76
174	Farm Buildings Set 2 containing Tractor, Plough and Harrow, plus two buildings, trees, figures and animals	1973–74

Further Reading

Cole, Peter, 2004, *Suspended Animation – An Unauthorised History of Herald & Britains Plastic Figures*. Plastic Warrior.

Joplin, Norman, 1993, *The Great Book of Hollow-Cast Figures*. New Cavendish Books.

Joplin, Norman and Dean, Philip, 2005, *Hollow-Cast Civilian Toy Figures*. Schiffer Publishing Ltd.

Maughan, Michael, (undated), *The A to Z of Timpo (2nd edition)*. Published by the author.

Newson, Robert, 1988-89, 'Timpo Toys', *Model Collector* Vol.2 nos.5 and 6 and Vol.3 no.1.

Newson, Robert, 1998, 'More Timpo Planes', *Model Collector* Vol.12 no.12.

Plath, Alfred, 2005, *Timpo Toys Die Goldenen Jahre Einer Schottischen Spielzeugfabrik*. Published by the author (German text).

Richardson, Sue, 1998, 'Timpo Planes', *Model Collector* Vol.12 no.10.

Timpo Toys 1952 Catalogue, reproduction 1978. Medcalf & Carlson publication (but not marked as a reproduction on the catalogue).

TIMPO MEANS HIGHEST QUALITY — IN HOLLOW CAST METAL TOYS

TIMPO MODEL FARM ANIMALS & FIGURES (Loose)

No.	Description	Packing: Per Box
MF 1029	Shepherd	1 doz.
MF 1030	Milkmaid, sitting	1 ,,
MF 1031	Milkmaid, walking	1 ,,
MF 1032	Woman, haymaking	1 ,,
MF 1033	Woman, feeding chicks	1 ,,
MF 1034	Scarecrow	1 ,,
MF 1035	Hurdle	3 ,,
MF 1036	Trough	6 ,,
MF 1037	Bench	4 ,,
MF 1038	Hedge and Bush	6 ,,
MF 1039	Tree	3 ,,
MF 1041	Bucket	2 ,,
MF 1042	Churn	2 ,,
MF 1043	Large Tree	2 ,,
MF 1044	Small Tree	3 ,,
MF 1045	Gander, Wild	1 ,,
MF 1046	Donkey	3 ,,
MF 1047	Farmer leading Bull	Single
MF 1048	Turkey Hen	1 doz.
MF 1049	Turkey Cock	1 ,,
Farmcart with Horse		Single
Watercart with Horse		Single
Logwagon with Horses		Single
Farm Roller with Horse and Farmer		Single
Farm Harrow with Horse and Farmer		Single

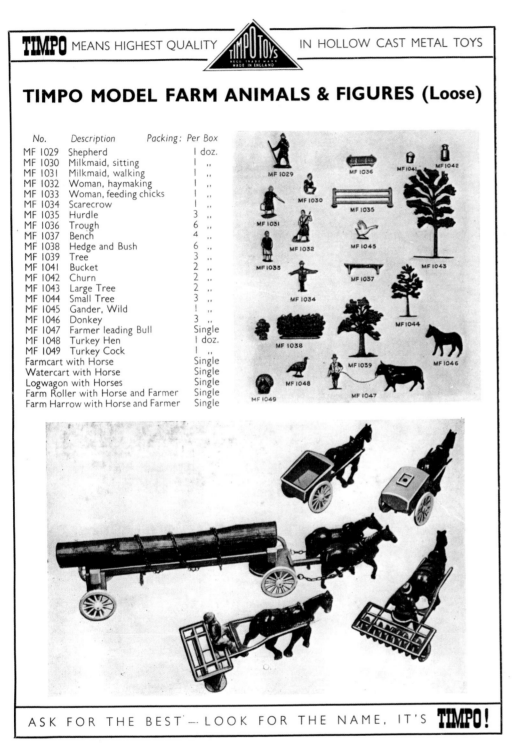

ASK FOR THE BEST — LOOK FOR THE NAME, IT'S **TIMPO!**

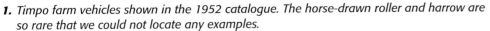

1. *Timpo farm vehicles shown in the 1952 catalogue. The horse-drawn roller and harrow are so rare that we could not locate any examples.*

TIMPO MEANS HIGHEST QUALITY — IN HOLLOW CAST METAL TOYS

TIMPO MODEL FARM ANIMALS & FIGURES
(Boxed sets)

All Sets packed in Covered Boxes with Multi-coloured Labels
Items sewn on Scored Covered Platforms

Measurements of Box: 14″ × 10″ × 1¼″

Bottom Tier · 154 · Top Tier

Two-Tier Box. Measurements: 14¾″ × 8½″ × 3″

ASK FOR THE BEST—LOOK FOR THE NAME, IT'S **TIMPO!**

2. *Another page of farm items from the 1952 catalogue.*

3. *Timpo Farmcart, early version without any lettering cast underneath.*

4. *Timpo Farmcart with horse, with* TIMPO TOYS *cast underneath (left), and the Benbros re-issue of the same model with a blank panel in place of the maker's name (right).*

5. *Timpo Farmcart with Horse, red with green chassis and yellow wheels, brown horse with no lettering cast on the horse (130 mm).*

6. *Timpo Farmcart with Horse, as (5) but cream wheels, horse with* TIMPO TOYS *on the right side and* MADE IN ENGLAND *on the left side.*

7. *Timpo Farmcart with Horse, as (6) but dark blue cart, horse with* TIMPO TOYS *on the left side and* MADE IN ENGLAND *on the right side.*

8. *Timpo Farmcart with Horse, as (7) but grey cart with yellow wheels.*

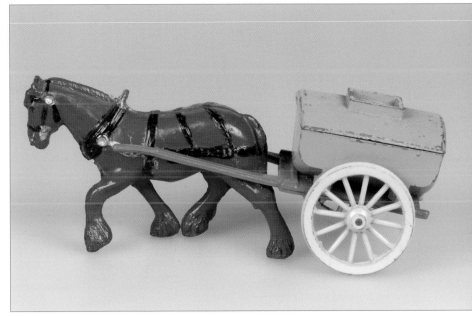

9. *Timpo Watercart with Horse, red with green chassis and yellow wheels, brown horse with* TIMPO TOYS *on the right side and* MADE IN ENGLAND *on the left side, end-opening* TIMPO VEHICLES *box with dark blue printing (130 mm).*

10. *Timpo Watercart with Horse, as (*9*) but light blue cart and cream wheels.*

11. *Timpo Watercart with Horse, as (*10*) but grey cart, horse with* TIMPO TOYS *on the left side and* MADE IN ENGLAND *on the right side.*

12. *Timpo Watercart with Horse, as (*11*) but olive-green wheels, horse with no lettering cast.*

13. *Timpo Logwagon with Horses, red with green turntable and cream wheels, brown horses. Black plastic horse collars are not original (290 mm).*

14. *Timpo Logwagon with Horses, as (13) but light blue wagon (horse collars missing).*

15. *Timpo Logwagon with Horses, as (14) but grey wagon (horse collars missing), plain card lift-off lid box with Union Jack label on the end.*

16. *Timpo Logwagon with Horses, as (15) but red rear bolster. Has the correct original black-painted metal horse collars.*

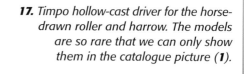

17. Timpo hollow-cast driver for the horse-drawn roller and harrow. The models are so rare that we can only show them in the catalogue picture (1).

18. *Timpo Irish Jaunting Car, produced as a tourist souvenir by Timpo Toys Ltd in Dublin. Black chassis, brown superstructure with red trim on the seats, dark brown wheels, dark brown horse, driver with bowler hat (both unique to this model),* FROM LIMERICK *transfers, plain card box with printed label (121 mm)* (photo by Philip Dean).

19. *Timpo Irish Jaunting Car, as (18) but red wheels, no transfers, dark brown horse as on the farm cart and water cart.*

20. *Timpo Irish Jaunting Car, as (19) but dark brown wheels, green trim on the seats,* SOUVENIR FROM IRELAND *transfers, horse has* MADE IN ENGLAND *cast but it may not be the original horse.*

21. *Timpo no.170 Farm Tractor, counter pack which would originally have held a dozen models (broken steering wheels and exhausts, drivers missing).*

23. *Timpo Tractor, later version without steering, red bonnet, mudguards and hubs, light blue body and seat, silver air cleaner, exhaust, steering wheel, engine and axles (90 mm).*

22. *Timpo Tractor, first version with operating steering, red bonnet, mudguards and hubs, light blue body and seat, black air cleaner, exhaust, steering wheel, engine and axles (90 mm).*

24. *Timpo Tractor, as (23) but red body, light blue hubs, black air cleaner, exhaust, engine and axles (steering wheel missing).*

25. *Timpo Tractor, as (24) but light blue mudguards and body, red seat and hubs, silver air cleaner, exhaust, steering wheel, engine and axles.*

26. *Timpo Tractor, as (25) but light blue bonnet and red body (steering wheel post broken and steering wheel glued in position).*

27. *Timpo Tractor, as (26) but light blue rear hubs.*

28. *Timpo no.171 Tractor and Trailer with hay bales, tractor as (27) but light blue front hubs, brown trailer, in a window box with* TRACTOR AND TRAILER REF.171 *on a stick-on label on one end.*

29. *Timpo no.171 Tractor and Trailer, tractor as (25), grey trailer with brown chassis and tailboard, missing the hay bales that were included in the set (tractor 90 mm, trailer 105 mm).*

30. *Timpo no.172 Tractor with Plough and Harrow, in a window box. A stick-on label on the other end of the box from that shown reads* Ref 172 Tractor, plough and harrow. *The clear plastic window is missing from the box.*

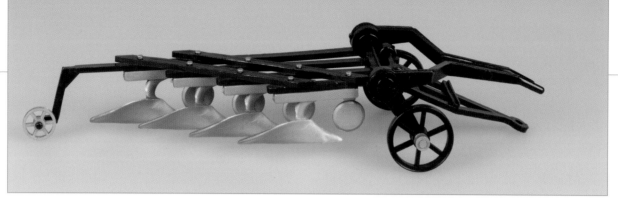

31. *Timpo Plough with the mouldboards in the raised position, black and silver plastic (140 mm).*

33. *Timpo Harrow in a different combination of yellow and brown colours from the one in (30) (50 mm).*

32. *Timpo Tractor (as (22)) and Plough (as (31)) with the mouldboards in the lowered position for ploughing.*

34. *Timpo Tractor (as (27)) and Harrow, with the harrow in a third combination of colours.*

35. *Timpo sets nos.171, 172, 173 and 174 shown in the 1974 catalogue.*

36. *Timpo no.174 Farm Buildings Set 2, in a window box, includes the Tractor, Plough and Harrow* (photo by Alfred Plath)*.*

37. Timpo no.160 Big Farm Set, earlier style of packaging, includes the Tractor (as (**25**)) and Livestock Trailer (as (**40**)) (playmat not shown).

38. Timpo no.160 Big Farm Set in the MEADOW FARM packaging used from 1977 to 1979.

39. *Timpo no.160 Big Farm Set, as (**38**) showing the contents of the set, includes the Tractor (as (**25**)) and Livestock Trailer (as (**41**) but grey chassis) (playmat not shown).*

40. *Timpo Livestock Trailer, brown body, chassis and tailgate, light brown net (108 mm).*

41. *Timpo Livestock Trailer, as (**40**) but grey body and tailgate.*

Company History

The Early Years

The history of Tri-ang started in the 1870s when two brothers, George and Joseph Lines, established the company of G. & J. Lines to manufacture children's wooden toys, including rocking horses, in the Caledonian Road area, Kings Cross, London.

Joseph Lines had four sons – George, William, Walter and Arthur – and as each left school he joined the family company. George Lines, one of the original founders, retired in the late 1890s, and in 1905 William and Walter became joint managing directors of G. & J. Lines. By 1910 the company was manufacturing a broad range of wooden products such as perambulators, children's horses and carts, doll's houses, pedal cars and rocking horses. The G. & J. Lines catalogue for 1914–15 advertised various ranges of wooden wheeled vehicles - Builders' Carts and Horses, Farm Carts and Horses,

Hansom Cabs and a Bread Cart and Horse. There was even a Dog Cart, a wooden horse pulling a wicker cart for a dog to sit in.

All the toys were relatively expensive; the cheapest builders' cart was 17 shillings, equivalent to a week's wages for a farm labourer in 1914. Some of the carts utilised the same metal-spoked, rubber-tyred wheels as the firm's perambulators, but most of the wooden parts were hand cut and formed, and the toys were almost coach-built. Although this method of wooden toy manufacture was common during the late nineteenth and early twentieth centuries, it was labour-intensive and would prove unsustainable in the long term. The catalogue stated that the Farm Cart and Horse was built of nicely polished elm, had removable tail and side-boards and was mounted on extra strong rubber wheels, but unfortunately we do not have an example of the cart.

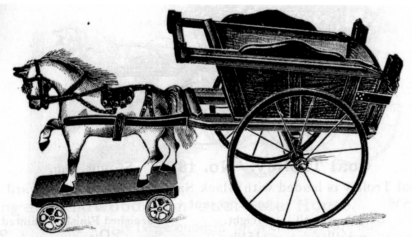

Best Builders' Carts and Horses. No. 47.

Constructed of Solid Elm and Beech.

All sizes tip. Wheels extra strong, with Rubber Tyres and Steel Axles.

No.	long over all		high						Price
1	27 in.	,,	12½in.	,, ...	...	...	...	...	17/-
,, 2	28½in.	,,	14 in.	,, ...	...	...	...	...	21/-
,, 3	33½in.	,,	16½in.	,, ...	...	...	...	...	27/-
,, 4	39 in.	,,	21 in.	,, ...	...	...	...	...	37/-
,, 5	44 in.	,,	24 in.	,, ...	...	...	...	...	47/-
,, 6	48 in.	,,	27 in.	,, ...	...	...	...	...	60/-

An excerpt from G. & J. Lines 1914–15 catalogue, showing Best Builder's Carts and Horses, No.47.

Best Strong Cart Horses. No. 146.

These are the same Horses as are put in the above Carts.
Iron Wheels on Horses.

No. 1	12 in.	high to top of Head,	11 in.	long	...	...	...	...	6/-
,, 2	13 in.	,,	,,	12½in.	,, ...	...	...	...	8/-
,, 3	15 in.	,,	,,	15 in.	,, ...	...	...	...	10/-
,, 4	17½in.	,,	,,	17 in.	,, ...	...	...	...	13/-
,, 5	22 in.	,,	,,	21 in.	,, ...	...	...	...	17/-
,, 6	23 in.	,,	,,	22 in.	,, ...	...	...	...	23/-

Farm Cart and Horse. No. 453.

Built of fine quality Elm. Cart nicely polished and mounted on extra strong Rubber Wheels. Our extra strong Cart Horse is fitted to this Cart. Tail-board and Side-boards are removable.

No.				Price
1	33in. long over all,	14 in. high,		20/-
2	35in. ,,	16 in. ,,		23/-
3	41in. ,,	18½in. ,,		29/-

From G. & J. Lines 1914–15 catalogue, showing Farm Cart and Horse, No.453.

G. & J. Lines registered the Thistle Brand in 1910 and the success of the company allowed it to move during 1914 to a purpose-built factory named the Thistle Works, in Tottenham in north London.

At the outbreak of the First World War, Walter and Arthur joined the British Armed Forces, the other brothers being too old for active duty. On their return, the pair decided to set up in business together and, with William, they formed Lines Bros Ltd which was registered in May 1919. In June 1919 the enterprise moved into an old sawmill in Ormside Street, just off the Old Kent Road in London SE15. The new company registered its trademark of three lines enclosing *L BROS* in 1920, while the toys were brand marked with the *TRIANGTOIS* logo, and the premises was renamed the Triangtois Works. Three joined straight lines make a triangle, and over the next seventy years or so, the Tri-ang name or logo would appear on a vast range of toys.

After the First World War

The new company initially produced a range of wooden toys very similar to that of G. & J. Lines, including rocking horses, pedal cars, scooters and large 'pull-along' toys such as horses and carts. In fact, in the Lines Bros catalogue for 1921-22, the 'BEST LARGE ELM TIP CARTS & HORSES' seemed to be almost identical to the G. & J. Lines Builders' Carts and Horses of 1915.

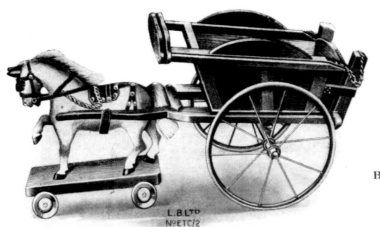

Lines' lines are selling lines.

Logos, left to right: G. & J. Lines Ltd, Triangtois and L. Bros Ltd.

BEST LARGE PINE TIP CARTS & HORSES

No. 1   Length overall   ...   ...   24 inches.
„   2   „   „   26   „
„   3   „   „   29   „

These Horses and Carts are of very superior quality. The Wheels are much stronger than usual, and the Horses have hard wood legs. All Carts tip.

BEST LARGE ELM TIP CARTS & HORSES.

No. ETC/2   Length overall   ...   26 inches.
„   ETC/3   „   „   ...   29   „
„   ETC/4   „   „   ...   34   „
„   ETC/5   „   „   ...   39   „
„   ETC/6   „   „   ...   44   „

These Toys are magnificently made, and practically unbreakable. They are fitted with rubber tyred wheels.

Best Large Elm and Pine Tip Carts and Horses as shown in the Lines Bros 1921-22 catalogue.

While Walter Lines was credited with the design of most of the early Lines Bros toys, it probably did not take much originality to plagiarise the products of his father's company. Ironically, this plagiarism right at the start of the company reflected a lack of originality which was to dog Lines Bros throughout its life. The 1921-22 catalogue showed a Pine Tip Cart with lath wheels, which we illustrate in the Models section of this chapter (**1**). Interestingly, in the catalogue the cart was attached to a standing horse, but we have seen no other examples of standing horses attached to carts in any Lines Bros literature. Perhaps the contradiction of a wheeled standing horse, 'pulling' a cart, was a poor seller and that version was dropped from the range. One other point worth mentioning is the strap line declaring that 'Lines' lines are selling lines' across the top of the catalogue page. Taking into consideration that at the time of the catalogue there were two firms with Lines as part of the company name, this encapsulated some of the muddled marketing which was to characterise much of Lines Bros' advertising and packaging.

Lines Bros exhibited at the British Industries Fair in 1920 where it was very successful, taking the equivalent of £1 million in orders although, in common with the general U.K. toy trade, orders at the next year's fair were considerably less. William was in charge of sales for Lines Bros, Arthur ran the factories, while Walter was the managing director and the most forward thinking of the three brothers. He understood that the future lay in mass production and division of labour, which allowed workers to specialise in specific areas of production and thus achieve higher output, while keeping the skills required, and therefore the wages, to a minimum.

Triangtois Works was organised into approximately thirty separate departments and investment was made in the latest production machinery such as capstan lathes, electric welders and leather cutters. A government report of 1924-25, originally requested by the perambulator trade to prove unfair competition, found that in 1923 each Lines Bros employee produced 625 doll's prams a year, compared with 181 at G. & J. Lines and 458 at Patterson Edwards, a local London competitor and one of the traditional pram makers (see the Leeway chapter). The report stated that the Lines Bros products were of the same quality as those of G. & J. Lines, so in the light of the huge disparity in productivity of the two workforces, it is not difficult to imagine the conversations between Walter and his father Joseph, prior to Walter starting his own company.

A showroom was opened in Fore Street, London EC2 in 1921, and in 1922 another factory was acquired near to the Triangtois Works, at 761 Old Kent Road, which was quickly adapted to Walter's ideas of how modern factories operated. By the end of 1923 it had become obvious that even larger premises were required, and within 18 months the company had moved into a purpose-built factory on a 27-acre site in Morden Road, Merton, London SW19. The factory was originally named the Triangtois Works, but was soon renamed the Tri-ang Works, when that trade mark was registered in 1927. Initially there were 500 employees on the site, but expansion meant that this had trebled by 1932, by which time the premises included a cricket pitch, two football pitches and a number of tennis courts. Lines Bros became a public company in 1933, and profits more than doubled between 1932 and 1937 from £37,367 to £79,717.

While offering an increased range of wooden products, during the late 1920s the company introduced large tinplate, clockwork and pressed-steel vehicles, and from mid-1932 this included a range of crawler tractors. The Pedigree name was registered in 1931, to be used for the plastic doll-making business, and the U.K. rights to K-boats, a range of steel-based toys, were acquired in 1934. Minic toys, a series of tinplate vehicles that filled a gap in the market between expensive large-scale pressed-steel models and diecast Dinky Toys were introduced in 1935. The increasing production and sales of soft toys were acknowledged when Pedigree Soft Toys was formed at the Merton factory in 1937. In addition to expansion of the Tri-ang range, success in the 1930s had also been driven by the acquisition of other toy makers. The company had bought the Unique and Unity Cycle Co. of Handsworth, Birmingham in 1927, and Hamleys, the chain of toy retailers, in 1931. Lines Bros gained full control of International Model Aircraft, the manufacturers of FROG scale model aircraft, in 1932, and bought G. & J. Lines Ltd when the company's founder, Joseph Lines, died.

The Second World War

Lines Bros' expertise in mass production was widely recognised, so the company was well-placed to help the war effort. The company employed around 7,000 workers to make magazines for fighter aircraft, shell cases, land mine cases, optical apparatus, four types of pilot-less and rocket-propelled glider and, after some redesign involving Walter Lines, over a million Sten guns. Toy production had continued in the early years of the war, until metal was no longer available for toys, and some of what are now the rarest Minic vehicles were created by painting standard items in camouflage colours.

Toy production quickly resumed after the war, and in 1945 an old munitions factory near Merthyr Tydfil in south Wales was bought, and Lines Bros (South Wales) Ltd was formed. In 1946, the Pedigree Works was opened in Belfast, Northern Ireland, initially for the production of soft toys, and later for wooden and diecast toys (see the Spot-On chapter). In 1951 Rovex Plastics Ltd of Richmond, west London, was bought and in 1954 was moved to a new factory in Margate, Kent, although Lines Bros retained the Richmond premises. Rovex was a successful manufacturer of electric train sets and, in common with a number of other purchases by Lines Bros, part of the reason for buying them was to save Lines Bros the problem of competing with them.

As well as acquiring new companies and factories in the U.K. during this period, Lines Bros also expanded abroad. While it had always had representation in the English-speaking nations around the world, the company now proceeded to buy factories and manufacturers, partly driven by the protectionism that sprang up after the Second World War. During the late 1940s and early 1950s, this led Lines Bros to buy Joy Toys Ltd of New Zealand, Cyclops and Moldex of Australia, and the Jabula Company in South Africa. Additionally, factories and sawmills were acquired in Canada, where a range of toys was designed, manufactured and marketed using the old G. & J. Lines Thistle trademark.

By 1950, the Tri-ang Works at Merton was considered to be the largest and most up-to-date toy factory in the world. An area of approximately 750,000 sq. ft. housed over 500 power presses, 1,000 hand presses, and had a half-mile of railway sidings, all manned by 4,000 employees. But expansion continued, and after Rovex moved to its new factory in 1954, the Richmond factory was used for Tri-ang's musical toys, later being used for the development of electronic toys. Minimodels, the maker of Scalextric racing systems, was bought in 1958. A maker of rubber toys, Young and Fogg, and a rival pram maker, Simpson, Fawcett & Co, were also bought.

By the end of the 1950s Lines Bros had 39 separate manufacturing facilities around the world, including a number of factories in France. Just in the U.K., the company produced metal toys, from doll's prams and pedal cars made of pressed steel to small tinplate tanks, ranges of plastic dolls, soft toys and diecast vehicles. It made the Tri-ang Railways range, doll's houses, wooden, steel or aluminium toy yachts, educational toys, construction toys and children's bicycles and tricycles. Under the FROG brand there was a series of scale model aircraft and under the Scalextric brand a range of plastic slot-car racing sets. The wooden toys made ranged from children's desks to pull-along carts, service stations to accompany Minic vehicles to children's play-stores complete with pretend groceries lining the shelves. Additionally there was a vast range of wooden vehicles, from floor trains to push-along and pull-along carts marketed under the names Toddler Toys, Tinkle Toys, Teachem Toys and Animated Toys.

The 1960s

On 1 January 1960 Lines Bros opened a new 14,000 sq. ft. showroom in Haymarket, London SW1. A German manufacturer of wooden toys, J. Schowanek GmbH, was bought in 1961, following which Shuresta (A. Mirecki) Ltd and Walker Industries Ltd, two baby carriage makers, were bought in 1963. Meccano Ltd and Dinky Toys were bought in 1964, and subsequently Meccano (France) and a 49 per cent share of Subbuteo Ltd were acquired in 1965. Further overseas expansion involved the purchase of Steelcraft Ltd, an Australian manufacturer of baby carriages, and a 50 per cent share of Regal Trading, at the time the largest toy wholesaler in South Africa.

Sindy dolls were introduced in 1962, and Tri-ang Publications Ltd was formed in 1964, in conjunction with the Thomson publishing organisation, to launch the Tri-ang magazine. The Hornby-Dublo range of toy railways was merged with Tri-ang's own railway system, and production moved to the Rovex factory at Margate. Lines Bros kept the Meccano factory in Liverpool and used the Meccano name to market

Play-Doh and the Cliki construction toy range. The maze of companies, controlling interests, brand names and trade marks used in the U.K. now included Arkitex, Cliki, Dinky, Dublo, FROG, Hornby, International Model Aircraft, K-boats, Lines, Meccano, Minic, Minimodels, Pedigree, Penguin, Play-Doh, Rovex, Scalextric, Schowanek, Sindy, Spot-On, Subbuteo, Tri-ang, Unique, G. & R. Wrenn and Hamleys.

Unfortunately, what on paper was one of the most successful U.K.-owned toy companies, proved to be a castle built on sand. Lines Bros had always struggled to bring original ideas to market, and most of its best-selling toys were products of the companies it had bought, such as FROG, Minimodels, Rovex, Subbuteo and Wrenn. Successful in-house products such as Spot-On and Sindy were actually copycats of Dinky and Barbie, respectively.

Tri-ang's biggest problem was that however big it grew, its only product was toys and any general downturn in sales of toys would affect it particularly badly. Additionally, its range of toys looked increasingly outdated. It is a supreme irony that one of Lines Bros' most collected and appreciated ranges, Minic tinplate toys, was introduced at a time when the rest of the toy world was moving into diecast models. Equally, although Lines Bros' wide range of wooden toys are now very collectable, they were competing with plastic toys that were lighter and thus more suitable for export. The plastic competitors were easier to mass-produce, could not rust and generally did not have sharp edges or contain any toxic paints.

In addition to its in-house problems, the Meccano acquisition soon proved to be an albatross, as it would be in later years for Airfix (see the Airfix chapter in Volume 1). There were good reasons why, in 1964, Lines Bros was able to buy Meccano for half its market value. Despite rationalising products and premises, and various financial measures, Lines Bros struggled to make acceptable levels of profit throughout the 1960s. The purchase of Meccano exacerbated the trend, as the company resisted Lines Bros' attempts to bring its working practices up to date.

The 1970s

Despite efforts to move away from its traditional toys such as prams and wooden vehicles, and into the newer toys such as diecast models, plastic toys and products aimed at a poorer and younger demographic, Lines Bros found trading increasingly difficult towards the end of the 1960s. Due to an over-valued pound, temporarily halted by devaluation in 1967, and falling home demand, the late 1960s and early 1970s were in general a difficult time for British industry. This was especially true for a company like Lines Bros that was already losing its way, and which relied heavily on overseas sales through its representatives in 63 countries. A predicted trading loss of £4.6 million for 1971 led to the collapse of Lines Bros in late 1971, and the various parts of the business were separated and sold.

Rovex Tri-ang Ltd, based in Margate, was sold to Dunbee-Combex-Marx, which was a short-term reprieve, as that conglomerate had closed by the end of the decade. G. & R. Wrenn continued independently; Meccano Ltd was sold to Airfix; Canterbury Bears took over the soft toy bear range, and the ex-production manager of Pedigree Soft Toys in Belfast set up a company to take on that range. Tri-ang Pedigree, based in Merton and Birmingham, became part of the Barclay Toy Group and was soon rationalised. The Morden Road factory was sold for development and the cycle business was sold to Raleigh Cycles. The wooden toys business went to Goodwood Toys, a manufacturer based near Chichester, Sussex that had been set up by Walter Lines upon his retirement from Lines Bros in the early 1960s. He died in November 1972, having masterminded the family business through its halcyon years.

Tri-ang Pedigree almost became part of Airfix Industries in 1975, but Airfix decided against the purchase. In 1983 Tri-ang was bought by Sharna Ware Ltd, its major competitor in the manufacture of large plastic toys such as pedal cars. The Tri-ang name is currently used to promote a range of large garden toys including plastic pedal cars such as tractor dumpers, loaders and wagons. Pedigree Toys and Dolls Ltd is now owned by Tamwade, a U.K.-based holding company, and still has the Sindy range of dolls amongst its products.

Models

Over the course of its half-century of business, Lines Bros made toy vehicles in all the popular and economical materials. Early toys were wooden, then pressed steel and tinplate were used, and gradually plastic parts were added. Post-Second World War, the company moved into diecast toys (see the Spot-On chapter), and increased its range of plastic toys, while still using wood and steel where it was appropriate to the toy and the target market.

Rather than list the farm models made by Lines Bros in a purely chronological order, which would have involved constantly moving between material, brand and toy range, it seemed more sensible to list them by material, and then chronologically within each section. We have also separated the early steel vehicles from the Minic range, as the Minic range was so broad and mostly of post-Second World War manufacture.

The triangular Lines Bros trademark was registered in 1920, and the Tri-ang name was registered in 1927. Subsequently, dependent on the space available, one or both appeared on every toy and piece of packaging made by Lines Bros. The L Bros triangle usually had *L.BROS LTD., LONDON* and *ENGLAND* around its three sides, and whenever *Tri-ang* was used it was accompanied by the words *REGD. TRADE MARK*, or later with the abbreviation ®. However, to save pointless repetition, we have not shown these when we have described any brand marking. The *Tri-ang* mark is normally found with the hyphen, but can sometimes be found as *Triang*, and we reflect these variations in the descriptions.

Wooden Toys

The earliest toys made by Lines Bros were wooden, and Lines Bros catalogues and advertising material from the 1920s promoted tip carts and horses made from either pine or elm. Both types of cart were available in four sizes and, dependent on the budget of the customer, the carts could be fitted with wooden wheels or metal wheels on rubber tyres, and the horses with or without manes and tails. Production of the carts seemed not to survive the Second World War.

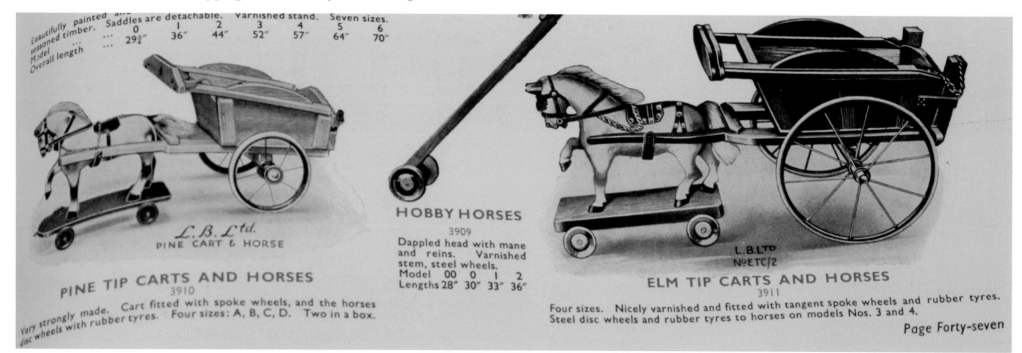

Lines Bros Elm and Pine Tip Carts and Horses from a late 1920s catalogue.

The excerpt from the Lines Bros catalogue shows the ranges of Pine Tip Carts and Horses, and Elm Tip Carts and Horses. The excerpt is as interesting for how it advertises the ranges as it is for what it advertises. Immediately beneath the illustration of the pine cart is 'L.B.Ltd.' and 'PINE CART & HORSE'. Immediately beneath the illustration of the elm cart is 'L.B.LTD' in a different, smaller type face, and 'No.ETC/2', i.e. Elm Tip Cart, size 2. Beneath the item descriptions are Lines Bros' internal numbers: 3910 for the pine cart and 3911 for the elm cart. The pine cart was available in four sizes: A, B, C and D. The elm cart was also available in four sizes. While superficially they were the same horses and carts, available in a choice of woods, on further inspection it can be seen that very few components, if any, are common to both of them. The elm carts were gradually phased out during the 1930s, although that was probably as much due to the availability of elm as it was to the fact that in the early 1930s Lines Bros were advertising a toy that was effectively the same as G. & J. Lines had made before the First World War.

In the 1930s Lines Bros gradually moved from making mainly wooden toys into metal toy production, and this accelerated after the Second World War. However, for younger children, the company continued to make some of the pre-Second World War wooden toys, at least until the mid-1950s. Sold under names such as Teachem Toys, Tinkle Toys and Toddler Toys, these toys consisted of carts being pulled by a range of animals, from elephants to geese, and were aimed at the pre-school market.

Pine Tip Cart and Horse

The cart was made of pine with a plywood base (**1**), and our example had a top-rail at the front, onto which was pinned a metal badge with the *TRIANGTOIS* name, thus dating the toy to between 1920 and 1927. Similar pine models were manufactured by Lines Bros at least until the start of the Second World War. As our version is 740 mm, it would seem to be an example of the largest, 29-inches-long model. The cart wheels were attached to a wooden platform that hinged on the axle. A piece of wood across the front of the platform, behind the horse, held the cart in the upright position and could be swivelled to make the cart tip. The cart had a separate tailgate, held by wooden pegs on chains, and each cart side carried what could almost be termed greedy boards.

The horse was made of wood, gessoed and painted a dapple grey. It had a horsehair mane and tail, and leather harness and pad. (Interestingly, in his privately published history of Lines Brothers Limited, Walter Lines recalls that 'My brothers and I were hard at it in the factory within a couple of days of leaving school. We carved pole horses and rocking horses and loaded vans, and by far the most horrible job of all, cleaned the raw cows tails before they were cured for nailing to the rear ends of rocking horses.') The horse had a leather belly band which was pinned to the shaft on one side, while the other end had a hook which fixed over a nail on the other shaft. A chain,

attached between the shafts, ran over the horse and kept the cart at the correct height, and smaller chains were attached between the horse's collar and the shafts each side to allow the horse to 'pull' the cart along.

The cart wheels were made completely of wood. They had eight spokes, and rims made from wooden lath. The horse stood on a wooden platform with four small solid-wooden wheels, and at the front of the platform was a small towing eye. Because both sets of wheels were wooden, it is likely that this was one of the cheaper toys within the cart range, as the more expensive versions had metal wheels with rubber tyres. In the Leeway chapter we show a similar, but slightly more expensive horse and cart. It was made by Patterson Edwards, one of Lines Bros' major competitors, whose products were generally known to be slightly inferior to those made by Lines Bros.

Simple Wooden Tractor

This was part of a range of utilitarian toys made by Lines Bros immediately after the Second World War, which used readily available, left-over components when quality materials were in short supply (**2**). It was a very simple toy, made from a small block of wood, painted green, with wooden wheels painted red. The tinplate front was from the Minic range, the aluminium *TRI-ANG* nameplate pinned each side seemed to have been left over from some other use, and the seat, exhaust, air cleaner and steering column were each made from an odd piece of metal. Only the stamped-tin steering wheel required specific manufacture and even that was possibly obtained from a spring washer. The front wheels were push-fitted onto a wire axle that was held to the body by two staples, and the larger rear wheels were push-fitted onto wooden pins, and could not rotate on our examples. The two examples shown have minor differences - one is missing the air cleaner at the rear, the other has a towing eye at the front, and the example on the right does not have the *TRI-ANG* name and triangle trademark above the radiator grille.

Tinkle Toys Pony Cart

This was part of a range of similar wooden carts with bells, pulled by different animals including a chick, a duck and an elephant. It was made from five pieces of wood nailed together, with a smaller block underneath (**3** and **4**), all painted in light pink. A nail and washer each side of the wooden block held the wooden wheels which were painted dark blue. The pony was cut out of plywood and then painted white with dark grey dappling, and details of the pony's tail, mane and facial features added in black. No harness was depicted and it was attached to the shafts by nails each side. Originally the pony would have had a piece of coloured string attached, with a wooden grip at the end to enable a child to pull the cart along. The pony was set on four small wooden wheels, painted yellow. Nailed to the top edge of the back of the cart was a thin strip of tinplate, and a tinplate bell with clanger was attached to the strip. However, it is likely that the bell was originally at the front of the cart on our example, as that was more common with toys in the Tinkle Toy range.

On the rear panel of the cart was the standard Lines Bros logo of *L BROS* in a triangle, with *TRI-ANG MADE IN ENGLAND* (**4**) underneath. Tinkle Toys were in the Tri-ang Toys range from the 1930s onwards, and it is likely that the example pictured was from the late 1940s and made at the Merton factory.

Tinkle Toys Horse Cart

Similar to the Pony Cart, this was made from five pieces of wooden board, with a smaller block underneath (**5** and **6**), all painted in light green. The light blue wooden wheels were held by a nail and washer each side of the wooden block. The horse was cut out of plywood and then painted white, with details of the horse's tail, mane, hooves, mouth, bridle and collar all spray-painted in black. Nailed to the top edge of the front of the cart was a thin strip of tinplate, and a tinplate bell with clanger was attached to the strip. However, a design flaw meant that when the horse and cart was on a flat surface, the horse's tail stopped the bell from ringing.

On the rear panel of the cart was the standard Lines Bros logo of *L BROS* in a triangle, but underneath it was *TRI-ANG MADE IN IRELAND* (**6**), indicating that the model was made in the Lines Bros Belfast factory in the late 1940s or early 1950s.

Early Metal Toys

Lines Brothers started to advertise its new range of large clockwork toys in mid-1932. According to an advertisement in the July 1932 edition of *Meccano Magazine*, the range consisted of three tractors – Nos.1, 2 and 3 – a Whippet Tank, a Tiger Tank, and three Magic Cars, a Racing Car, a Sports Car and a Saloon Car. In fact the 'Magic Autos' were not yet available, the advert stating that, 'These three cars will be ready in August'. As this implied that the other five models were available, it allows us to date them quite precisely. The tractors were similar in concept to the Animate Climbing Tractor, a toy made in America by the Animate Toy Co. that had been advertised in *Meccano Magazine* from 1929. Gamages retailed the Animate products and later the Tri-ang equivalents, and used similar adverts – of the models climbing a pile of books – for both! While there was no doubt that Tri-ang was inspired by Animate, the U.K.-made models were significantly cheaper than the American-made originals.

Elsewhere in this book we have referred to pressed steel under the umbrella term 'tinplate'. However, Tri-ang did refer to some of their models as steel or pressed steel, and so in this chapter, where appropriate, we too have used the term pressed steel. All the toys had Tri-ang internal reference numbers that were not always present on advertising material

but seem to have been used more consistently as the 1930s progressed. In the Model List later in this chapter we show the internal reference numbers, where known, in addition to the references used on advertising material. While it is not known precisely when the following toys disappeared from the Tri-ang range, they were not produced after the Second World War, though some residual stock might have still been available for sale.

Tractor No.1 (Nippy)

This crawler tractor was the simplest and cheapest model in the new range (**7** and **8**), initially advertised for 1/3d each, whereas the Magic Car, which was the most expensive model, was 12/6d. Along with the Whippet tank that was based on it, Tractor no.1 was made from tinplate, whereas all the other models were made from pressed steel, a heavier gauge of sheet metal. The tractor was red with green metal rollers around which ran white rubber tracks. The clockwork motor drove the two front rollers, and there was a fixed key for the motor on the left side of the tractor body. Details of the engine compartment were stamped into the tinplate and were enhanced with black print.

Printed above the front grille was the Lines Brothers logo of *L BROS* set within a triangle, and beneath was *TRIANG TRACTOR No.1 Made in England*. The driver was a slightly militaristic, three-dimensional tinplate figure, holding an integral steering wheel which fitted through a hole cut in the rear of the engine compartment. Another hole was cut out of the back panel of the tractor, which allowed a trailer to be towed. The driver's seat was made by cutting and forming the platform at the rear of the engine compartment, and there was an on/off switch to the right of the driver.

Almost immediately, the cheaper toys within the new range were subject to price reductions by Lines Bros. The July advertisement was repeated in the August 1932 edition of *Meccano Magazine*, but now 'TRACTOR no.1' was only 1/-. By November 1934 it had been renamed 'NIPPY TRACTOR' and was only 6d. To facilitate the price reductions, the model was modified so that it was cheaper to produce. The black printing was gone, and the detail and lettering on the front were now achieved by stamping them into the tinplate (**9**). The driver, complete with his seat, had become a two-dimensional piece of printed tinplate which slotted into two cuts in the tinplate at the rear of the tractor, and the hole for towing was now a very crude towing eye that was punched in the tinplate.

Farm Tractor No.1

At some point during mid-1933 Lines Bros adapted Tractor no.1 by removing the metal rollers and tracks, and adding vulcanised rubber wheels (**10**). The wheeled model became Farm Tractor No.1 while the original tracked model was renamed the Nippy Tractor. The new model had larger wheels at the rear than the front, and on our example the *L BROS* above the front grille was no longer present, probably because the stamp to make the front had worn. Otherwise it was the same model as before, and in the *Meccano Magazine* advertisement for November 1934 it was advertised for 9d, or 1/- complete with an all-steel trailer. Unfortunately we do not have an example of the trailer, but from the illustration in the advert, it was similar to the no.2 Trailer (**15**) but shorter and shallower. We have been unable to find a box for Tractor no.1 or its variants, with or without trailer, and it is likely that it was sold in trade packs as was common for cheaper toys at the time, rather than individually boxed.

Tractor No.2

Made of pressed steel and 8½ inches (215 mm) long (**11** and **12**), this was a more substantial model crawler tractor than no.1 (note that the black tyres on the wheels in (**11**) are not original). It had metal rollers the same colour as the body, which was described as 'cellulose red' in a Gamages advertisement for December 1932, and there was no additional paintwork. The fixed key for the clockwork motor was on the right side of the body, the motor drove the front rollers and the on/off switch was on the left beside the driver. It had white rubber tracks which unfortunately are no longer present on our examples. The steering wheel was formed from part of the rear panel of the engine compartment and a small towing hook was formed from the rear of the base.

The driver, which was almost three-dimensional, was made of two pieces of tin clipped together and slotted into three holes at the rear of the tractor. The driver figure was always made in the same colours of blue trousers, red jumper with yellow shirt, brown boots and white hat with a blue band, although the shades of all of the colours varied. The details of the engine compartment and grille were pressed into the steel. There was a large grille effect on the front panel, with the *L BROS* logo in a triangle at the top and *TRIANG TRACTOR No.2* beneath. On the rear panel, behind the driver's seat, was *TRIANG TRACTOR No.2 Made in England by L.B. Ltd London*.

In the July 1932 *Meccano Magazine* advertisement Tractor no.2 was 4/11d, but a month later it was 3/11d. The price continued to drop, and by November 1934 the model was advertised in *Meccano Magazine* at 2/6d, an almost 50 per cent reduction on its opening price only 28 months previously. In contrast to Tractor no.1, there seemed to be no change in quality or cost of production to accompany these reductions. Possibly Lines Bros had simply got the original price wrong or, of course, had been taken by surprise by the severity of the 1930s economic upheavals. Alternatively, the range was so successful that the company was able to reduce the price to encourage further sales.

Farm Tractor No.2 and Trailer

The September 1934 *Meccano Magazine* advertisement introduced 'FARM TRACTOR No.2 AND TRAILER'. The farm tractor was produced by replacing the tracks and rollers of Tractor no.2 with wheels of vulcanised rubber (**13** and **14**), which were larger at the rear than the front, while the trailer was a new item, also made from pressed steel (**15**). Unlike the original Tractor no.2, the new model tractor was available in a variety of colours, and we have found a blue tractor (**16**), an orange tractor (**13** and **14**) and a yellow tractor (not illustrated). In all other respects, the Farm Tractor no.2 was the same model as Tractor no.2.

The trailer was simply made from two pieces of pressed steel, one for the trailer and the other for the chassis, a metal axle with white rubber wheels push-fitted onto the axle ends, and our example was light green (**15**). At the front of the chassis the metal had been cut and formed to create a crude spring to hold the body in place. When the spring was released, the body could tilt on two metal supports at the rear of the chassis. *TRIANG TOYS PAT PENDING* was marked around the side wall of the wheels, and this patent actually referred to the rubber wheel.

Farm Tractor no.2 was sold both with and without a trailer. Unfortunately we have been unable to find a box for Tractor no.2 or its variants, with or without trailer, and we do not know how the toy was packaged.

Tractor No.3 (6 wheel)

As the name denotes, this crawler tractor had six wheels of equal size, but the front two on each side were rollers, around which ran white rubber tracks (**17** and **18**). The two rear wheels had pieces of the rubber track stuck around their rims, and additionally the two rear axles had cogs fixed between the wheel and the side of the tractor, and a chain ran around the cogs. The clockwork motor drove the middle wheels, which powered the rear wheels via the chain and the front wheels via the tracks. The winder for the motor was on the left side of the body, just in front of the middle wheel and beneath the track and, unlike the previous tractors, required a separate key.

This tractor had two levers at the rear, one either side of the driver which was the same two-piece figure as used on Tractor no.2. The lever on the right side had the words *STOP* and *START* pressed into the metal, and the left-side lever had *FORWARD* and *BACK*.

In many ways, Tractor no.3 resembled Tractor no.2, but with a longer bonnet, and many of the same press tools would have been used to make the two vehicles. Similar to Tractor no.2, the steering wheel was formed from the rear of the engine compartment, and a small towing hook was formed from the rear of the base. The same grille effect was on the front panel, with the *L BROS* logo in a triangle at the top and *TRIANG TRACTOR No.3* beneath. On the rear panel, behind the driver's seat, was *TRIANG TRACTOR No.3 Made in England by L.B. Ltd London*. We have found the tractor body unpainted (**17**, **18** and **20**) and painted green (**19**), and these were possibly the only finishes available. We have not found any trailers for Tractor no.3 to pull other than the trailer that was advertised with Tractor no.2.

As befitted what was a very expensive toy when first sold, Tractor no.3 was supplied individually boxed (**20**). On its four faces the box had an illustration of the toy traversing rough ground, with ploughed fields in the background, while being driven by a schoolboy. Interestingly, the illustration showed the toy with the key in place, but only mentioned in passing, on the end panels, that the toy was in fact clockwork. At the top left of each face was the Lines Bros logo of *L BROS* in a triangle, and beneath was *TRI-ANG TRACTOR, NUMBER 3* in a roundel, and *REGD. TRADE MARK* beneath the Tri-ang name. On each end flap, at the top, was *TRI-ANG TRACTOR NUMBER 3* and *REGD. TRADE MARK*, with two of the Lines Bros logos, with *MADE IN ENGLAND by Lines Bros. Ltd. London* beneath. On the bottom flap was *The TRI-ANG series comprises many fine clockwork and mechanical models. Have you seen them all?*

Tractor no.3 was initially advertised at 10/6d, and only the Magic series of cars was more expensive within the Tri-ang clockwork range introduced in 1932. While retaining its high price during 1932 and 1933, in the November 1934 *Meccano Magazine* advertisement the toy was only 7/6d, and was now 'TRACTOR no.3C'. The possible implication of the number amendment was that C was short for continuous track, and that Tri-ang intended to introduce a wheeled farm version. We have no evidence that a wheeled version was ever introduced and, in fact, the lack of clearance between the winder and the wheels already used by Tri-ang, would have necessitated an expensive new tool.

Traction Engine

In the September 1934 edition of *Meccano Magazine* Tri-ang advertised three sizes of Traction Engine for 9d, 1/3d and 1/11d. We have found only one size of traction engine (**21** and **22**), although the red model in (**22**) had a pole to hold a light bulb and a battery compartment underneath, so perhaps this was one of the more expensive versions. The toy was made of pressed steel with vulcanised rubber wheels and a wooden fly wheel. When pushed, the left rear wheel would turn the flywheel that rested on it. In turn, the flywheel would action a pushrod connected to the flywheel via an eccentric axle. The pushrod would move in and out of the cylinder box, thus simulating the action of a real traction engine. In (**22**) the circumference of the flywheel has been increased with a thin strip of rubber.

Pressed into the steel base at the rear was *TRI-ANG TOYS MADE IN ENGLAND.*

Wagon and Horses

It is not known exactly when this toy was first introduced. In the 1935-36 Tri-ang catalogue it appeared as part of the non-mechanical steel toys range, which consisted of farm trucks, tipping lorries and breakdown lorries in addition to the wagon and horses. Each model was available in three sizes, 0, 00 and 000, 0 being the largest size of each, and in the catalogue all the dimensions were shown in both metric as well as imperial measurements. The three available lengths of wagon and horses were 35.9 cm, 27.9 cm and 21 cm, and the catalogue also showed the Tri-ang internal reference number for each size, 1867, 1868 and 1869. The example illustrated is size 00 (**23**).

In later Tri-ang advertisements, the non-mechanical steel toys were listed as part of the 000 series, and although the range of non-mechanical steel toys had been extended, the number of sizes available had been decreased. The expanded range included a motor coach, a crane, a forward control lorry and a coupe, but the wagon and horses was only available in one size and the farm truck – now listed as market lorry – in only two sizes. The later advertisement for the wagon and horses listed it with a Tri-ang internal reference number of 3762, and it was supplied with a load of coloured bricks.

The wagon and horses was of all-steel construction with four rubber wheels (**23**), similar to those on the trailer for Farm Tractor no.2. The wagon had a steel support clipped to the body at the rear which held the back axle, and the front axle was attached to the turntable. The turntable was all-white and had an eye at the rear which attached it to the wagon via a hook, cut and formed from the wagon bed. The turntable had to be detached to enable the toy to be packaged in its individual box. The horses were two-dimensional profiles attached to the turntable, with detailing on their outer faces. The driver clipped into a slot in the front seat and was also two-dimensional but with detailing on both sides.

The wagon and horses were totally unmarked, other than on the side wall of the wheels. In contrast, the box had almost too much information on it. Printed in red on a white background, each of the four faces had an illustration of the contents, and *Tri-ang* with *REGD*, at the top left, and *WAGON & HORSES* at top right. On the left side was the *L BROS* logo, plus *MADE IN ENGLAND Lines Bros Ltd. LONDON, S.W.19*. Under the illustration was *HEAVY GAUGE STEEL* and to the right was *SOLID RUBBER WHEELS*.

Minic Models

The Minic trade mark was registered in January 1935, and the first advertisement for the range appeared in the June 1935 edition of *Meccano Magazine*. The 14 models that comprised the first series were advertised as 'ALL TO SCALE CLOCKWORK TOYS'. The advert stated that the toys were of 'STRONG CONSTRUCTION', and that they had 'POWERFUL LONG RUNNING MECHANISM' with 'FRONT WHEEL DRIVE'. All the models were made of tinplate and, with the exception of a tractor, a tank and a steam roller, they were all road vehicles that ran on tinplate wheels with rubber tyres. Minics were sold in a variety of packaging, some of which was generic, some depicting the contents and some differentiating the more expensive models in the range from the others. The expensive models, including the tractor and trailer set, were packaged in either red or green cardboard boxes with lift-off lids. The two tracked models had their own packaging, as did the caravan and steam roller, but the majority of models were packed in thin cardboard boxes with end flaps illustrated with British street scenes involving vehicles from the Minic range.

Although the Minic name – created from **MINI**ature and **C**lockwork – is synonymous with tinplate clockwork toys, the range of toys sold using the trade mark was wide. It ranged from the pre-Second World War tinplate models to much larger post-war pressed steel vehicles, and from large plastic clockwork or battery-driven toys, through small plastic push-n-go model vehicles, to miniature electric motorways, racetracks and accessories.

Tri-ang introduced a series of larger-size clockwork models in the early 1950s, and the models were advertised as Minic no.II series. There was little consistency to the range, which included vehicles with gears, novelty items such as a musical car with built-in musical box, a car with a working horn and a stop-on Police car which would automatically avoid falling from a table top.

No.11M Tractor

Model no.11M, Minic Tractor (**24**) was a crawler tractor that was part of the Minic range from the outset and, although it underwent a number of minor changes, it continued in the range until clockwork Minics were discontinued in the early 1950s.

The first versions of no.11M had white rubber tracks which ran around wooden rollers, although the original tracks are now very brittle and consequently have quite often been replaced, sometimes with black tracks. The tractor body, made completely of tinplate, stamped and formed and then held together with tabs, appeared in various shades of green (**24** and **25**). The steering wheel and seat were separate pieces held in place with their own tabs and there was a tow hook formed from the tractor base. The seat is often found with slots to hold a driver, but none was ever issued. Underneath, the tractor was marked *TRI-ANG MINIC TOYS MADE IN ENGLAND*. The winder was on the left side and the motor drove the front wheels. The front of the tractor was made from a piece of tinplate that was unpainted but which had been anodised against rust. A filling cap was stamped into the top, and a grille, with *TRI* and *ANG* either side of a triangle, was stamped on the front.

The first type of box was made of thin card. It was illustrated on two faces, had advertising text on the other two and on the end flaps was *TRI-ANG MINIC* above a panel listing the contents. The illustration was of a country lane with Tractor no.11M heading towards a fence, and on the other side

of the fence was a Light Tank no.20M in battle (**24** to **27**). One side panel had the *L BROS* triangular logo, and *TO COLLECTORS MINIC MINIATURE CLOCKWORK SCALE MODELS are obtainable in all Good Toy Shops in many types representing all the traffic seen in Town or Country*. Beneath was *Made in England by LINES BROS. LTD. Morden Rd. LONDON, S.W.* On the other was *New Types will be available from time to time of the TRI-ANG MINIC SCALE MODEL CLOCKWORK TOYS*. Beneath that was *Ask your Toy Dealer for the latest* and beneath that *Made in England by Lines Bros.Ltd. London. S.W.19*.

In 1940, a military, camouflage-coloured version was issued, catalogue no.11MCF, but these are now very hard to find. Soon after production recommenced post-Second World War, the rollers were changed to diecast metal and the tracks were changed to black rubber (**26** to **29**), although it is possible to find black tracks on wooden rollers (**25**). The tractor was now made in various colours – the ubiquitous green, red (**26**), or blue (**29**) – and the box changed to one specific to the tractor (**29** to **31**). The two large panels had an illustration of the tractor in a country setting, with *Tri-ang MINIC TRACTOR* above the illustration and *SCALE MODELS WITH POWERFUL CLOCKWORK MOTORS* beneath. On the two other side-panels was an illustration of the tractor in profile and the Lines Bros logo. Above the illustration was *Tri-ang MINIC TRACTOR* and *MINIC TOYS – Made in England by LINES BROS. Ltd. Merton. London. S.W.19* was beneath. At some point during the post-war period the steering wheel changed slightly to one made from a thin piece of tinplate held in by a small nail (**28**). The last variation, introduced during the 1950s, replaced the metal rollers with similar, though slightly smaller, black plastic versions (**30** and **31**).

No.83M Farm Tractor

This model, first advertised in November 1950, was based on the pre-war Tractor no.11M but it had been modified with mudguards at the rear, and the tracks were replaced by wheels (**32** and **33**). The wheels had rubber tyres on metal hubs, later changed to black plastic hubs (**34** and **35**), and the rear wheels were larger and had treads, but the motor still drove the front wheels. An additional hole was cut either side at the rear of the body to take the higher clearance required for the rear axle. There was a simple exhaust made from a pop rivet pushed through a hole in the rear left side of the bonnet.

The farm tractor has been found in green (**32** and **35**) and red (**33** and **34**), and although it was always marked underneath exactly the same as no.11M, the filler cap, Tri-ang name and logo were no longer stamped on the front. In a number of the advertisements for this vehicle, including the strip catalogues given away with Minics in the 1950s, no.83M is illustrated with two pipes, perhaps meant to represent an air cleaner in addition to the exhaust. This would seem to be artistic licence, as we have never seen a version of no.83M with a second pipe.

Model no.83M was only in the Minic range for a few years until the tinplate clockwork toys were discontinued in the mid-1950s. Its thin cardboard box stayed the same during its life, and instead of being multi-coloured like the pre-war boxes, the box was illustrated using only black and red against a white background (**32** to **35**). In all the illustrations of the tractor it is shown with the Tri-ang name and logo at the top of the front grille, which were not present on the model. Two of the box's faces showed the tractor in what seemed to be a shed. Above the illustration was *Tri-ang MINIC FARM TRACTOR* with *CLOCKWORK SCALE MODEL* beneath. The other two faces were illustrated with the tractor and a driver to the right of the panel with *A True Scale Model* in the top right corner and *Tri-ang MINIC CLOCKWORK FARM TRACTOR* to the left of the illustration. *MINIC TOYS – Made in England by LINES BROS. Ltd. Merton. London. S.W.19* ran along the bottom of the panel.

On one end flap was *Tri-ang MINIC CLOCKWORK TOYS* above a panel listing the contents as *FARM TRACTOR*. On the other end flap was *Tri-ang MINIC TOYS To Collectors A large range of MINIC clockwork toys is available, representing all types of road vehicles. These are illustrated in the full colour brochure enclosed. MINIC Toys are obtainable from all good Toy Shops.*

No.26M Tractor & Trailer with Cases

This set combined Tractor no.11M with a simple four-wheeled trailer, and was probably issued as part of the second series of Minic models in 1936 (**36** and **37**). The tractor was the same as supplied individually as no.11M but was available in other colours. The trailer body was stamped and formed from one piece of tinplate with simple brackets either end to hold metal axles, and the wheels were the same wooden rollers used on the tractor, but with part of a rubber track stuck on to act as a tyre. The tow bar was made from the same heavy-gauge wire as the axles, with its ends bent around the front axle. We have found the trailer in blue (**36**), red (**37**) or light brown (**40**), and none of the examples was marked. Nine wooden packing cases were on the trailer, implying that it was intended to represent an industrial rather than an agricultural vehicle. The cases were variously ink-stamped on all faces with the Lines Bros name, triangles, squares, rhombuses, *MINIC, LONDON* and *CALCUTTA*.

Before the Second World War, Set no.26M was packaged in a plain red box with a label listing the contents, *MINIC TRACTOR & TRAILER* above the *L BROS* logo, and *MADE IN ENGLAND by L. B. Ltd. LONDON* beneath (**36** and **37**).

No.26M Farm Tractor with Trailer, with or without Cases.

The post-1950 version of set no.26M combined Farm Tractor no.83M with a new version of the trailer (**38** and **39**). The simple axle brackets of the pre-war version were replaced by a two-piece axle 'platform', the same at both ends, which retained the heavy-gauge wire drawbar. The new type of axle platform

was similar to the arrangement used on other Minic articulated trailers. The post-war trailer initially had metal hubs and black rubber tyres but, in line with the Farm Tractor, the hubs changed to black plastic. Photo (**40**), shows the evolution of the trailers. From left to right, first is the original trailer with white rubber tracks stuck around wooden hubs, in the middle is the new type of trailer with rubber tyres on metal hubs, and lastly is the second version of the new trailer with rubber tyres on plastic hubs.

There were two versions of the post-1950 set, with or without cases, though in advertisements they both had the 26M set number. The version with cases was sold in a multi-coloured box which depicted a red tractor pulling a green trailer on each face (**38**). The tractor illustrated on the larger panels had a driver, whereas on the smaller panels it did not. In fact no driver was ever issued for the no.26M tractor or any of its variants. Above the illustration was *Tri-ang* in large white letters, *MINIC* in large yellow letters, with *CLOCKWORK POWERED · SCALE MODELS* in smaller green letters, and *FARM TRACTOR WITH TRAILER* in white lettering, with *Complete with CASES* in smaller yellow letters beneath. *MINIC TOYS – Made in England by LINES. BROS. Ltd. Merton. London. S.W.19* was under the illustration in white.

The two smaller panels had *Tri-ang* in an oval, the same description of the contents but with *CLOCKWORK* above the illustration, and *SCALE MODELS WITH POWERFUL CLOCKWORK MOTORS* beneath. One end panel had *Tri-ang MINIC CLOCKWORK TOYS* and *FARM TRACTOR WITH TRAILER* beneath. The other had the, by now, standard Tri-ang words, informing collectors that a large range of Minic clockwork toys was available from the nearest good toy shop.

The version without cases was sold in the same box, with similar artwork and words (**39**), except that it used only red and white colours, there was no driver illustrated, no words referring to the cases, and no cases were depicted on the trailer (**41**).

No.211M Nuffield Tractor

Introduced in 1950, and sometimes listed and referred to as Minic no.2 Nuffield Tractor, its correct number was no.211M. One of the largest vehicles within the Minic no.II series, this was a totally new model made mainly of plastic, which is always found in red with green mudguards (**42** to **45**). The sides of the engine and the front grille area were painted black to enhance the detail of the plastic moulding. The body of the tractor was constructed from two pieces of plastic which were stuck together down the central axis. It had a plastic seat with pips underneath that push-fitted into two holes in the tractor body, and was supplied with a separate plastic driver with a hole underneath that fitted over a pip in the seat. The driver was made of flesh-coloured plastic, and always had a blue shirt and brown overalls with black boots.

The tractor had a tinplate baseplate which was held to the body with small screws, and small tinplate mudguards which were riveted to the rear of the tractor. The front axle mounting with stub axles was made of plastic the same colour as the tractor. A metal steering mechanism was riveted to the axle, which allowed the steering wheel to turn the front wheels via a rod inside the tractor body. The steering wheel was diecast metal, and there was a belt pulley, exhaust pipe and air cleaner all made of black plastic and push-fitted into holes in the tractor.

The tractor had a powerful clockwork mechanism which was wound on the right side, using holes that were provided through the rear tyre (**43** to **45**). There was a lever on the left side of the tractor, just in front of the belt pulley (**42**), which was actually the on/off switch, and at the very rear of the model was a gear lever with forward, reverse and neutral positions. Neither lever was marked in any way. The sides of the bonnet had transfers with *No.2 MINIC TRACTOR*, and beneath the grille at the front the tractor had a transfer with the registration number *APL42*. Underneath, the tractor was marked *TRI-ANG MINIC TOYS MADE IN ENGLAND*.

To cap it all, the model had an *Automatic "exhaust-smoke" puffing device*. The boxed tractor came supplied with white powder in a bag and a bulb-filler. The bulb-filler was used to suck in and hold some of the white powder, by compressing the rubber bulb of the filler and thus creating a vacuum into which the powder was drawn when the bulb was released. The exhaust was removed, powder was applied to its base, and then the exhaust was replaced. Air, set in motion by the clockwork motor inside the tractor, would puff out the powder through the hollow exhaust pipe, creating a smoke effect just like the real thing.

The wheels had metal hubs with rubber tyres, and push-fitted onto the axles. All the tyres had substantial treads and *TRI-ANG MADE IN ENGLAND L B LTD* running around the side wall. The rear hubs had *L. BROS. LTD* cast into the inner surface, and the rear tyres had six holes cut equidistantly around the hub which allowed access to the motor winder.

The model was initially supplied in a substantial corrugated cardboard box with illustrated labels stuck to three sides (**43**). The fourth side was left blank, as were the end flaps. Each of the three illustrated sides had a different aspect of the tractor on it. The main side had a scene of the tractor being driven right to left across it. To the left and above the tractor was *Tri-ang MINIC* with *Series II* in tiny letters beneath the Minic name and *CLOCKWORK POWERED* on the top right. In large letters in the bottom right corner was *NUFFIELD TRACTOR*.

One of the other two sides had a rear view of the tractor, showing the towing mechanism (**44**), while on the third side the tractor was being driven towards the viewer. Both these rear and front illustrations took up just over half the side, and above the illustrations was *Tri-ang MINIC Series II* and beneath the illustration was *CLOCKWORK POWERED*. To the right of the illustration was a panel with *NUFFIELD TRACTOR* at the bottom and the following points emphasising the toy's play value:

- *A true scale model with fully detailed body and chassis.*
- *Powerful, specially built, clockwork power unit with reversing gear and hand brake.*
- *Automatic "exhaust-smoke" puffing device.*
- *Full steering operated from wheel.*
- *Large section rubber tyres.*

The box contained a small paper bag with a key, and a general instruction sheet, in addition to the bag containing white powder and the bulb filler. For ease of packing the model into its box, the driver and front wheels were supplied loose.

Subsequently the box changed to a slightly larger one made of thinner cardboard. The illustrations and words were unchanged, but they were now printed directly onto all four sides (**44** and **45**). The large end flaps were also printed, with *Tri-ang MINIC CLOCKWORK POWERED NUFFIELD TRACTOR* above a small line drawing of the tractor at work.

Although no.211M Nuffield Tractor was only in the Tri-ang range for a few years, it underwent a number of minor but significant changes. There was still nothing for it to tow, and the towing mechanism on the rear was changed to a simple towing eye. The wheels were changed to one-piece, without separate hubs, and made completely of rubber (**44** and **45**). A larger driver was used, made of a harder plastic and more in scale with the tractor, and the painted detailing of the engine was a deeper black. The most significant change was that the on/off switch was no longer a lever on the left side, but was replaced by a smaller second lever at the rear, on the opposite side to the gear lever.

No.M250 Clockwork Marvel Tractor

During the early 1960s, Tri-ang introduced a series of vehicles under the name 'CLOCKWORK MARVEL' (**46** to **48**). The series initially consisted of a Traction Engine with 'CHUGITY-CHUG' noise, a Veteran Car with 'WOBBLE-WHEELS', and a tractor with 'CLICKITY-CLICK' noise. Subsequently a fire engine with wobble wheels was added to the series.

The tractor was made in various shades of blue plastic (**46** and **48**), with chromed plastic engine, exhaust, grille and air cleaner. There was a plastic push-fit steering wheel, normally white but sometimes black. The wheels were red plastic with metal caps on the axle ends, and the front axle swivelled, though there was no steering. The clockwork motor was wound via a large handle on the right side of the tractor, and the motor drove the rear wheels and made a clicking noise as the wheels revolved. Underneath, the tractor was marked *M250 MADE IN ENGLAND BY MINIC LIMITED*, and there was a towing eye at the rear. Although the tractor seats were shaped to take a driver, no driver has been found. The only modification to the model during its life was that the winder changed from a circular knurled wheel to a hexagonal handle (**46** and **48**).

The three vehicles were initially packaged in a generic orange window box which showed the contents and which had a photograph of the vehicles on the back (**47**). Above the photo of the series was *MARVELS are marvellous!* and below the photo was *The more MARVELS the merrier!* On the ends was *POWERFUL CLOCKWORK MOTOR FIXED WINDING KEY* and *ENGINEERED FOR STRENGTH* above *a MARVELLOUS toy!* Beneath was a blank panel, presumably intended to list the contents, and below that was *MADE IN ENGLAND BY MINIC LIMITED* and *A Tri-ang PRODUCT*.

At some point during its life, presumably after the rationalisation of the Tri-ang, Rovex and Hornby product lines in 1967, each Clockwork Marvel vehicle was repackaged into its own box (**48**). The new boxes were slightly smaller than the previous version, no longer had a cellophane window, and were in blue with photographs of the contents on all sides. The two main sides had a large photograph of the contents with *CLOCKWORK MARVEL* above. The bottom panel had a smaller photo to one side with *CLOCKWORK MARVEL* and *a Pedigree Playtime clockwork toy* above, and • *Powerful wind-up motor* • *Big easy-to-turn key - cannot be removed* • *Wheels will not come off* • *Tough plastic construction* and • *Steerable front wheels* on the left side of the panel. The top of the box had a similar photograph, *CLOCKWORK MARVEL* running along the top, the flags of eleven European countries beneath that, and the remaining space taken up with the same promotional phrases regarding the virtues of the toy, but in small script in five European languages.

The end flaps had *CLOCKWORK MARVEL a Pedigree Playtime clockwork toy TRACTOR with clicking engine noise!* and *Another safe play-tested Tri-ang product made in Great Britain by Rovex Industries Ltd*. At the bottom right was a small panel with *M250*, the item number.

No.P717 Pedigree Playtime Nuffield Tractor with Tipping Trailer

These models were inherited by Rovex after the Lines Bros group bought Raphael Lipkin Ltd in 1969 (see the Lipkin chapter for a detailed description of this model) (**49**). Some of the words on the box – *FOR CHILDREN OF ALL AGES* – and the photograph of the contents being played with showed that Pedigree Playtime toys were aimed at early school-age children. The multiplicity of brand names on the box – *Pedigree, Playtime, Rovex, Tri-ang* and *Nuffield* – seemed to indicate a firm that was losing its way (**50**). The model was given the Pedigree item number of *P717*.

Hi-way Series of Models

In about 1966 Tri-ang introduced the Hi-way series of model vehicles. Most of the vehicles seemed to be for sale to a pre-school age group and stayed in the Tri-ang range until the company's demise in the 1970s.

In common with a lot of Tri-ang's later products, the range was ill-defined, ranging from generic racing cars made of diecast metal to large pressed-steel cranes, and from lorries that looked like contemporary vehicles, such as ERFs or Scammells, to tractors that looked like nothing you were likely to see on a farm.

Jumbo Tractor

The design of this vehicle seemed to be based around simplicity and cost of manufacture rather than any notion of what a real tractor looked like (**51**). However, in its favour was the fact that it was so robustly made that it must have given hours of play value even in the harshest of environments! The tractor was made from diecast metal, one piece for the tractor body, mudguards and engine including exhaust pipe, with a two-piece base. All the pieces had scratch-resistant paintwork. There was working steering for the front wheels, and the steering wheel was made of silver-coloured plated metal. On most examples the tractor baseplate was painted the same colour as the body, but we have seen one on which the base was made of the same unpainted, plated metal as the steering wheel and front axle. The baseplate was riveted to the tractor and had a large bracket at the front, and a tow hook at the rear. On top of the tractor, behind the driver's seat, was another towing eye. The vehicle was marked underneath with *Tri-ang MADE IN ENGLAND*.

All the wheels were made of two pieces with red plastic hubs and black plastic tyres, and were held to the body via metal axles with turned ends. Engine detail was provided by identical stickers, one each side, which depicted the engine, with the word *TUGSTER* above it. Another sticker, running around the front bumper, had *TYPE 37A* on each side and the number plate *CHE 4531* on the front, with chevrons and small side lights also depicted. A similar sticker was on the rear panel, and on it were chevrons and *CHE 4531* plus side and indicator lights. Two headlights and the engine grille were detailed in the casting at the front and these were painted in silver.

The box was as functional as the toy, and was made of corrugated cardboard with printing on all six faces (**51**). On two sides was an illustration of the tractor, complete with its stickers, in profile, with *TRACTOR* in large letters above the illustration and *JUMBO die-cast metal series by Tri-ang* beneath. On two other sides was *JUMBO die-cast metal series by Tri-ang* above *TRACTOR REALISTIC STEERING DIE-CAST METAL BODY BEAUTIFULLY DETAILED CUSHION TYRES* and *NO SHARP EDGES*. On both ends was *JUMBO series by Tri-ang TRACTOR*. All the printing was in black or red, against the white background of the box. What was odd was that the box did not mention the Hi-way range at all.

Although we have seen no casting variations of the model, the box changed to simpler, shrink-wrapped and open-sided packaging, like sleeves, which displayed the contents (**52**). On the top was *Tri-ang TOYS*

Jumbo TRACTOR A DETAILED MODEL WITH REALISTIC STEERING & FITTED WITH CUSHION TYRES with a silhouette of the tractor. On the base were silhouettes of a hay trailer, a lorry and an earth mover, with the words *OTHER MODELS IN THIS SERIES INCLUDE THE JUMBO HAY TRAILER DESIGNED FOR TOWING BY THIS TRACTOR also EARTH MOVERS LORRIES and CRANES.* On one end was *Jumbo TRACTOR TRI-ANG PEDIGREE LTD. LONDON S.W.19*, with the tractor silhouette, and on the other *Jumbo DIE·CAST METAL TRACTOR MADE IN ENGLAND Tri-ang TOYS.* On the plinth holding the tractor inside the packaging was *SECURE WHEEL FIXING* and *NO SHARP EDGES*, again illustrating that the product was expected to be bought for young children.

Jumbo Hay Trailer with Hurdles

Made for use with the Jumbo Tractor, this was a robust model simply made from one piece of diecast metal, with a very heavy-duty axle with turned ends holding on the two wheels (**53**). The wheels used the same black plastic tyres as the front of the Jumbo Tractor, although the red plastic hubs were different, and aluminium spacers kept the tyres away from the sides of the trailer. At the front of the trailer was a drawbar with a towing eye that fixed over the hook at the rear of the tractor. Underneath, the trailer was just marked *Tri-ang*, and we have only found it in red.

The trailer was supplied with four greedy boards, called hurdles by Tri-ang, which were made of diecast metal and always coloured yellow. There were two sizes, with the longer ones for the sides. Each hurdle had two eyes which clipped onto pips around the top of the trailer.

The packaging was very similar to the later packaging of the Jumbo Tractor, consisting of a simple, shrink-wrapped and open-sided sleeve which displayed the contents (**54**). On the top was *Tri-ang TOYS Jumbo HAY TRAILER with HURDLES* and a silhouette of the trailer attached to the tractor and with the hurdles erected. On the base was *MODELS IN THIS SERIES INCLUDE*, then silhouettes of three lorries and an earth mover, followed by *ALSO J.C.B. EARTH MOVER CRANES.* On one end was *Jumbo HAY TRAILER TRI-ANG TOYS LTD. LONDON S.W.19*, and on the other *Jumbo HAY TRAILER TRI-ANG MADE IN ENGLAND.* On the plinth holding the trailer inside the packaging was *DESIGNED FOR TOWING BY THE JUMBO TRACTOR* and *THE HURDLES MAY ALSO BE USED TO FORM A SHEEP PEN.*

Jumbo Tractor Dumper

This was a set which married the standard Jumbo Tractor with a dumper, and the pair could be used for either a farm or a road works scene (**55**). The dumper had a diecast metal chassis which rose at the front into a tow bar with pin that slotted into the towing eye at the rear of the driver's seat on the tractor. The large dumper body was made of plastic, and hinged on

the axle. A metal ram attached to the bottom of the dumper slid into a plastic sleeve attached to the chassis and allowed the dumper to tip and hold. The wheels on the dumper were the same as those used on the front of the tractor, and we have only found the dumper in orange plastic. On both sides of the dumper was a sticker with *MAXIMUM CAPACITY 3 cu. yds.*, and around the outer face of the tow bar at the front of the chassis was a sticker with *ROAD-WORKS Ltd.* on either side and *26* in a circle on the front. We presume that 26 was the Tri-ang item number for the Dumper, but we have not seen one individually boxed.

The box, made of corrugated cardboard held together with heavy-duty staples, was white with black and red printing. The two main faces had an illustration of the contents with its selling points highlighted, *FANTASTIC DETAIL REALISTIC STEERING DETACHABLE DUMPER TIPPING ACTION* and *NO SHARP EDGES.* Above the illustration was *JUMBO die-cast metal toys by Tri-ang TRACTOR DUMPER* which was repeated on both ends, which also had *TRI-ANG TOYS LTD MORDEN RD, MERTON S.W.19* on them. On the top was just *JUMBO TRACTOR DUMPER by Tri-ang.*

Jumbo Buck Rake Tractor

Tri-ang achieved this model by attaching a buck rake to the front of the Jumbo Tractor (**56**). The buck rake was made of three pieces of diecast metal riveted together, and it was attached to the tractor by a bolt that ran through holes cut in the tractor sides. On the left side of the buck rake a handle with black rubber tip allowed it to be raised and lowered, and a crude aluminium clip could be used to hold the rake in the elevated position. The tractor was painted yellow, but otherwise all the details of stickers etc., were the same as the standard Jumbo Tractor.

The packaging was similar to the corrugated cardboard, sleeve-type boxes of the Tractor and the Hay Trailer. The top and the two ends had an illustration of the model with the buck rake raised and lowered. On the top, the model was described as *WITH RAISING AND LOWERING BUCKRAKE* and *STRONG DIECAST TOYS WITH MOULDED PLASTIC PARTS.* Also on the top was *Jumbo BUCK RAKE TRACTOR by Tri-ang TOYS.* On both ends was *Jumbo BUCK RAKE TRACTOR*, but one end also had *TRI-ANG BRITISH ENGINEERED* while the other had *MADE IN ENGLAND by TRI-ANG PEDIGREE LTD.* On the base were silhouettes of six lorries and above the silhouettes was *OTHER MODELS IN THE JUMBO SERIES INCLUDE* while beneath was *AND OTHER EXCITING MODELS ARE ADDED TO THE RANGE REGULARLY*, and on the plinth holding the model inside the packaging was *RESILIENT TYRES.*

Hi-way Jumbo Farm Gift Set

This large boxed set comprised a Jumbo Tractor, a Hay Trailer and a Dumper, with a Hi-way series lorry that carried three pigs in the back (**57** and **58**). The three farm items were all standard models as described above. The lorry is outside the scope of this book, so will not be described in detail, but was one of those shown in profile on the packaging for the Jumbo Tractor and Hay Trailer models, but which did not mention the Hi-way series. The lorry had the same stickered registration number, *CHE 4531*, as the tractor, so the set would seem to have been a clever repackaging exercise by Tri-ang.

The box was the by-now standard corrugated cardboard box with inner plinth, all made from one piece but cleverly folded so that the lid could be arranged so as to display the contents (**58**). On the top of the lid was *Tri-ang TOYS HI-WAY Jumbo FARM GIFT SET* with individual illustrations of the contents, above a farm scene which included the four models and farm buildings. *Tri-ang TOYS HI-WAY Jumbo FARM GIFT SET* was also on the two sides of the plinth, but across the front was *BRITISH ENGINEERED WITH BUILT-IN STRENGTH*. The bottom of the box was blank, but around the three sides were *HI-WAY Jumbo FARM GIFT SET FOR TREMENDOUS PLAY VALUE Tri-ang TOYS HI-WAY Jumbo DIE-CAST TOYS WITH PARTS IN DETAILED MOULDED PLASTIC* and *FARM GIFT SET MADE IN ENGLAND BY TRI-ANG-PEDIGREE LTD TRI-ANG WORKS, MORDEN ROAD, LONDON, S.W.19*, a plethora of brand names and makers.

Hi-way Mighty Mini Tractor

This was part of a range of much smaller toys introduced in the early 1970s, still within the Hi-way series, aimed at pre-school children (**59**). The tractor was made of yellow plastic with a red pressed-steel shell riveted to it, although we have seen one with a light-blue shell (**60**). It had a red plastic steering wheel which push-fitted into a hole in the tractor, and an exhaust which poked through a hole punched in the top of the metal shell. It had four plastic two-piece wheels, red hubs with black tyres, which were made with the same size hubs throughout, but with larger tyres at the rear than the front. The tyres were all marked, the smaller ones with *Tri-ang Grip* and the large ones with *Tri-ang Mighty Grip*, and they all push-fitted onto the knurled ends of heavy-duty axles. The axles on the tractor clipped into slots in the plastic body, and the tractor had a tow hook at the rear.

The box was simply made from thin cardboard with a cellophane front (**59**). Above the cellophane window was *Tri-ang TOYS LTD Mighty Mini*, and beneath was *Tri-ang TOYS LTD BRITISH ENGINEERED WITH BUILT-IN STRENGTH*. On both ends was *Tri-ang TOYS LTD.* and a blank panel to list the contents with *HI-WAY* underneath. The back of the box had illustrations of the other models in the series which were a low-loader, a tractor with boom and dozer, crawler tractor with boom and the Junior Tractor. The bottom of the box illustrated the tractor and listed its selling points: *PRESSED STEEL BODY-NO SHARP EDGES PARTS IN MOULDED PLASTIC! RESILIENT TYRES CAN'T SCRATCH SURFACES! ACTION PACKED FOR MAXIMUM PLAY VALUE!*

Hi-way Mighty Mini Tractor & Trailer

This contained the Mighty Mini Tractor combined with a trailer (**60**). The tractor was made of yellow plastic with a light-blue pressed steel shell riveted to it. Otherwise it was the same as the Mighty Mini Tractor described previously in (**59**). The trailer was made of pressed steel painted yellow, and it had a drawbar clamped beneath and an opening tailboard. The wheels on the trailer were the same as those on the front of the tractor, with black tyres on red hubs. The axle was attached to the trailer through holes in the drawbar. A towing eye at the front of the trailer fitted over a hook at the rear of the tractor body.

The box was similar to the individual box for the Mighty Mini Tractor, and would originally have had a cellophane front (**60**). Above the cellophane window was *Tri-ang TOYS LTD Mighty Mini* and beneath was *Tri-ang TOYS LTD BRITISH ENGINEERED WITH BUILT-IN STRENGTH*. On both ends was *Tri-ang TOYS LTD. TRACTOR & TRAILER HI-WAY*. The back of the box had illustrations of the other models in the series, which were a low loader, a tractor with boom and dozer, crawler tractor with boom and the Junior Tractor. The bottom of the box had the tractor and trailer separately illustrated and listed their selling points: *PRESSED STEEL BODY-NO SHARP EDGES PARTS IN MOULDED PLASTIC! RESILIENT TYRES CAN'T SCRATCH SURFACES! ACTION PACKED FORMAXIMUM PLAY VALUE! and DETACHABLE TRAILER WITH MOVABLE TAILBOARD!*

Hi-way Mighty Mini Gift Set

The Mighty Mini Tractor also appeared in a set combined with a low loader, and additionally in no.TM6680 Mighty Mini Gift Set, along with four other Mighty Mini vehicles.

Model List

Tri-ang Internal Reference No.	Model	Dates
Wooden Toys		
3910	PTC/x* Pine Tip Cart and Horse	1920–39
3911	ETC/x* Elm Tip Cart and Horse	1920–1930s
	Simple Wooden Tractor	1945– ?
5TT	Tinkle Toys Pony Cart	Late 1930s–early 1950s
	Tinkle Toys Horse Cart	Late 1940s–early 1950s

*Tri-ang varied the sizes of the carts and the reference numbers for each size over the life of the range.

Metal Toys		
2881	Tractor No.1 (Nippy)	1932–39?
2885	Farm Tractor No.1	1934–39?
2886	Farm Tractor No.1 and Trailer	1934–39?
2882	Tractor No.2	1932–39?
2887	Farm Tractor No.2	1934–39?
2888	Farm Tractor No.2 and Trailer	1934–39?
	Tractor No.3 (6 wheel) later No.3C	1932–39?
	Traction Engine	1934–39?
1867	Wagon and Horses size 0	Mid-1930s-1939?
1868	Wagon and Horses size 00	Mid-1930s-1939?
1869	Wagon and Horses size 000	Mid-1930s-1939?
Minic Models		
2561	11M Tractor	1935–54?
	83M Farm Tractor	1950–54?
2573	26M Tractor & Trailer with Cases	1936–50?
	26M Farm Tractor with Trailer, with or without Cases	1950–54?
	211M Nuffield Tractor	1950–late 1950s
	M250 Clockwork Marvel Tractor	Early to late 1960s
Plastic Model		
P717	Pedigree Playtime Nuffield Tractor with Tipping Trailer	1969–78?
Hi-way Models		
	Jumbo Tractor	1966–78?
	Jumbo Hay Trailer with Hurdles	1966–78?
	Jumbo Tractor Dumper	1966–78?
	Jumbo Buck Rake Tractor	1966–78?
	Jumbo Farm Gift Set	1966–78?
TM6xxx	Mighty Mini Tractor	1970–78?
TM6xxx	Mighty Mini Tractor & Trailer	1970–78?
TM6680	Mighty Mini Gift Set	1970–78?

Further Reading

Bartok, Peter, 1987, *The Minic Book*. New Cavendish Books.

Brown, Kenneth D., 1996, *The British Toy Business, A History since 1700*. The Hambledon Press.

Fawdry, Marguerite, 1990, *British Tin Toys*. New Cavendish Books.

Meccano Magazines website: www.meccanoindex.co.uk.

Richardson, Sue, 1981, *Minic Lines Bros. Tinplate Vehicles*. Mikansue Publishing.

Salter, Brian, 2013, *The Ultimate Book of Spot-On Models Ltd.* 'In House' Publications.

Tri-ang large road vehicles website: www.triang.nl.

V & A Museum of Childhood website: www.museumofchildhood.org.uk/collections.

1. *Triangtois Pine Tip Cart and Horse. Wooden cart pulled by a painted wooden horse, all on wooden wheels (740 mm).*

2. *Tri-ang simple wooden tractors, green body with red wheels and tinplate grille (88 mm).*

3. *Tri-ang wooden Tinkle Toys Pony Cart, light pink cart on dark blue solid wooden wheels, with black-and-white horse (318 mm).*

4. *Rear view of (3) showing TRI-ANG logo.*

6. *Rear view of (**5**) showing TRI-ANG MADE IN IRELAND label.*

5. Tri-ang wooden Tinkle Toys Horse Cart, light green cart on light blue solid wooden wheels, with black-and-white horse (300 mm).

8. *Tri-ang Tractor no.1 (Nippy) in (**7**) but right side view. Note broken track.*

7. Tri-ang Tractor no.1 (Nippy) red-and-black tinplate clockwork crawler tractor, with green rollers and white tracks (130 mm).

9. *Tri-ang Tractor no.1 (Nippy) as (**7**) but later version with no printed detail and simpler driver.*

10. Tri-ang Farm Tractor no.1, red tinplate tractor with black vulcanised rubber wheels (missing the driver) (215 mm).

11. Tri-ang Tractor no.2, red pressed-steel clockwork crawler tractor, with red rollers, missing white rubber tracks and with replacement black rubber tyres (210 mm).

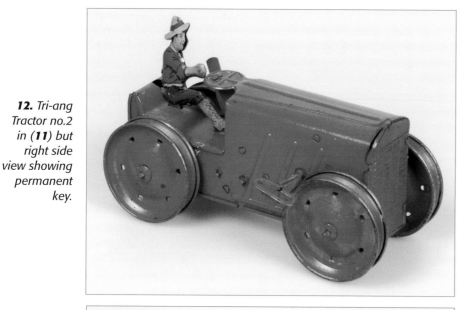

12. Tri-ang Tractor no.2 in (**11**) but right side view showing permanent key.

14. Tri-ang Farm Tractor no.2 in (**13**) but rear three-quarter view, showing permanent key on side, tractor and maker's names pressed into rear panel, and tow hook.

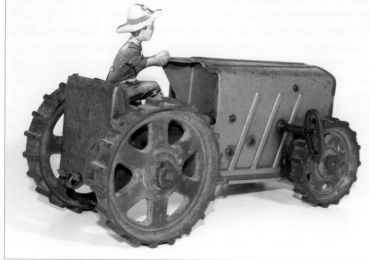

13. Tri-ang Farm Tractor no.2, orange pressed-steel clockwork tractor with black vulcanised rubber wheels (215 mm).

16. Tri-ang Farm Tractor no.2 as (**13**) but blue.

15. Tri-ang pressed-steel green farm trailer from Farm Tractor No.2 and Trailer set (210mm).

18. Tri-ang Tractor no.3 (6 wheel) in (**17**), but left side showing winder.

17. Tri-ang Tractor no.3 (6 wheel), unpainted pressed-steel clockwork crawler tractor with six red pressed-steel rollers and white rubber tracks. Right side view showing chain connecting the two rear axles (310 mm).

19. Tri-ang Tractor no.3 as (**17**) but green.

20. Tri-ang Tractor no.3 as (**17**) but showing box (note tractor is missing its tracks).

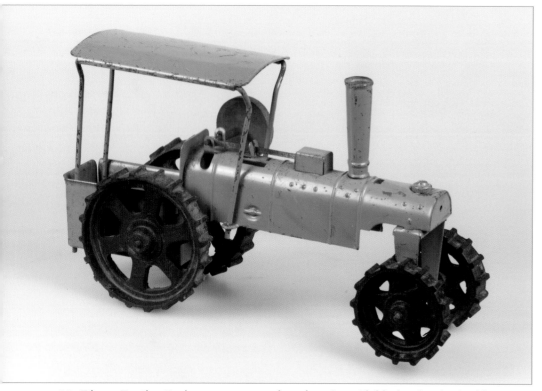

21. *Tri-ang Traction Engine, green pressed-steel engine with black vulcanised rubber wheels and wooden flywheel (250 mm).*

22. *Tri-ang Traction Engine as (21) but red and with pole to hold light bulb.*

23. *Tri-ang Wagon and Horses, all-steel red wagon on white rubber wheels, with two-dimensional white horses (280 mm).*

25. *Tri-ang Minic no.11M Tractor as (24) but darker green and with black rubber tracks.*

24. *Tri-ang Minic no.11M Tractor, green tinplate crawler tractor with red wooden rollers and white rubber tracks, showing first type of box (75 mm).*

28. *Tri-ang Minic no.11M Tractor as (**27**), showing steering wheel made from a nail and stamped tin.*

26. *Tri-ang Minic no.11M Tractor as (**25**) but red and with diecast metal rollers.*

27. *Tri-ang Minic no.11M Tractor as (**26**) but green. Rubber tracks are probably replacements.*

29. *Tri-ang Minic no.11M Tractor as (**28**) but blue and showing second type of box.*

30. *Tri-ang Minic no.11M Tractor as (**29**), but green and with black plastic rollers.*

31. *Tri-ang Minic no.11M Tractor as (**30**) but red.*

32. *Tri-ang Minic no.83M Farm Tractor, green tinplate tractor with black rubber tyres on diecast metal hubs (85 mm).*

33. *Tri-ang Minic no.83M Farm Tractor as (**32**), but red.*

34. *Tri-ang Minic no.83M Farm Tractor as (**33**) but black plastic hubs.*

35. *Tri-ang Minic no.83M Farm Tractor as (**34**) but green.*

36. *Tri-ang Minic no.26M Tractor & Trailer with Cases. No.11M Tractor with blue tinplate trailer on red wheels with pieces of track for tyres, carrying cases (180 mm).*

37. *Tri-ang Minic no.26M Tractor and Trailer with Cases as (**36**) but green tractor and red trailer (note tractor is missing its tracks).*

38. *Tri-ang Minic no.26M Farm Tractor with Trailer, with Cases. No.83M Farm Tractor as (**33**), with second type of trailer, with black rubber tyres on diecast metal hubs, and carrying cases. Showing multi-coloured box, for version with cases. (187 mm.)*

39. *Tri-ang Minic no.26M Farm Tractor with Trailer, without Cases. No.83M Farm Tractor as (**34**), with black plastic hubs, and third type of trailer, in green with black rubber tyres on black plastic hubs. Showing less colourful box for version without cases.*

40. Tri-ang Minic Farm Trailers showing evolution of versions, first version on left, last version on right.

41. Two types of box for Tri-ang Minic no.26M Farm Tractor with Trailer, with and without Cases.

42. Tri-ang Minic no.211M Nuffield Tractor. Red plastic, clockwork tractor with green tinplate mudguards, on black rubber wheels with diecast metal hubs and plastic driver. Left side showing start/stop lever (200 mm).

43. Tri-ang Minic no.211M Nuffield Tractor in (**42**), but right side of tractor showing winder, and first type of box.

45. *Tri-ang Minic no.211M Nuffield Tractor in (**44**), showing the other face of the box.*

44. *Tri-ang Minic no.211M Nuffield Tractor as (**42**), but second version with one-piece black black rubber wheels and second type of box.*

47. *Tri-ang Minic no.M250 Clockwork Marvel Tractor as (**46**) but dark blue tractor, white steering wheel and showing early generic-type orange box.*

46. *Tri-ang Minic no.M250 Clockwork Marvel Tractor in light blue plastic with silver-chromed plastic engine block, exhaust and air cleaner, and red plastic wheels. Early knurled, circular winding knob (156 mm).*

48. *Tri-ang Minic no.M250 Clockwork Marvel Tractor as (**47**), but right side showing hexagonal winding knob and later panelled box.*

49. *Pedigree Playtime no.P717 Nuffield Tractor with Tipping Trailer, in original box (tractor 215 mm, trailer 280 mm).*

50. *Pedigree Playtime no.P717 Nuffield Tractor with Tipping Trailer, box end showing* Pedigree Playtime, ROVEX, Tri-ang *and* nuffield .

52. *Tri-ang Hi-way series Jumbo Tractor as (**51**) but blue, showing second type of box with open sides.*

51. *Tri-ang Hi-way series Jumbo Tractor, green diecast tractor, with black plastic wheels on red plastic hubs, and original corrugated cardboard box, (160 mm).*

53. *Tri-ang Hi-way series, Jumbo Hay Trailer. Red diecast trailer with yellow hurdles and black plastic tyres on red plastic hubs (180 mm).*

54. *Tri-ang Hi-way series, Jumbo Hay Trailer, as (53) but showing original packaging.*

55. *Tri-ang Hi-way series, Jumbo Tractor Dumper. Tractor, as (51) but light blue, towing orange plastic dumper on light blue diecast chassis. Showing original corrugated cardboard box (overall length 298 mm).*

56. *Tri-ang Pedigree Jumbo Buck Rake Tractor. Tractor as (51) but yellow and with buckrake attachment on front, in box with open sides (200 mm).*

57. *Tri-ang Hi-way series Jumbo Farm Gift Set, showing illustrated top of box.*

58. *Tri-ang Hi-way series Jumbo Farm Gift Set, showing contents of Lorry, Jumbo Tractor and Jumbo Dumper as (55), and Jumbo Hay Trailer as (53).*

59. *Tri-ang Hi-Way series Mighty Mini Tractor. Tractor in yellow plastic with red pressed-steel shell, on plastic, two-piece wheels (78 mm).*

60. *Tri-ang Hi-Way series Mighty Mini Tractor & Trailer. Tractor in yellow plastic with light blue pressed-steel shell and yellow pressed-steel trailer, both on plastic two-piece wheels (172 mm).*

Company History

Tudor Rose was one of the brand names used by Rosedale Associated Manufacturers Ltd. The first reference we have found for Rosedale was in telephone directories for 1941 to 1946 in which the company was listed as 'Rosedale Associated Manufacturers Ltd, Imitation Jewellery', of Treforest Trading Estate, Glamorgan, and then as 'Rosedale Associated Manufacturers Ltd, Plastic Moulders', at various addresses on the same trading estate from 1946 to 1971. As plastic moulders they were also listed at premises in Gloucestershire between 1946 and 1968.

Rosedale was owned by Norman Rosedale. Because the earliest directory entry described the company as makers of imitation jewellery, we assume that was the initial business of Rosedale Associated Manufacturers before they moved into the production of toys. Interestingly, O. & M. Kleemann, the manufacturers of Kleeware products (see the Kleeware chapter), which also had a considerable business in imitation jewellery, considered Rosedale to be their biggest competitors. This competition is normally taken to be in reference to Rosedale's toy business, but was possibly actually referring to the rival imitation jewellery businesses. Early Tudor Rose products used cellulose acetate and polystyrene-based plastics, both of which would also have been used in the manufacture of imitation jewellery, but the company moved into the manufacture of polythene toys when the material became commercially available.

Rosedale was one of the firms associated with Islyn Thomas (see the T.N. Thomas chapter), who in later years considered Norman Rosedale to be a personal friend. During the immediate post-Second World War period, Thomas established and part-owned a plastic products manufacturing company based on another industrial estate in Glamorgan. So perhaps it was no coincidence that Rosedale started to manufacture toys contemporaneously with Thomas's company. In common with the products of a number of other U.K. plastic toy companies, many of the early toys sold under the Tudor Rose name were copies of American-made toys, possibly using the original moulds, and this was partly because of the influence of Thomas. Tudor Rose toys such as North American cowboys and Indians were similar to the products of Beton Toys of the U.S. and also of Airfix in the U.K. (see the Airfix chapter in Volume 1), and some of its vehicles seemed to be based on the same moulds as products from Poplar Plastics (see the Poplar Plastics chapter). As early as 1953, Tudor Rose were making a range of space figures and vehicles that were identical to the products of American plastic companies Pyro Plastics and Archer Plastics, and with some items it is impossible to determine who copied whom.

In 1959 Rosedale purchased the Kleeware business of O. & M. Kleemann,

transferring the production to Rosedale's Glamorgan factory in the mid-1960s. The company now produced a broad range of plastic toys, from small-scale doll's house accessories to a three foot long Greyhound bus. The vehicles made covered most aspects of civilian life: London black cabs and double-decker buses, sports cars such as a Jaguar XK140 Roadster, steam rollers, construction vehicles, lorries, farm vehicles, ships and a range of Old Timer cars. The toys could be found in a variety of scales, and most were made in primary colours with the same colours used interchangeably over the life of the toy. Pedal toys such as a motorcycle and sidecar, a jeep and various tractors and trailers were made, as were garden toys like wheelbarrows, watering cans and junior cricket sets. The majority of these late 1960s and early 1970s toys were made of polythene, often by the blow-moulding method of production, which allowed larger toys to be made without a commensurate increase in weight. Blow moulding had been introduced after the Second World War but was really only used for bulk manufacture from the 1960s.

The company seemed to have a successful business during the late 1960s. It annually sold around £1,000,000 worth of toys to F.W. Woolworth in the U.K., and exported to many European countries including Spain, where some Tudor Rose toys have been found packaged under the Nacoral brand. But in 1970 Rosedale Associated Manufacturers Ltd lost £186,000, and it was bought by a finance company named Heenan Beddow. The name changed to Rosedale Industries, and in an advertisement in the October 1971 edition of *Games & Toys*, the new company stated 'We admit it. Nine months ago our production, delivery, packaging and marketing just weren't up to scratch.'

Rosedale Industries was sold on to Mettoy in 1978, which in turn sold it to another finance company, Tamwade, in 1983. The last telephone directory record for Rosedale is as a plastic moulder, in Glamorgan trade directories for 1971, although we know that in 1977 a version of the Tudor Rose Routemaster bus was issued to take advantage of H.M. The Queen's Silver Jubilee. The box was marked *A Rosedale product* and *Made in England*, but there are no clues as to which firm actually manufactured the toy.

In the 1990s a company called Springwell Mouldings issued some former Tudor Rose toys in small quantities. These are usually found in polythene bags with thin cardboard headers which had a generic illustration of toys including a Jeep, construction vehicles and a double-decker bus, but without any detail of the actual toy contained in the polythene bag. In addition to the toys illustrated on the packaging, the range reissued by Springwell included a traction engine and a small boat, but it is not known what happened to the majority of the Tudor Rose moulds.

Models

Unfortunately we have been unable to find any original Rosedale catalogues, though we believe the Tudor Rose brand was first used in 1952. In addition to the farm items listed below, there is a Tudor Rose Irish Jaunting Car, which we have not included, but we mention it because we did include a Jaunting Car in the Timpo chapter. The farm items are listed in what we believe is the chronological order of their manufacture, taking the following points into consideration.

In general, toys made from brittle plastic such as cellulose acetate are from an earlier date than polystyrene-based toys and those in turn are likely to have been made earlier than toys which include components made of polythene. However, this is not foolproof, because toys can be found that use both polystyrene and polythene-based plastics, as the different properties of different types of plastic meant that the most suitable type for a particular task would be used whenever it was available.

Packaging marked Rosedale is later than packaging bearing only the Tudor Rose brand, and packaging illustrated with a photograph of the contents is later than packaging illustrated with artwork of the item.

Most Tudor Rose toys were marked with the Tudor Rose name, the logo of a Tudor rose and *Made in England*, sometimes in a roundel. Additionally, some of the toys can be found marked Rosedale. However, unfortunately not all Tudor Rose toys are marked, so precise identification can be difficult.

To our knowledge, the toys reissued by Springwell Mouldings were completely unmarked.

Small Rowcrop Tractor

This small tractor, only 75 mm long, was simply made from a single piece of hard plastic complete with integral driver, with four separate wheels each on its own stub axle. It can be found in a myriad of colours, and those found to date are a swirly blue colour body (achieved when the molten plastic mix is not stirred sufficiently) with yellow wheels (**1** and **2**), dark green body with off-white wheels (**3**), red body with off-white wheels (**4**), bright green body with either yellow or red wheels (**5** and **6**), red body with yellow wheels and yellow body with red wheels (**7**). The model has been found fully marked with the Tudor Rose logo, *MADE IN ENGLAND TUDOR ROSE*, and *ROSEDALE*, but has also been found completely unmarked. It was almost definitely copied from an American original, as very similar models were made by a number of American companies, including Banner Plastics.

Medium Rowcrop Tractor

This toy tractor was probably also copied from an American original, as the toy was modelled on an American rowcrop tractor made by Massey-Harris. It was made from a single piece of plastic complete with integral driver, but with some minor variations to the markings, wheel arrangements and the types of plastic used over the life of the toy. What was probably the first version was made from a hard cellulose-acetate-based plastic. The tractor body was unmarked, but *TUDOR ROSE*, a Tudor rose and *MADE IN ENGLAND* were marked around the outer wall of the rear wheels, which had the tread across the tyres (**8**). This version had stub axles on the rear onto which the wheels pushed, which probably led to breakages as cellulose acetate is quite brittle.

The second version was marked underneath with both *MADE IN ENGLAND* and *TUDOR ROSE* set in roundels, and the Tudor rose logo was set within the *TUDOR ROSE* name. This version was made from a softer plastic, and while it also had stub axles and the tread across the tyres, the rear wheels were unmarked (**9** and **10**).

The third version had the same markings on the body but the unmarked rear wheels now had v-shaped treads, and there was an independent axle at the rear, clipped into slots beneath the tractor body, which allowed the wheels to turn independently of each other (**11** to **16**). On all three versions the front wheels and axle were moulded in one piece and clipped into a slot at the front of the tractor.

The model has been found in a broad range of colours, including yellow body with green wheels and green body with yellow wheels (**8** and **15**); blue body with red wheels (**9**); red body with green wheels (**10** and **11**), blue wheels (**12**) and a combination of those wheel colours (**13**); yellow body and blue wheels (**14**); and red body with yellow wheels (**16**). While we do not know who made the original toy, we have found a model very similar to the first Tudor Rose version, but made in Australia by Thomas Hore & Co., and another similar to the third version but marked *MADE IN HONG KONG*.

Trailers for Medium Rowcrop Tractor

The rowcrop tractor was made with a towing eye (**10**), which allowed implements to be attached, and although to date we have found no implements that we can attribute to Tudor Rose, we have found two types of four-wheeled trailer. The first type was a flatbed wagon, with wheels and axles similar to those used on railway wagons. We have found an example in red with yellow wheels and side bars (**16**).

The second type had larger wagon-type wheels set on stub axles, and a chassis with an integral drawbar and trailer ends, and the two sides were separately moulded pieces which slotted between the end pieces (**17** to **19**). Both end pieces had clips on their top edge, possibly to hold tools such as hoes or rakes, but we have not found any trailers with items attached to the clips. The second type of trailer has been found moulded in blue plastic with either yellow or red wheels (**17** and **18**), and in yellow with red wheels (**19**). Both types of trailer were marked underneath with *TUDOR ROSE MADE IN ENGLAN*D in a rectangular panel.

Tractor Trailer Hoe & Rake

The tractor in this set was based on an International Harvester Farmall C rowcrop tractor, made in America after the Second World War (**20**), and was the largest of the three model rowcrop tractors made by Tudor Rose. The model had possibly been made previously by both Ideal Toy & Novelty Corporation and Thomas Manufacturing Corporation in America and a similar model was made by Jouef in France. A version was also made by T.N. Thomas in the U.K. (see the T.N. Thomas chapter). The Tudor Rose model was marked underneath with *TUDOR ROSE*, the Tudor rose logo and *MADE IN ENGLAND*, and the letter *C* at the front on each side. It differed from the T.N. Thomas model in having a metal axle at the rear and because it had an integrally moulded driver's seat. The tractor was yellow with green wheels and steering wheel, and the trailer was a red one-piece moulding with two small green wheels on a metal axle. In our example the hoe and rake that formed part of the boxed set are missing.

The set was contained in a bright yellow box which had illustrations of its contents on all four sides. *TRACTOR TRAILER HOE & RAKE* appeared in large letters on every side, two sides had *Flexible Polythene* and the *Tudor Rose* name and logo, while the other two sides had *THIS IS A Tudor Rose Flexible Polythene TOY* and *MADE IN ENGLAND*, which perhaps implied that the toy was made at Rosedale's Gloucestershire factory. From the box illustration we assume that the two missing farm tools were out of scale with the farm vehicles.

Heavy Duty Farm Tractor, clockwork version

There were two versions of the Heavy Duty Tractor, with and without clockwork motor. We have found one boxed example of the clockwork version, which was a substantial toy with an integral motor to drive the rear wheels, the key for which was permanently fixed to the mechanism under the tractor (**21**). Although the tractor was large, at 270 mm long, it was cleverly made from three pieces, with separate accessories which push-fitted into holes or onto raised pips on the tractor chassis.

The clockwork version had a silver engine and front grille, which matched the colour of the engine and grille on the tractor illustrated on the box, and

the seat, steering wheel, headlights, exhaust, air cleaner and belt pulley were also silver. The bonnet, front axle and wheel hubs were dark blue; the body – which included the mudguards and tow-hook – was red, and it had yellow tyres and a cream driver. Both sets of wheels were attached to the body via metal axles which would have made the toy more robust, as stub axles were an area of weakness on all plastic toys. The front axle was proud of the hubs on each side, while the rear wheels push-fitted onto the axle.

The box for the Heavy Duty Farm Tractor was impressively illustrated, with three different scenes of the tractor in use across its six faces. Two sides showed the tractor pulling a plough across a field, with farm buildings in the background. The box stated *HEAVY DUTY FARM TRACTOR* in large letters. *EVALUATED POLYTHENE HYGIENIC AND ALMOST INDESTRUCTIBLE*, in smaller letters, reassured purchasers that the recipient would get plenty of play value from what would have been a relatively expensive toy and that they would be in no danger from the new-fangled polythene from which it was made. In surprisingly smaller script *The Tudor Rose* and the Tudor Rose logo half-heartedly promoted the maker, and in the top right corner, in even smaller letters, was *POWERED BY ROBUST CLOCK-WORK MOTOR*. Two other sides had exactly the same arrangement of words but with an illustration of the tractor on its own, and the ends also had the same words, albeit with *The Tudor Rose* more prominent and an illustration of the tractor pulling a muck-spreader.

Heavy Duty Farm Tractor, non-clockwork version

The non-clockwork version of the Heavy Duty Tractor used all the same plastic components as the clockwork version. We have found three examples which used parts in a whole array of colours which seem to have been combined at random (**22** to **24**). However, on further analysis it can be seen that the tractor must have been made in large quantities on a well-organised batch-production line, and that the colours had been cleverly combined to differentiate the toy from the more expensive clockwork version. Unfortunately we have not found any packaging for the non-clockwork version of the Heavy Duty Tractor, but it is likely that this version was marketed to a different demographic from the clockwork version, probably packaged in headed plastic bags which allowed the bright colours to 'sell' the toy, and which are now missing.

Throughout the non-clockwork examples, the seat, steering wheel, driver, headlights, exhaust, air cleaner and belt pulley all matched the colour of the engine, and the wheel hubs matched the bonnet, although in two examples (**23** and **24**), the hubs were a slightly darker shade of blue. They all had black tyres, and on this version the belt pulley was attached the opposite way round to that on the clockwork model, although we have no idea if that was significant.

The colours of the non-clockwork tractors were: red bonnet, wheel hubs and front axle, yellow body and all other pieces in light blue (**22**); the same yellow body, retaining the red front axle, but for all other items the red and light blue colours were reversed (**23**); and red body with light blue bonnet, front axle and hubs, with all other items in yellow (**24**).

We have found no separate implements or trailers which this tractor might have pulled, although it was fitted with a tow-hook as part of the rear of the body. However, it was sold in a set with a tipping trailer (see below).

Heavy Duty Tractor with Tipping Trailer

This set combined a non-clockwork version of the Heavy Duty Tractor with a simple two-piece trailer (**25**). Although the tractor in the set was non-clockwork, it was the same colours as the boxed clockwork version. The tractor had a red body and yellow tyres on dark blue hubs, but the engine and grille were in silver which matched the colours of the tractor illustrated on the box, as did the bonnet and front axle in dark blue. All the other components – seat, steering wheel, headlights, exhaust, air cleaner and belt pulley – were also in silver. Similar to the clockwork model, the belt pulley was attached the opposite way round to the unboxed examples.

On the box, the trailer was red and was shown with the body held in a tipped-up position by a telescopic support, but although the model in the box was red, the support was not present on the toy which merely tilted on two pips at the rear of the chassis. The trailer sat on two small yellow wheels, push-fitted onto a metal axle, and had a drawbar with a towing eye which fitted over a hook at the rear of the tractor.

Although the tractor did not have a clockwork mechanism, it must have been a relatively expensive toy, and, similar to the clockwork tractor, this was reflected in its box. In fact, in the example photographed, the original price of 17/9 was written on the box. Two sides of the box showed a front view of the tractor and trailer, with the body tilted to empty its contents. The box stated *HEAVY DUTY TRACTOR with TIPPING TRAILER* in large letters. *EVALUATED POLYTHENE HYGIENIC & ALMOST INDESTRUCTIBLE* was in smaller letters, as were *The Tudor Rose*, and the Tudor Rose logo and *MADE IN ENGLAND* in even smaller lettering. Two other sides had exactly the same arrangement of words but with an illustration of the tractor and trailer from the rear, carrying a load of sacks.

Steam Traction Engine

Although not dedicated farm vehicles, traction engines were used extensively on farms to provide greater and more sustained power than horses. The example shown in (**26**) had a black body with red wheels, and yellow canopy, front axle, chimney, flywheel and steering wheel. The canopy and its four supports were moulded in one piece. The model had a heavy metal rear axle and a tow hook, though we are not aware of any

Tudor Rose implements for it to tow. It was made of polythene and was marked with three roundels underneath with the words *THIS ITEM IS A TUDOR ROSE DESIGN MADE IN ENGLAND* and *TUDOR ROSE PLASTIC PRODUCTS* surrounding the Tudor Rose logo.

We have seen many examples of the traction engine, with various colour combinations of body, canopy, wheels, axle, flywheel, steering wheel and chimney. The traction engine in (**27**) also had a black body, but had yellow wheels, canopy and steering wheel, and red axle, flywheel and chimney. The examples in (**28** and **29**) both had red bodies, but all other parts were either blue or yellow. Note that the chimney is missing from both vehicles and the example in (**29**) is missing its steering wheel. It is likely that other combinations of colours were also used.

The box was superbly illustrated on all four sides (**28**). The two main faces had an idyllic farm scene of a traction engine powering a thresher, with farm workers and the engine driver all present in the illustration. *STEAM TRACTION ENGINE* in large letters, *HYGIENIC AND ALMOST INDESTRUCTIBLE EVALUATED POLYTHENE* in smaller letters and *Tudor Rose* and the Tudor Rose logo were all in a panel beneath the illustration. The other two faces of the box were illustrated with a scene of the engine as it towed the thresher, and *STEAM TRACTION ENGINE EVALUATED POLYTHENE HYGIENIC AND ALMOST INDESTRUCTIBLE Tudor Rose* and the Tudor Rose logo in a panel alongside. The end panels had similar words but also had a smaller, slightly inaccurate drawing of the steam traction engine.

This form of packaging was very similar to that of the Heavy Duty Farm Tractor, so we assume that the two toys date from the same period.

Rosedale Heavy Duty Tractor

This model was another clockwork tractor which used all the same components as the earlier Tudor Rose clockwork Heavy Duty Farm Tractor, but with fewer colours (**30**). The driver was in blue – perhaps suggesting the increased use of overalls on the modern farm? The tractor body, wheel hubs and front axle were all in yellow, and all the other separate components – the seat, steering wheel, headlights, exhaust, air cleaner, belt pulley, tyres, engine block and grille – were made of black plastic. The wheels were attached to the tractor by metal axles.

The box was a very subdued affair, in a neutral all-over green colour, with a photograph of the actual contents on each side, the words *Heavy Duty TRACTOR* prominent at the top left, and with a Tudor rose and *A Rosedale Product Made in Tough, Non-Toxic Polythene Powered by Clockwork Motor Age: 2 to 8 years*. The two ends had no illustration, just the words *Heavy Duty TRACTOR* and 'clockwork tractor' translated into four foreign languages.

The subdued nature of the box might have reflected changing consumer legislation but also the compromises needed when the same product was to be sold in many countries.   Unfortunately we do not know at what point Rosedale stopped using the Tudor Rose name, but it is likely that this version was made in the last years of Rosedale's production and is consequently rare.

Model List

Model
Small Rowcrop Tractor
Medium Rowcrop Tractor
Flatbed Trailer
High-sided Trailer
Tractor, Trailer, Hoe & Rake
Heavy Duty Farm Tractor, clockwork version
Heavy Duty Farm Tractor, non-clockwork version
Heavy Duty Farm Tractor, non-clockwork version with Tipping Trailer
Steam Traction Engine
Rosedale Heavy Duty Tractor

Further Reading

Brown, Kenneth D., 1996, *The British Toy Business, A History since 1700*. The Hambledon Press.

Plastic Warrior, Dec 2003, *The Plastic Warrior Guide to UK Makers of Plastic Toy Figures*. Plastic Warrior Publications.

Various editions of *Plastic Warrior* magazine.

Young, S. Mark, et al., 2001, *BLAST OFF! Rockets, Robots, Ray Guns and Rarities from the Golden Age of Space Toys!* Dark Horse Comics, Inc.

1. *Tudor Rose Small Rowcrop Tractor in swirly blue plastic with yellow wheels (75 mm).*

2. *Tudor Rose Small Rowcrop Tractor in (**1**), rear view.*

3. *Tudor Rose Small Rowcrop Tractor, as (**1**) but dark green with off-white wheels.*

4. *Tudor Rose Small Rowcrop Tractor, as (**1**) but red with off-white wheels*

5. *Tudor Rose Small Rowcrop Tractor, as (**1**) but bright green with yellow wheels.*

6. *Tudor Rose Small Rowcrop Tractor, as (1) but bright green with red wheels.*

7. *Tudor Rose Small Rowcrop Tractors, both as (1) but one version red with yellow wheels and the other yellow with red wheels.*

9. *Tudor Rose Medium Rowcrop Tractor, as (8) but blue with red unmarked wheels.*

8. *Tudor Rose Medium Rowcrop Tractor, in yellow plastic with green wheels marked with Tudor Rose name and logo, and tread across the rear tyres (133 mm).*

10. *Tudor Rose Medium Rowcrop Tractor, as (9) but red with green wheels. Rear view showing towing eye.*

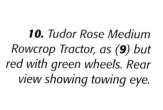

12. *Tudor Rose Medium Rowcrop Tractor, as (**11**) but blue wheels.*

11. *Tudor Rose Medium Rowcrop Tractor, as (**10**) but tyres with v-shaped treads.*

13. *Tudor Rose Medium Rowcrop Tractor, as (**12**) but with green front wheels.*

14. *Tudor Rose Medium Rowcrop Tractor, as (**13**) but yellow with blue wheels.*

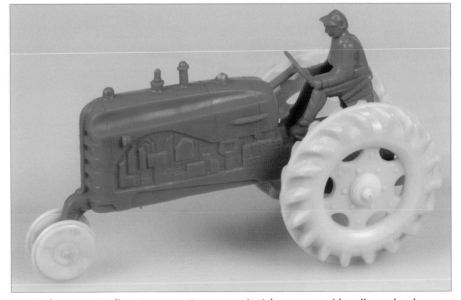

15. *Tudor Rose Medium Rowcrop Tractor, as (**14**) but green with yellow wheels.*

16. Tudor Rose Medium Rowcrop Tractor pulling flatbed Trailer. Tractor as (**15**) but red with yellow wheels, trailer in red with yellow wheels and side bars (241 mm).

17. Tudor Rose four-wheeled Trailer in blue with yellow wheels (110 mm).

19. Tudor Rose four-wheeled Trailer, as (**17**) but yellow with red wheels.

18. Tudor Rose four-wheeled Trailer, as (**17**) but with red wheels.

22. *Tudor Rose Heavy Duty Farm Tractor, non-clockwork version. Tractor with yellow body, red bonnet, hubs and front axle and all other components light blue (270 mm).*

20. *Tudor Rose Tractor Trailer Hoe & Rake. Yellow rowcrop tractor with green wheels and steering wheel pulling a red trailer with green wheels, showing original box. Hoe and rake are missing from photograph (480 mm).*

23. *Tudor Rose Heavy Duty Farm Tractor, non-clockwork version, as (22) but yellow body, light blue bonnet and hubs and all other components red.*

21. *Tudor Rose Heavy Duty Farm Tractor, clockwork version, showing box. Tractor in silver, red, and blue with yellow tyres and cream driver (270 mm).*

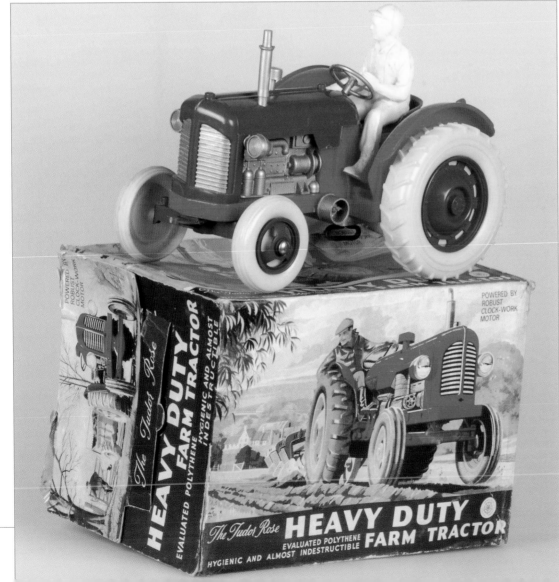

24. *Tudor Rose Heavy Duty Farm Tractor, non-clockwork version, as (22) but red body, light blue bonnet, hubs and front axle and all other components yellow.*

25. Tudor Rose Heavy Duty Tractor with Tipping Trailer, showing box. Tractor as (**21**), but without clockwork motor; red trailer with yellow wheels (610 mm).

26. Tudor Rose Steam Traction Engine, with black engine, red wheels, and yellow canopy, front axle, chimney, flywheel and steering wheel (178 mm).

27. Tudor Rose Steam Traction Engine, as (**26**) but with yellow wheels and red axle, chimney and flywheel.

28. *Tudor Rose Steam Traction Engine, showing box. Traction engine as (**26**) but with red body, yellow canopy and flywheel and blue wheels, axle and steering wheel. Note the chimney is missing.*

29. *Tudor Rose Steam Traction Engine, as (**28**) but with blue canopy and flywheel, and yellow wheels and axle. Note the chimney and steering wheel are missing.*

30. *Rosedale Heavy Duty Tractor, showing box. Tractor with yellow body, bonnet and hubs, blue driver and all other components black (280 mm).*

Unidentified Heavy Metal Tractors

The toys in this section are all solid cast metal, and were certainly not made with any regard to economy of raw materials. The resulting heavy castings are sometimes called 'fairground toys', because they would be made by fairground stallholders as cheap prizes for their sideshows. Some such toys were simply made by pushing a master pattern (which could be a Dinky Toy, for example) into wet sand and then filling the indentation with molten lead and attaching wheels and axles. These four tractor models are a little more sophisticated and seem to have been made from proper moulds, but the results were still very crude. Three of the models are copies of the Dinky Toy Fordson tractor no.22e, and therefore may date from the late 1930s. Shortage and rationing of raw materials make it unlikely that this sort of toy was made in the 1940s, but there are certainly later examples and the tractor in photo (**6**) could be a 1950s style vehicle with its square lines. Whether or not they were actually fairground prizes, these models had very limited production and distribution and are all rare. No maker's name or other identification was cast on any of them.

Models

Tractor (copy of Dinky no.22e)

This is probably the nicest of this group of models, being the closest copy of the original. Two different colour schemes have been found, either a blue upper casting with green wheels (**1**) or a turquoise upper casting with light blue wheels (**2**). Both had a red-brown chassis, engine and mudguards. One of these models came from Denmark, and that does suggest the possibility that it was a Danish-made toy. There were several makers of lead toy cars in Denmark in the 1930s and their output is detailed in an interesting book (Johansen & Hedegård, 2002). However the tractor is not mentioned in the book, so we have still included it here as possibly a British toy.

Tractor (copy of Dinky no.22e)

This tractor was similar to the above, but a zinc casting (note the fatigue cracks in the model photographed) and with less detail. It had a yellow upper casting, blue chassis, engine and mudguards, and unpainted wheels (**3**).

Tractor (copy of Dinky no.22e)

This was a narrower and more upright model with a separate casting for the steering wheel. It had an orange upper casting with silver grille and black seat, yellow steering wheel, green engine block and mudguards and dark red wheels with black hubs (4 and 5).

Tractor

A very angular design, this had a blue body with black steering wheel and black wheels (**6**).

Further Reading

Johansen, Dorte and Hedegård, Hans, 2002, *Danske Modelbiler.* Samlerbørsen (Danish text).

1. Heavy lead copy of Dinky Toy no.22e, possibly Danish, blue and red-brown with green wheels (65 mm).

2. Heavy lead copy of Dinky Toy no.22e, as (1) but turquoise and red-brown with light blue wheels.

744

3. *Heavy zinc copy of Dinky Toy no.22e, a different casting from the tractor in (1) and (2), yellow and blue with unpainted wheels (the wheels are replacements, except the front right) (65 mm).*

4. *Heavy lead copy of Dinky Toy no.22e, narrower and more upright than the others, orange and green with dark red wheels (66 mm).*

5. *Heavy lead copy of Dinky Toy no.22e, rear view of the tractor in (4).*

6. *Heavy lead tractor of angular design, blue with black wheels (55 mm).*

Unknown Manufacturers

In any comprehensive survey like this a small number of models will come to light for which no manufacturer can be identified. Their origins are unknown, so we cannot be sure they were made pre-1980 or that they were made in Britain, and for these reasons we have rejected several models, especially those made of plastic. Only two items illustrated (**8** and **9**) carry the claim *MADE IN ENGLAND*. Some, especially the wooden items, may have been home-made using purchased parts such as the tinplate wheels on (**1**), or made in small-scale craft workshops (**7**). Those we are most confident to include are illustrated below. They are described according to the materials used, in the order wood, tinplate and card.

Near the end of the sequence is a remarkable hay wagon which we suspect was home-made (**12**). Although it is therefore actually beyond the scope of this book, it is a useful example of what can be achieved at home with sufficient metal-working skills.

2. Red wooden steerable tractor with black wooden seat, black plastic steering wheel, wooden lights with metal fronts and a bare wooden strip nailed along each side of the engine block probably to identify the manufacturer, but now blank. A narrow yellow plastic strip was nailed along the top of the bonnet and down the centre of the grille. The front hubs were white plastic and the rear hubs red plastic. The tyres were black plastic (195 mm).

1. Light blue wooden tractor with red tinplate wheels, a nail for a long gear lever, a metal seat and a tow hook; the steering wheel is missing (72 mm.)

3. Red wooden cart with bare metal tinplate wheels, wire shafts and a rather delicate beige tinplate horse with a light brown saddle. There are traces of a metal seat half way along the insides of the cart (150 mm).

4. *A large blue and green wooden tractor with a silver-painted metal mesh grille set within a metal surround nailed onto the front. The silver paint on the grille extended onto the woodwork and wrapped around the sides and on the top of the blue bonnet. It did not have a steering wheel. The wheels were each cleverly turned from a single piece of wood with the hubs painted red and the tyres black. Pressed steel mudguards in green were nailed onto the wooden body, and there was a metal drawbar (315 mm).*

5. *A green wooden trailer to match (4) with turned wooden wheels and an articulated front axle. The drawbar was made from soldered heavy wire forming a 'V' and a towing eye for (4).*

6. *A wooden tractor with red and blue wooden wheels, a metal rear tow hook and a hole in the top for a driver (missing) (145 mm).*

7. *A wood and metal tractor with a red wooden body and blue fuel tank. The yellow and black wheels, yellow grille, baseplate, exhaust, steering wheel and seat were metal. The front axle pivoted (100 mm).*

8. *A tinplate clockwork cart with driver and a pair of horses with wheels on the rear hooves. The cranked axle was joined to the shafts so that horses rose and fell as the cart travelled. The key and the spring mechanism to drive the axle were located in a small metal box under the cart. The cart was red with yellow interior, with red wheels and blue shafts. The multiple colours on the driver and horses hardly look suitable for farm work, but it was a cheerful piece which no doubt had a wide range of play purposes. Underneath the girth strap on both horses in very small print was* MADE IN ENGLAND *(185 mm).*

9. *A tinplate horse and cart as (*8*) but with blue wheels and red shafts; the driver and reins are missing.*

10. *A tinplate clockwork traction engine without a roof painted mainly cream and green with red and black wheels. The clockwork mechanism was in the rear section wound by a key pushed through a hole in the left-hand rear wheel (230 mm).*

11. *Tinplate clockwork tractor in red with black tinplate wheels and black exhaust. At the top of the grille was a transfer JBE 452, rather confirming that it was British-made (135 mm).*

12. *Hay wagon with two Britains horses, probably home-made. The body of the wagon was cut from 22-gauge copper sheet, bent up at the edges to form the sides and then soldered at the corners. The top edges were rolled over and strips applied to the sides to form the wagon ribs. The raves were framed from narrow strips of copper or bronze and filled with a mesh of 16-gauge copper-coated mild steel welding rods hooked through loops in the floor of the wagon. The shafts were fabricated from soldered strips of brass or copper, hinged onto the front chassis with a pin made from the spoke of a bicycle wheel. Although there were*

chains over the saddles, the horses were actually held firmly in place with two pins which went through both horses. The front chassis and turntable were fabricated from soldered brass or bronze, and the rear chassis was cut out of thin sheet steel. The hollow front wheels were assembled in two halves, one slightly smaller than the other, hammered into shape over a circular former. By contrast, the larger rear wheels were solid, cut out of sheet steel. The model was then covered with a thick layer of paint, probably red oxide, and a very dark blue trim. This project must have involved an enormous amount of work (263 mm).

13. *Card Leyland 804 tractor. The model, including wheels, axles and exhaust, was made of pre-printed cut-out sections of thin card all glued together (178 mm).*

Company History

Wells-Brimtoy was formed in 1932 when Brimtoy Ltd was bought by A.Wells & Co. Ltd. The Brimtoy brand name, with its trademark of Nelson's Column and '"BRIMTOY" BRAND BRITISH MAKE', was registered in December 1914 by a company called The British Metal and Toy Manufacturers Ltd (BMTM), BRIMTOY being achieved by BRI, M and TOY from British, Metal and Toy.

BMTM made a range of quality tinplate toys, including pull-along carpet toys, kitchen stoves and other doll's house accessories as well as the first clockwork train set made entirely in England (previous examples had been made in collaboration with German toy makers). For obvious reasons, German toys were not available during the years of the First World War, and initially toys made in

America and Japan took the available market. However, in 1916 the British government announced restrictions on the importation of games and toys, and BMTM were one of the U.K. companies that sought to capitalise on the situation. Brimtoy made its first clockwork vehicle in 1918, part of a range that soon contained lorries, delivery vans, open tourers and landaus, ocean liners and a model steam tractor. However, post-war success was short-lived and BMTM closed in 1921.

Within two years a new company, Brimtoy Ltd, had been registered at the old BMTM factory in Riverside Works, Highbury Quadrant, London N5. The new firm made an extended range of tinplate toys based on the old BMTM products and sold through the old BMTM showrooms in Holborn, London EC1.

A. Wells was started in 1919 by Alfred W. J. Wells, and while the company started in a single basement room at 28 Dame Street, Islington, London N1, it quickly expanded into bigger premises at Shepherdess Walk, London N1, and then moved again to a larger factory a few miles away in Somers Road, Walthamstow, London E17, which Wells named the Progress Works. The 'Wells o' London' trade mark was registered in 1924, and A. Wells & Co. Ltd was incorporated as a limited company on 21 August 1930.

Alfred Wells, an engineer by profession, kept his company up to date with the latest technology, introducing one of the first conveyor belts in Britain in 1928-29. Part of Wells & Co.'s success stemmed from the fact that the company made all the components it required itself, using German machinery that Alfred Wells personally adapted, whereas even its German competitors used some bought-in components in their product assembly.

The company now employed its own travellers to sell its products direct to retailers around the U.K., and started to export. Advertisements from the late 1920s listed African, Australian, Belgian and Dutch agents, and its continued success allowed the company to move in 1930 to a new factory at Stirling Road, Walthamstow, London E17, which was again named the Progress Works. The Stirling Road factory contained its own machine shop, tool room with toolmakers, paint shop, diecasting shop, packing room and main factory floor. Here, up to 150 females undertook the assembly of the parts made by their male co-workers in the other departments. The Wells o' London trademark was synonymous with some of the higher-quality tinplate and clockwork toys from the pre-Second World War era. Train sets, toy clocks, racing cars, fire engines, a whole range of lorries and delivery vans, as well as some of the finest British-made tinplate cars were all produced during this period.

A.Wells & Co. Ltd and Brimtoy Ltd

In 1932 A. Wells & Co. Ltd took over Brimtoy Ltd. The Brimtoy factory at Highbury was kept, as was the Holborn showroom. The combined companies employed around 500 people, and the two ranges of toys – A. Wells & Co. and Brimtoy – were sold and promoted alongside each other.

Up until 1941, tin was not rationed in the U.K., so limited toy production aimed at export sales continued much as it had prior to the war, but this was curtailed when both the toy factories at Highbury and Somers Road were badly damaged in the London Blitz. In the late 1930s new factories had been acquired at Holyhead in north Wales and in Barnet in north London where a subsidiary company, Welnut, was based. These two premises, along with the main factory at Stirling Road, Walthamstow, continued to produce for the war effort. Working round the clock, more than 1,000 employees made items such as gas masks, Sten gun magazines and anti-vibration equipment which protected sensitive aviation instruments.

Immediately after the end of the Second World War a new larger factory, also named the Progress Works, was opened in Holyhead, and a new company, Anglesey Instrument and Clock Company, started to mass-produce alarm clocks. From 1947 until 1952, a workforce of between 400 and 500 produced up to 15,000 clocks a week using tools designed and made by the toolmakers back in Walthamstow. While the new factory made clocks – sold under the Weltime brand name – Welnut moved to the old Holyhead factory which reverted to toy production, making simple tin whistles and noise makers for what was known as the 'toy-filling trade', as well as clockwork mechanisms which were used in mechanical toys by the London factories at Barnet and Walthamstow.

Wells-Brimtoy Distributors Ltd

In 1949 a new sales organisation, Wells-Brimtoy Distributors Ltd, was formed. An injection moulding plant was acquired in 1952, and alongside its traditional tinplate mechanical toys the company started to produce a new range using plastic components. Under the Brimtoy brand, the Pocketoy series of vehicles was launched in 1948, using diecast, tinplate and plastic components. As the name suggested, the range was targeted at children by being both smaller and cheaper than previous clockwork tinplate toys made by the company.

The 1951 catalogue showed many mechanical walking animals and figures, large clockwork vehicles such as speed boats, lorries, buses and coaches, racing cars, trains, road rollers and aeroplanes, and of course train sets and accessories. Alongside push-along musical toys were simple tinplate racquets and doll's house stoves as well as a whole range tied in with the release of Cinderella by Walt Disney in 1950. These consisted of tinplate toy carpet sweepers and brush-and-pan sets, as well as plastic waltzing figures of Cinderella and Prince Charming which incorporated a clockwork mechanism inside Cinderella's figure. Additionally, four pages were devoted to the 15 vehicles that constituted the newly introduced Pocketoy series. Although not shown in the catalogue, the company also made a small range of wooden toys, some of which also used metal components and, in keeping with previous practice, all the components were made in-house.

Confusingly, the 1954 Pocketoy leaflet had the names 'A WELLS & CO. LTD.', then 'BRIMTOY, LTD' and 'WELLS-BRIMTOY DISTRIBUTORS LTD.' on it, while many of the commercial vehicles carried the Pocketoy name or displayed either the Wells & Co or Brimtoy logos. However, the only address given for the enterprise was 'Progress Works – Stirling Road, Walthamstow, London E17'.

To further complicate the situation, in 1955 the 'Welsotoys' trademark was registered, based on Wells & Co.'s pre-Second World War telegraphic address

of 'Welsotoys, Walt, London'. The new trademark was used across the different mechanical and non-mechanical, tinplate or plastic toy ranges, some of which featured television and film characters such as Mickey Mouse, Fred Flintstone and Superman. It appeared on advertising for various sizes of friction-drive, gyro-powered vehicles, with the smaller versions becoming Pocketoys, complementing the clockwork vehicles already in that range.

But the company was finding trading more difficult, and in 1965 closed the London factories, while moving all production to the newer factory in Holyhead. The company founder, Alfred Wells, died later that year at the age of 77, being replaced as chairman by his son, also named Alfred.

C.M.T. Wells Kelo Ltd

In 1970, Wells-Brimtoy was bought by an industrial products company called Central Manufacturing and Trading Group Ltd, and merged with Keith Lowe Ltd, one of their subsidiaries. Keith Lowe made large metal toys such as scooters, rocking horses and pogo sticks, all sold under the 'Kelo' brand name. The new toy company became C.M.T. Wells Kelo Ltd in 1975, based at the Holyhead factory and the business enjoyed some success in the 1970s, continuing to make large metal toys such as mini-gardener sets, prams and toy ironing boards, cleaning sets and sweepers, beach toys, bath toys and pogo sticks as well as a range of clockwork tinplate train sets. Marguerite Fawdry, in her book *British Tin Toys* (1990), shows the front cover of the C.M.T. Wells Kelo 1984 catalogue. The company logo incorporated a well, harking back to the original Wells & Co. logo of 1924.

The industrial products side of C.M.T. included anti-vibration equipment partly developed from the original Wells business of the 1940s and it was the non-toy side of the company which led to its being bought by Caparo Industries Ltd in 1981. After a period of trading from within Caparo Industries, in September 1988 a private limited company called Wells

Kelo Ltd was incorporated in Southend, Essex for the manufacture of games and toys, but this company was dissolved in March 1998.

(The authors are grateful for the research and co-operation of Jim Lindsay in the compilation of this history.)

Models

Despite the company being in business for over 80 years in various guises, very few farm-related items seem to have been made. Alfred Wells, the founder, was a London boy and indeed the vast majority of commercial vehicles made by his companies were based on vehicles that would have passed the doors of his premises every day. Unfortunately, we do not have examples of some of the farm-related toys, but we have included them here for the sake of completeness.

The Pocketoy series was introduced in 1948, and the catalogue for 1951 described all the models in the series as mechanical. The Pocketoy leaflet for 1954 introduced the clockwork tractors. However, the 1955 Wells-Brimtoy catalogue described the final series of Pocketoys as 'Friction Drive Pocketoys' and although many of the original clockwork vehicles had been modified into friction-drive models, the tractors were no longer part of the Pocketoys series.

Wells-Brimtoy were still advertising the clockwork tractor models in the late 1950s, but it is not known how long they continued in the Wells-Brimtoy range.

Brimtoy Model

No.87 Tractor

This was sold by BMTM under the Brimtoy brand name, and described in the 1920 catalogue as 'Tractor. Clockwork, slow running. Catalogue No.87. Size 7"' (**1**). (Seven inches is almost 180 mm.) The toy would have been of all-tinplate construction, possibly with a brass top to the chimney and steam dome. The original vehicle was more likely to have been seen on the streets of London than on a farm. Our illustration is taken from Fawdry (1990).

Wells-Brimtoy Models

No.66 Clockwork Tractor

Described as 'Tractor. Red and yellow. Clockwork. Catalogue No. 66. Size 5¾" [146 mm]' by Wells-Brimtoy, this model is dated to 1934 in Marguerite Fawdry's book. Unfortunately we do not have an illustration of it, but we believe it was a crawler tractor with rubber tracks and a tinplate driver. It differed from the pre-war crawlers by Marx and Tri-ang by having

lithographed engine detail on the sides. It was pictured in a 1937 wholesaler's catalogue from Kleiner of Houndsditch, London EC3, which is reproduced in Hertz, 1970, p.167.

No.549 Mechanical Tractor with Wheels

This model, and two others based on it, was introduced in the Pocketoy catalogue for 1954 (**2**). The tractor body was simply made from two pieces of tinplate, with part of the top piece cut and bent to form the seat. It had separate mudguards that were held in place with tabs. The large rear wheels were made of tinplate; it had rubber front wheels and a cast metal steering wheel which was coloured black on the example shown (**3** and **4**). The rear wheels were coloured grey and black with red hubs, and the tractor was red and yellow, with lithography providing all the detail of the engine, grille and rear wheels. The winder was accessed from the right side of the vehicle, and the clockwork motor drove the front wheels. At the rear was a towing eye, though we know of no implements for it to tow, and across the rear panel of the tractor was *MADE IN GT.BRITAIN*.

The box was a standard Brimtoy Pocketoy box made of thin card, which used just black and red colours on a white background. On two faces was a motorway scene showing vehicles from the Pocketoy range, with *Brimtoy POCKETOY* in the bottom left corner. On one face was a drawing of a boy with a speech bubble asking, *HAVE YOU SEEN THE BRIMTOY POCKETOY SCALE MODEL SERIES?* The Brimtoy logo including Nelson's Column was in the middle of the fourth face, with *LOOK FORWARD TO FURTHER MODELS IN THIS FASCINATING SERIES OF TOYS* and *ASK YOUR LOCAL DEALER TO RESERVE THESE MODELS FOR YOU* either side. *New models are on the way* was beneath the logo, and beneath those words was *MADE IN GT. BRITAIN by BRIMTOY, STIRLING ROAD, WALTHAMSTOW*. At the bottom was *ALL SALES ENQUIRIES WELLS BRIMTOY DISTRIBUTORS LTD*. The end flaps listed the contents, *9/549 MECHANICAL TRACTOR with Wheels*, the 9 prefix being an internal company number, as the catalogue number was just 549.

No.550 Caterpillar Tractor

This model used the same tinplate body and mudguards as no.549, and at the rear part of the body of some examples were two holes at different heights to take the axles for either the mechanical (**3**) or caterpillar tractors (**5** and **6**). The model had four red plastic rollers around which ran white rubber tracks, and the two front rollers were driven by the motor. The steering wheel was the same cast metal item as on the tractor, though unpainted in our examples.

It was packaged in what was essentially the same box as no.549, with similar illustrations and promotional words on all faces. The only differences were that in the motorway scene some of the vehicles were coloured light blue and *No. 550. CATERPILLAR TRACTOR* was on both end flaps.

The basic tractor was also used by Wells-Brimtoy for another two models. In the 1954 Pocketoy leaflet, model no.551 was a bulldozer that was made by attaching a blade to the Mechanical Tractor. A 1950s advertisement (7), showed that no.551 was also used for a bulldozer tractor that was made by attaching the bulldozer blade to the caterpillar tractor. No.551 Bulldozer Tractor was described with the words 'With rubber track and lifting blade. Fitted with clockwork motor and sound effect as No.549. Singly boxed. No.550. Also available as caterpillar tractor without bulldozer blade.'

The same advertisement described the Mechanical Tractor as 'No. 549 CLOCKWORK TRACTOR WITH WHEELS. Fitted with long-running clockwork motor geared to run slowly and make realistic exhaust noise.' The realistic exhaust noise was generated by a simple hammer tripped by one of the cogs of the clockwork mechanism, as was the lifting blade of the bulldozer, no.551.

Further Reading

Fawdry, Marguerite, 1990, *British Tin Toys*. New Cavendish Books.

Hertz, Louis H., 1970, *The Complete Book of Building & Collecting Model Automobiles*. Crown Publishers Inc.

David Rodgers' blog: www.rodgersantiques.co.uk.

3. *Wells-Brimtoy Pocketoy no.549 Mechanical Tractor with Wheels. Clockwork tinplate tractor with lithographed detail in red and yellow, tinplate rear wheels and rubber front wheels. Right side showing winder for clockwork motor (85 mm).*

1. *Brimtoy Model Tractor no.87 (illustration taken from* British Tin Toys *by Marguerite Fawdry).*

Brimtoy model tractor No.87

550 CATERPILLAR TRACTOR
4 × 1½ × 1⅛ ins.

546 SIX-WHEEL ARTICULATED MILK TANKER
6 × 2¼ × 1¼ ins.

545 ARTICULATED COAL LORRY
6 × 1½ × 2¼ ins.

551 BULLDOZER
with Automatic Lifting Scoop
4½ × 2¼ × 1½ ins.

ARTICULATED AIRCRAFT CARRIER

549 TRACTOR

2. *Excerpt from Wells-Brimtoy Pocketoy catalogue showing no.549 Tractor, no.550 Caterpillar Tractor, and no.551 Bulldozer.*

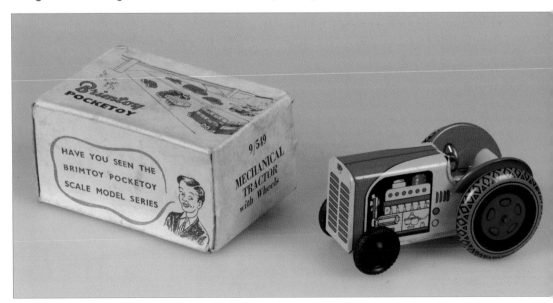

4. *Wells-Brimtoy Pocketoy no.549 Mechanical Tractor with Wheels as (*3*), but left side and showing original box.*

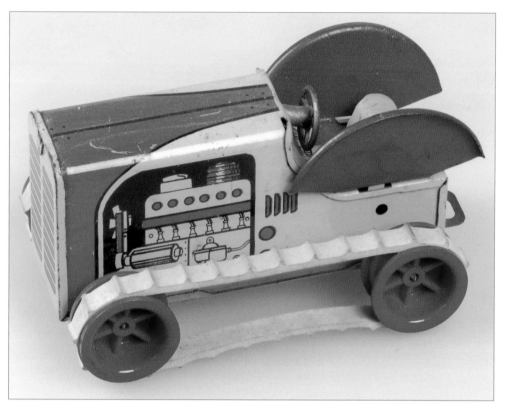

5. *Wells-Brimtoy Pocketoy no.550 Caterpillar Tractor. Clockwork tinplate tractor with lithographed detail in red and yellow, plastic rollers and rubber tracks (85 mm).*

6. *Wells-Brimtoy Pocketoy no.550 Caterpillar Tractor as (**5**), but showing original box.*

7. *Excerpt from 1950s Wells-Brimtoy advertisement showing no.549 Clockwork Tractor with Wheels and no.551 Bulldozer Tractor.*

Wend-al

Company History

Wend-al Toys Ltd, which specialised in making 'unbreakable' solid-cast aluminium toys, was formed in 1947 in Blandford Forum, Dorset. The founder, Edgar Kehoe, had become aware of the success of a French company, Quiralu, which produced toys from aluminium. Aluminium was not used for toy manufacture in the U.K. prior to the Second World War, probably due to its high cost. But while post-war U.K. toy companies used metals such as lead or Mazak, there was already a significant use of aluminium by French toy companies. Kehoe sought the assistance of Quiralu to set up a U.K.-based aluminium toy manufacturer, and was lent moulds, given sources of materials and lent staff to help him create a successful company.

Kehoe's original company was called Wendan Manufacturing Company Ltd (incorporated on 21 May 1946), Wendan deriving from the christian names of the works manager's daughters, Wendy and Ann. In common with French toy figure companies (Quiralu, Aludo, Mignalu), the material chosen for the products was alluded to in the chosen brand name of Wend-al for the company set up to produce aluminium toys.

Although Wendan Manufacturing was in business until the early 1980s, aluminium toy production ended in 1956, while the remaining stock of aluminium toys was sold for a few years afterwards. After ending the production of aluminium toys, the company sold a range of flock-covered plastic zoo animals through most U.K., and some European, zoos.

Because of the relationship between Wend-al and Quiralu it is often impossible to distinguish the products of one company from the other. Most of the early Wend-al figures had a baseplate and the base had a paper sticker, which stated *WEND-AL UNBREAKABLE TOYS MADE IN ENGLAND*. Quiralu products sometimes had a sticker with *Jouet Français Quiralu incassable* (French Toy Quiralu unbreakable). However, in order to decrease the costs of material used, later Wend-al figures were made without bases, and consequently with no identification, so it is virtually impossible to identify some products accurately. With the military figures, precise colour schemes had to be followed, but with civilian figures the painters were given a relatively free hand. This would explain the very wide range of colour variations on some items. The models were usually packed and dispatched either in boxes with lids – carrying the company logo of a soldier at present arms, wearing a red uniform with a busby – or individually in polythene bags.

A useful account of the company can be found in *Wend-al of Blandford, the place that made toys* by Philip Dean, published in 2005. This contains information based on a series of interviews conducted by Philip Dean with Edgar Kehoe and various ex-employees. It also contains a facsimile of the only known catalogue, for 1952.

There are two books on Quiralu: Jean Michel Tisne's *Les Figurines Quiralu* (undated) and *Les Jouets Quiralu 1933-1964* by R. Alazet, J. Borsarello and H. Giroud (1995). The latter has a particularly helpful farm section on pages 173 to 201. By comparing the outputs of the two companies it is possible to identify the models that were unique to Wend-al. These books allow comparisons of painting styles, but there is no definitive evidence of date of manufacture.

The factory in Shaftesbury Lane, Blandford was made by stripping out an old isolation hospital, and at its height the company employed over 50 people: 7 men on moulding, 8 to 10 in the fettle shop finishing off the rough castings, 24 girls and 8 outworkers hand-painting the figures, and another 7 or so in packing. The aluminium toys were made by the sand-casting method (see the Glossary in Volume 1), which was rather slow and labour-intensive compared with zinc diecasting, or injection moulding of plastics.

The boxes describe the company as being in Blandford and London, but there is no sign of the company in London telephone directories, so this probably refers to a London agent. However, the company's success was short-lived. Although it was allocated twenty tons of aluminium a year under the U.K. government's regulations to support exporting companies, this was insufficient to allow the business to expand. In addition, export sales to the United States were relatively low margin.

The adverse effects of a shortage of aluminium, leading to increased cost, and the outdated sand-casting production process meant that Wend-al could not compete with the arrival of plastic toy figures (from around 1952) that were also marketed as 'unbreakable'. This led to the end of aluminium toy production in 1956.

The biggest prizes for the collector are the presentation boxed sets with lift-off lids and drop-down fronts which reveal dioramas of figures fixed down to the base card, against a backdrop of a farmyard or a nativity scene (**1**, **2**, **3** and **20**). The models may not have the same level of detail as Britains hollow-cast lead, but they do have their own charm.

Models

The farm implements which Wend-al produced were all horse-drawn, even though Quiralu made a farm tractor. It is tempting to wonder if

Wend-al ever tried, or intended, to produce one, because in the background farmyard scene in the presentation boxes with drop-down fronts there was a Fordson tractor in an open cart shed (**1**, **2** and **20**). Britains were bringing out their Fordson (see the Britains chapter), and Meccano had also produced a Dinky Fordson (see the Dinky chapter), so it is puzzling that Wend-al apparently did not try, unless it was the shortage of aluminium which made a tractor uneconomic.

The horse-drawn vehicles and implements which we can be certain that Wend-al manufactured were the harrow, hay cutter, hay rake, plough and tumbrel. The hay cutter is the easiest to find. Farm wagons were only made by Quiralu, and items with rubber wheels were probably all made by Quiralu. We state this because we have not seen a tumbrel or implement with rubber wheels in a Wend-al box or with a Wend-al sticker. It is interesting that in the 1952 catalogue the only horse-drawn farm implement listed was the harrow, so we suspect that Wend-al tried to sell the tumbrel and other implements in the early years and then decided to focus on figures because they probably offered more play potential for their price.

Wend-al used only one basic model of a cart horse with its head down and in a normal stride. When the horse was painted white, the harness moulded onto the horse was painted either black or brown. On the brown horses it was always painted black, and on the black horses the harness was painted brown. Sometimes there was gold trim on the harness. There were loops cast on the horse collars to take traces or shafts, a much more realistic approach than that taken by other manufacturers who simply had holes in the sides of the horses. This basic model can be found with or without a base, with offset or parallel ears, with a docked tail or, less frequently, with a full tail. Sometimes the bridle had holes for reins and sometimes these were absent. The horses without any holes in the bridles were probably to be sold as a generic cart horse.

It is clear that the company did allocate code numbers to their models, although most codes are now lost and, as already mentioned, only one catalogue is known to survive, that for 1952 illustrated in Philip Dean's book (2005). The harrow listed in that catalogue was sold as part of Farm Set FD3. Other codes can now only be pieced together where they can be discerned hand-written on circular labels on the ends of boxes. On the boxed items we illustrate, *71C* was on a presentation box of farm animals (**1** and **2**), *FB9* was on a hay rake (**14**), and *6FA* on a tumbrel (**24**).

Harrow

The harrow was sold either in the drop-down-front presentation box (**3**) or in a flat box in a set with farm animals and two figures all strung down on a plain backing card (**4**). The harrow was either dark red (**4**) or light red (**5**). The traces were of thin fuse wire (**4**) or thicker wire with hooks at each end (**5**). The horse was in cart harness although it was just working in traces.

Hay Cutter

The hay cutters we have seen were painted red with a yellow cutter bar (**6**), green with a red cutter bar (**7** to **9** and **13**), green with a blue cutter bar (**10**) or yellow with a red cutter bar (**11** and **12**), and there were further variations for all three colours, with short or long levers to the right of the driver and different colours of the seat, lever and footplate. Undoubtedly there were still more variations than we have managed to illustrate here. The wheels on this and on the hay rake had ten spokes. The shafts were made of thick wire and the reins of thin fuse wire. A pin in the driver's bottom fitted into a hole in the seat. The driver had a light blue, brown, grey or green shirt, and brown, black, dark blue or grey trousers, and wore a grey, brown or yellow hat. The horse was white, black or brown with or without a base.

An unusual factory variation to the standard design can be seen in (**13**) where the housing for the shafts was fixed in error upside down to the frame of the cutter.

Hay Rake

The only Wend-al hay rakes we have seen have been red with yellow tines (**14**) or yellow with red tines (**15**) and with a light blue (**14**) or red (**15**) seat. Again, the shafts were made of thick wire and the reins of thin wire.

Plough

The plough with frame, handles, coulter and mouldboard was cast as one piece with the addition of two wheels. Sticking up from the frame was the black top of the coulter, which was otherwise bare metal, and also a spike which we interpret as a depth-control lever. The traces from the horse to the whipple tree were either heavy wire bent to form hooks at each end, or thin wire.

There was a wide range of colours: yellow with red handles and red stripes along the sides (**16**), all-yellow with black stripes (**17**), all-red with yellow stripes (**18**), all-red with light blue stripes (**19**), all-green with red stripes (**20**) and green with red handles and red stripes (**21**). The mouldboard was always bare metal. The ploughman was the same figure that Wend-al used for its gardener with wheelbarrow. He usually had a light blue shirt, brown trousers and a waistcoat with a grey front and a black back. We have also found one variation of an unusual beige waistcoat and red shirt in (**20**).

Tumbrel

The tumbrel was available in yellow (**22**), brown (**23**), red (**24**, **26** to **28**) or dark green (**25**), and some had their wire shafts painted to match the body of the cart (**22**, **24** and **26**), while others were bare metal (**23** and **25**). All models had light blue hand-painted detailing applied on all four sides. The tumbrels were supplied with metal wheels with the edges painted silver to represent the steel outer rims. These were larger wheels than those on the hay cutter and hay rake, and with only six spokes.

While the horse implements so far described appear to have been produced unchanged during the period they were manufactured, the tumbrel went through two, and possibly three, casting changes. There is some uncertainty about the third type, since the examples we have seen had rubber wheels and so may have been made by Quiralu. The differences relate mainly to the release mechanism for tipping the tumbrel, described here as Types 1 to 3.

Type 1 is probably the original version, produced in by far the largest numbers. A bent wire pin ran through holes in two flanges fixed to the chassis and a third projecting from the tumbrel body. Removal of the pin allowed the body to tip, but this was a fiddly device, and the pin was easily lost. We have seen this version in yellow (**22**), brown (**23**), red (**24**) and dark green (**25**), with the original pin surviving in (**22**) and (**24**).

Quiralu tumbrels usually had an identical release mechanism which suggests that this was the original French-designed version.

All the type 1 tumbrels had an axle made from a rod which was threaded throughout its length, which allowed nuts to be fixed at each end to keep the wheels in place.

Type 2 is quite rare, and we have seen it in only one example (**26** to **28**). This had a long wire catch held in place under tension behind the left-hand wing of the front panel, and when this catch was released the lower

part of the wire was freed from a projecting knob on the cart body so it could then tip backwards (**26** and **27**). There is no doubt that this was a Wend-al product because of the original stickers underneath (**28**). Although this version involved an expensive change to the casting, the resulting model used less material. The axle on this version was smooth, and the wheels were held in place by crimping the ends of the axles.

Type 3. For no obvious logical reason this third version had a square-sectioned column cast onto the front of the tumbrel body, extending from the wire catch up to the top of the front panel. We have seen this only on tumbrels with rubber wheels and thicker smooth axles with crimped ends in dark green (**29**) and yellow (**30**). Neither of these examples had a manufacturer's sticker, so, as mentioned earlier, the rubber wheels raise the possibility that the models were made by Quiralu.

Model List

Model
Harrow
Hay Cutter
Hay Rake
Plough
Tumbrel

Further Reading

Alazet, R., Borsarello, J. and Giroud, H., 1995, *Les Jouets Quiralu 1933-1964*. Grancher.

Dean, Philip, 2005, *Wend-al of Blandford, the place that made toys*. Avalonimages.com.

Auction catalogue for the Philip Dean Collection sold on 1 December 2005. Christie's.

Tisne, Jean Michel, undated, *Les Figurines Quiralu*.

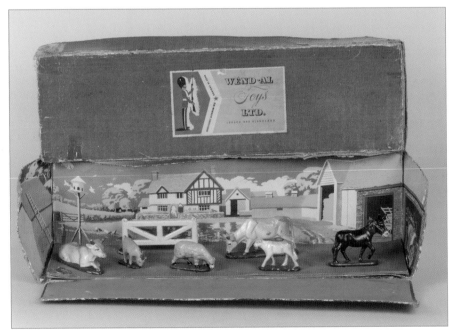

1. Wend-al presentation box with drop-down front for a group of farm animals. This was the longer version of the box with an extended scene to the left (compare with (3)). The label on the lid had WEND-AL Toys LTD. LONDON AND BLANDFORD UNBREAKABLE ALUMINIUM TOYS MADE IN ENGLAND. *A circular end label had* WEND-AL Toys LTD. MADE IN ENGLAND, *and* 71C *was hand-written in an orange panel on the label.*

2. Wend-al, close-up of the presentation box in (1) showing a Fordson tractor and a wagon in the open shed to the right.

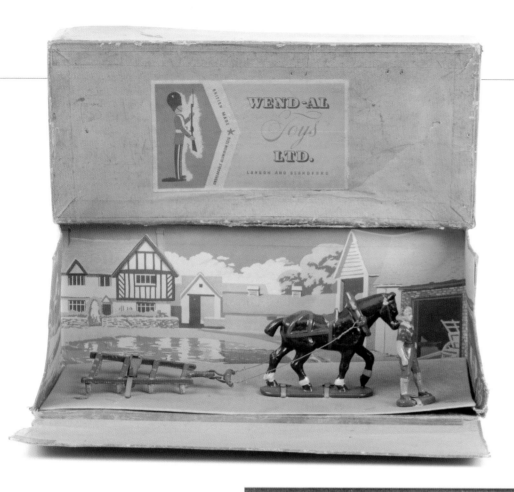

3. *Wend-al, a shorter version of the presentation box with a horse and harrow. The label on the lid was as on* (**1**), *except it said* BRITISH MADE *and not* MADE IN ENGLAND. *No details of the circular end label are available* (photo courtesy of Vectis Auctions).

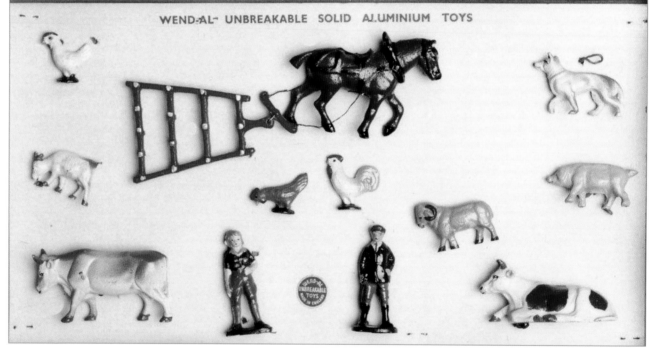

4. *Wend-al harrow and whipple tree in dark red in a boxed farm set. The backing card had* WEND-AL UNBREAKABLE SOLID ALUMINIUM TOYS *at the top and below was the standard small circular sticker found under the baseplates of many items, as on* (**28**), *which read* WEND-AL UNBREAKABLE TOYS MADE IN ENGLAND (photo by Philip Dean).

5. *Wend-al harrow and whipple tree in light red. Note that on this example the horse's harness and fetlocks have been retouched (240 mm).*

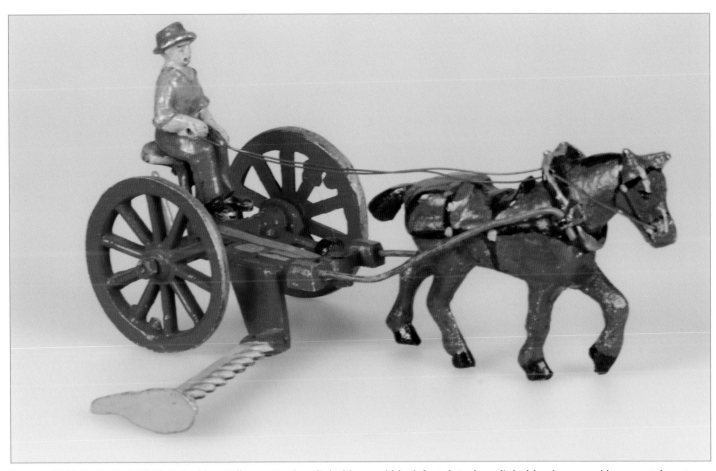

6. Wend-al hay cutter in dark red with a yellow cutter bar, light blue and black footplate, long light blue lever, and bare metal seat
(155 mm).

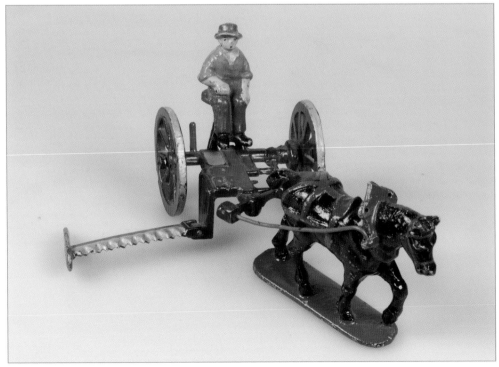

7. Wend-al hay cutter as (**6**) but in dark green with a red cutter bar, red and light blue
footplate, short red lever and red seat.

8. Wend-al hay cutter as (**7**) but with a long red lever.

9. *Wend-al hay cutter as (**6**) but in light green with a dark green and grey footplate (missing shafts and a horse).*

10. *Wend-al hay cutter as (**6**) but in mid green with a dark blue cutter bar, red and silver footplate, short red lever and red seat.*

12. *Wend-al hay cutter as (**11**) but with a long red lever.*

11. *Wend-al hay cutter as (**6**) but in yellow with a red cutter bar, light blue and black footplate, long light blue lever and bare metal seat in a dark blue box. The label on the lid was as on (**1**). On the end of the box was a circular label also as on (**1**), and hand-written in the orange panel there was lettering no longer legible.*

13. *Wend-al hay cutter as (**8**) but with a factory error, where the housing for the shafts was attached upside down to the frame of the hay cutter.*

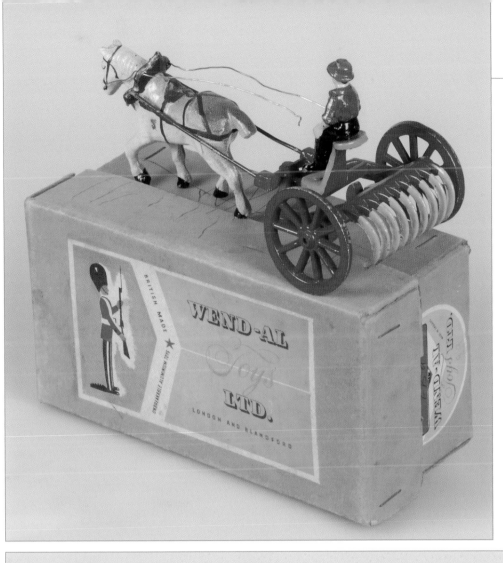

14. *Wend-al hay rake in red with yellow tines and light blue seat in a plain brown card box with the label on the lid as on (**11**). The circular label on the end was as on (**11**) but FB9 was hand-written in the orange panel (173 mm).*

15. *Wend-al hay rake as (**14**) but in yellow with red tines and red seat.*

16. *Wend-al plough in yellow with red handles, red stripes and red height-adjustment lever (132 mm).*

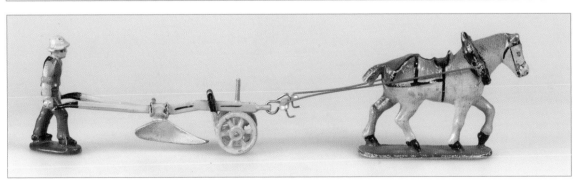

17. *Wend-al plough as (**16**) but with black stripes, yellow handles and light blue height-adjustment lever.*

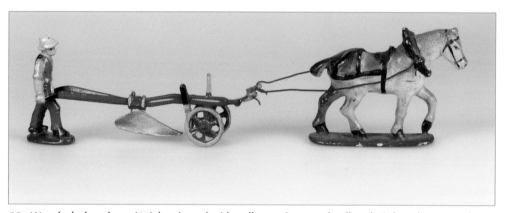

18. *Wend-al plough as (***17***) but in red with yellow stripes and yellow height-adjustment lever.*

19. *Wend-al plough as (***18***) but with light blue stripes and light blue height-adjustment lever (missing whipple tree, traces and a horse).*

20. *Wend-al plough as (***19***) but in mid green with red stripes and height-adjustment lever, in a presentation box* (photo by Philip Dean).

21. *Wend-al plough as (***20***) but in dark green with red handles and red stripes.*

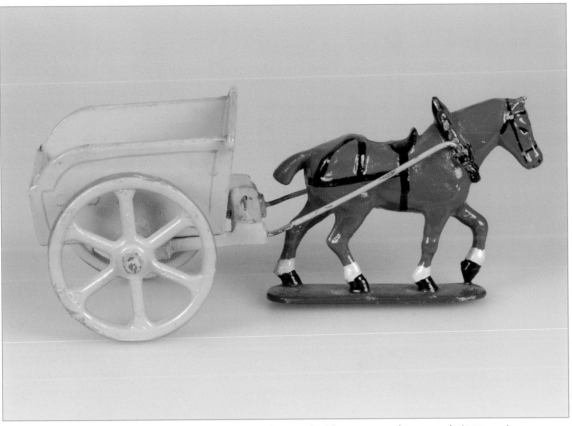

22. *Wend-al tumbrel in yellow with light blue detailing and with a type 1 release catch (165 mm).*

23. *Wend-al tumbrel as (22) but in brown (with pin missing on release catch).*

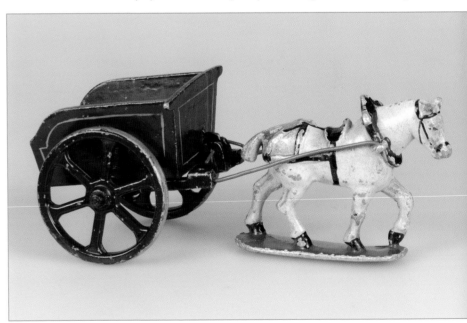

25. *Wend-al tumbrel as (22) but in dark green with light blue detailing (with a replaced red pin).*

24. *Wend-al tumbrel as (22) but in red (with bent shafts) still strapped to a backing card. On the backing card the sticker in the front right-hand corner had WEND-AL UNBREAKABLE TOYS, and the hand-written price on the sticker was 9/6. With this model came an incomplete box, and the circular label on the end was as on (1) but 6FA was hand-written in the orange panel.*

26. *Wend-al tumbrel as (**24**) but with a type 2 release catch.*

27. *Wend-al tumbrel in (**26**) showing a close-up view of the type 2 catch.*

28. *Wend-al tumbrel in (**26**) showing the Wend-al stickers underneath both tumbrel and horse.*

29. *Wend-al or Quiralu tumbrel in green with dark blue detailing and with a type 3 release catch and rubber wheels.*

30. *Wend-al or Quiralu tumbrel as (**29**) but in yellow with red detailing and rubber wheels (shafts missing).*

Addenda to Volume 1

We are grateful to the following people who contributed information to the Addenda: Chris Clemons, Gary Galvin, Cliff Maddock, Andrew Ralston, Nigel and James Robertshaw.

Introduction

(see Volume 1 pp.7-15)

On page 8, we referred to the Timpo Irish Jaunting Cart. However, the correct term is a Jaunting Car, as stated on the box for the Timpo model (see the Timpo chapter in this volume).

On page 12, we stated that the Dinky no.22e Tractor was at first made from lead, and later from zinc alloy. This was incorrect; in fact, the tractor was always lead, except for the front wheels which were sometimes zinc (see the Dinky chapter in this volume).

Benbros Pure Rubber Products Ferguson Tractor

(see Volume 1 p.31)

The tractor driver also exists in brown with green trim on her boots and hat.

Benbros Pure Rubber Products Timber Trailer

(see Volume 1 p. 31)

The Timber Trailer came with a diecast clip (painted in the body colour) which had two projections to fit into the holes in the chassis. The clip could be fitted so as to stop the bolster slipping off the end of the chassis pole and to adjust for the length of the log.

Benbros Pure Rubber Products Hay Rake

(see Volume 1 p.31)

The rake operator also exists in brown with green boots, the same colours as the tractor driver mentioned above.

Benbros Horse-drawn Water Cart

(see Volume 1 p.32)

The Water Cart has also been found in red with black wheels, the same colours as the Farm Cart in photo (**24**) on p.42.

Benbros Farm Hay Cart

(see Volume 1 p.33)

The Benbros tumbrel also exists with the body and shafts painted orange, red wheels and dark brown horse. The colours are similar to the Benbros timber wagon shown in photo (**55**) on p.48.

Benbros Large Tractor

(see Volume 1 p.34)

The Caterpillar crawler tractor also exists in orange with yellow outer wheels, the reverse colours of the example shown in photo (**70**) on p.53.

Britains Kenyan and Argentine variations

Since the Britains chapter was completed for Volume 1, two very unusual box types for horse-drawn models normally found in standard green boxes have been found. One was made of corrugated cardboard containing models originally sold in Nairobi, Kenya (**1** and **2**). The other was made of plain brown card containing models from Argentina (**7** and **8**). The Argentine boxes were part of a large consignment sent to England for auction at Bonhams Oxford on 5 December 2013 (**3** to **10**). The collection had some striking paintwork which we believe to be genuine although not normally seen on U.K. examples. The use of red paint on the wagon in (**4**) and on the mouldboard of the plough in (**5**) was extraordinary by Britains' usual exacting requirements. The unusual cream paint on (**6**), (**7**) and (**9**) has been seen previously on items from Argentina. Although lead figures are beyond the scope of this book, the painting on some of the farm figures from the Argentine collection is distinctive and unusual. The stable lads (**9** and **10**) had green shirts, when they were usually white. The tractor drivers in (**10**) had black or light blue belts, and the face painting in (**10**) had its own style. These colours and the unusual boxes underline the real possibility that agents abroad controlled some of the painting to suit local tastes as well as the packaging process.

Britains Clockwork Series leaflet

(see Volume 1 p.60)

We have also found a second very small leaflet for inserting into boxes; this one is entitled *BRITAINS CLOCKWORK SERIES* (**11**). The farm items illustrated on one side suggest that it was printed some time between 1951 and 1959.

Britains no.8F Horse Rake

(see Volume 1 p.66)

Although light blue examples of this rake are usual, this dark blue diecast version is seldom found (**12**).

Britains no.40F Farm Horse and Cart

(see Volume 1 p.68)

Two more no.40F Farm Horse and Carts (**13** and **14**) widen the colour range of this version of the cart which had large slender wheels more usually seen on the no.20F Farmer's Gig.

Britains no.128F Fordson Major Tractor

(see Volume 1 p.70)

In Volume 1 we illustrated the rare green box designed for the no.127F Fordson tractor to be laid on its side (photo (**110**) on p.111). We now have an example of a similar box for the no.128F tractor with rubber tyres (**15**).

Britains no.9525 Fordson New Performance Super Major Tractor

(see Volume 1 p.71)

When the Britains chapter was being compiled we were unable to illustrate the rarest of the blue Fordson tractors, the New Performance Super Major with red *FORDSON SUPER MAJOR* transfers. This has now been rectified (**16**). On p.71, the reference to photo (**281**) is incorrect and should read (**278**).

Britains Lilliput no.LV604 Fordson Tractor

(see Volume 1 p.83)

We illustrated the Lilliput no.LV604 Fordson tractor in the usual white card box and the window box (photos (**303**) and (**304**) on p.158), but it now appears that it was also sold for a short while in a picture box (**17**). This was a generic box that was also used for other Lilliput items.

Charbens pre-war no.505 Hay Cart

(see Volume 1 p.196)

Another variation has been found of the box label, similar to the example in photo (**11**) on p.209 but with a frame around the illustration (**18**).

Charbens pre-war no.513 Tree Wagon

(see Volume 1 p.196)

In Volume 1 we said that we had never actually seen the pre-war Tree Wagon with two different horses, but this does exist, as might be expected from the catalogue and box illustration. An example of the leading horse (without a saddle) is pictured in the Guide to Harnessed Horses in this volume (photo (**20**) on page 781), complete with the wire traces that hooked into the ends of the shafts.

A further type of box has been found for this model, having a small label which wrapped onto the end of the box, with *THE TREE WAGON MADE IN ENGLAND* printed (**19**). This label is very similar to the Farm Wagon box label in photo (**25**) on p.213.

Also, we have seen another example of the box pictured in photo (**23**) on p.212. This has an undamaged label, and *THE TREE WAGON 513* is printed on the label where it wraps onto the end of the box.

Charbens pre-war no.503 Farm Wagon

(see Volume 1 p.196)

It was perhaps over-confident to say that this model was first issued with a figure of a walking farmer with a stick. This statement was based on seeing just one boxed example, shown in photo (**25**) on p.213, and ideally we would want to see more original examples to be sure. Of course, now that we have published the photograph in Volume 1, collectors may be tempted to put this farmer figure with the wagon even if the combination was not original – so apparently confirming the point of which we are uncertain! The walking farmer may not even have been made by Charbens, and Norman Joplin put it in his 'Unidentified' chapter (Joplin, 1993, p.277) but captioned it 'possibly John Hill'. There is another colour variation of the figure, with a cream smock and green base rather than the yellow smock and brown base shown in Volume 1 (**20**). The representation of farm workers wearing smocks was archaic even in the 1930s, and is another example of nostalgia for a lost rural way of life that was depicted by the urban toy manufacturers.

Charbens no.2 Horse and Roller

(see Volume 1 p.199)

The yellow-painted model (first casting) also exists with a blue roller.

Charbens no.5 Farm Wagon

(see Volume 1 p.200)

A late version of the box has been found, similar to the boxes in photos (**88**) and (**89**) on p.231, but covered in pale green paper and with blue printing on the label. The model in this case was painted orange with a hollow-cast feather-legged horse, as photo (**85**) on p.230 (**21**).

Charbens no.6 Tractor

(see Volume 1 p.200)

The type B tractor also exists painted blue with yellow hubs (**22**).

Also the straight-backed driver, with hands on his hips, has been found in hollow-cast lead rather than zinc diecast, painted light blue.

Charbens no.17 Tractor and Tree Wagon

(see Volume 1 p.201)

The type B wagon also exists in cream with red wheels.

Charbens no.19 Tractor and Reaper

(see Volume 1 p.202)

The reaper also exists with no lettering cast underneath.

Charbens Hay Cart and Horse (plastic)

(see Volume 1 p.203)

An early version of the plastic cart has been found, with yellow-painted metal wheels as used on the metal hay cart. The cart is in light blue plastic, with red shafts and a white horse. On the front of the headboard is moulded *CHARBENS FARM*, and there is no *MADE IN ENGLAND* underneath. Also, another version of the red cart has been found, as in photo (**140**) on p.244 but with no lettering on the model at all.

Charbens Reaper (plastic)

(see Volume 1 p.203)

The reaper also exists in light blue plastic with red sails. Unfortunately the wheels are missing from the example seen, so we are not sure whether they were also red to match the sails, or yellow like the models in Volume 1.

Childs & Smith and Wagstaff Tractor-drawn Roller

(see Volume 1 p. 251)

In our discussion of the Childs & Smith tinplate tractors we said that no implements for them had been found. Since then, a tractor-drawn roller with the *E.W.WAGSTAFF LTD* crest has turned up just in time to be included here. The scale of the roller fits well with the tractors, and the loop of the drawbar sits easily on the towing pin of the tractor (**23**). The roll consisted of four sections of black Bakelite on a wooden frame, and the drawbar was all aluminium, with a tubular support to hold up the drawbar when it was not hooked onto the tractor (**24**). This roller opens up the real prospect that a whole range of implements for these tractors awaits discovery.

Crescent no.1808 Platform Trailer

(see Volume 1 p.295)

This model also came in the end-opening box with illustrations of farming scenes on the sides, as in photos (**39**) and (**40**) on p.309. This is not surprising as we already recorded the other versions of the trailer (nos.1809, 1810 and 1811) in this style of box. *NO 1808 PLATFORM TRAILER MADE IN ENGLAND* was printed on a label on one end of the box.

Crescent no.1822 Caterpillar Tractor

(see Volume 1 p.296)

The illustration shows a new colour variation of this model. It differs from the example in photo (**57**) on p.314 by having a yellow body with silver trim on the grille, red base and green control levers (**25**).

Crescent 1967 Catalogue

(see Volume 1 pp.292-299)

We have now seen a copy of the 1967 Crescent catalogue. This included the four-wheeled trailers, nos.148 and 149, and did not show the earlier two-wheeled trailers, nos.1810 and 1811, which were replaced by nos.148 and 149. In Volume 1 we were guided by Ray Strutt's research into Crescent catalogue dates, published in Ramsay, 1988, which showed the four-wheeled trailers as introduced in 1968; however, it appears that Ray was mistaken on this point. The four-wheeled trailers were not even shown as 'New' in 1967, so without having seen the 1966 or earlier catalogues, we are now unsure about the year that the change of design of the trailers occurred. It was certainly after 1963 (see the 1963 catalogue page in photo (**96**) on p.324) and before 1968.

The Plough, no.150, was not in the 1967 catalogue, confirming the 1968 introduction date.

The 1967 catalogue showed the Dexta tractor (no.1809) in the shorter box (as in photo (**90**) on p.322) which we previously dated to 1968 because it illustrates the four-wheeled trailers (see p.298).

Gift set no.1203 was still available in 1967. The set shown in the catalogue was very similar to photo (**97**) on p.325, except the tractor had no driver, and the window box had a design of six stars and the words *HAND PAINTED BY CRESCENT* on the front face.

Gift set no.9213 was also in the 1967 catalogue. The box was identical to the example in photos (**98**) and (**99**) on p.325. The contents were similar, except the tractor was pictured with a driver and there were minor differences in the figures included in the set.

Finally, the 1967 catalogue showed no.1803 Dexta Tractor and Trailer as 'New' in the box with a plinth, as in photo (**100**) on p.326. Again, we previously dated this box to 1968 (see p.298).

Lone Star Roadmaster Tractor

(see Volume 1 pp.330 and 331)

We said that the Lone Star Roadmaster Tractor was a copy of the American Tootsietoy Ford Tractor of 1956; however, this date is incorrect. There were in fact two post-war Tootsietoy tractors in around 1:32 scale. The first was a model of the 1947 Ford 8N tractor, introduced by Tootsietoy in 1951. It was later replaced by a model of the 1953 Ford NAA Golden Jubilee tractor, and it is this model that was copied by Lone Star. The Tootsietoy Ford 8N was still shown in a 1957 leaflet, but the Ford NAA must have appeared in 1958 or 1959, because it was shown in the 1959 catalogue. We took the date of 1956 from Wieland & Force, 1980, but that book does not distinguish between the two Ford tractors, and the 1956 date seems to be unreliable as it does not relate to the introduction or withdrawal of either model.

We are now able to illustrate the Lone Star tractor in its original yellow and red blister pack, numbered 1293 (see photo (**5**) on p.336). Note the absence of labels on this early model, and the silver steering wheel rather than the later black (**26**).

Deltoys Wagon

(see Volume 1 p.342)

Another example has been found of the wagon without the Deltoys brand name. This is as photo (**5**) on p.345, but painted dark green with red shafts and interior.

References

Joplin, Norman, 1993, *The Great Book of Hollow-Cast Figures*. New Cavendish Books.

Ramsay, John, 1988, *John Ramsay's Catalogue of British Diecast Model Toys, 3rd edition*. Swapmeet Toys and Models Ltd publication.

Wieland, James and Force, Edward, 1980, *Tootsietoys, World's First Diecast Models*. Motorbooks International.

*1. Britains no.4F Tumbrel in a short corrugated card box, with a long label as in photo (**5**) on p.88 but wrapped around both ends. Tumbrel as photo (**17**) on p.90, but with cream raves.*

*3. Britains no.5F Farm Waggon in a green box with label as photos (**24**) and (**25**) on p.92. Wagon as photo (**22**) but very dark green with dark grey floor.*

*2. Britains no.5F Farm Waggon in a corrugated cardboard box, with a label as photos (**24**) and (**25**) on p.92. Wagon as photo (**25**) but with pins to fix the horses to the shafts.*

*4. Britains no.5F Farm Waggon in a green box with a label as photos (**24**) and (**25**) on p.92. Wagon as photo (**25**) but with pins to fix the horses to the shafts and without red tips to the shafts, and with red paint on the tops of the side boards and on the ribs.*

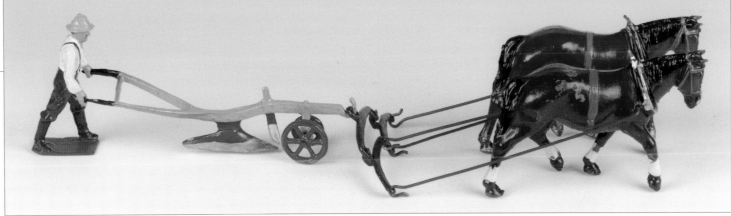

5. Britains no.6F General Purpose Plough, as photo (**36**) on p.95 but with the upper parts of the mouldboard and the middle of the coulter painted red.

6. Britains no.8F Horse Rake in a long box, with a label as photo (**44**) on p.97 but not wrapped around the end. Rake as (**44**) but with a cream frame and tines.

7. Britains no.9F Horse Roller in a plain brown card box, with label as photo (**56**) on p.99. Roller as (**56**) but with a cream frame.

8. Britains no.9F Horse Roller in a box as in (**7**). Roller as photos (**54**) and (**55**) on p.99.

9. *Britains no.12F Timber Carriage in box as photo (**72**) on p.102. Carriage as (**72**) but painted cream, and the stable lad with a green shirt.*

10. *Britains lead farm figures with some unusual painting, particularly the belts on the drivers and the stable lad with a green shirt.*

12. *Britains no.8F Horse Rake in a green box with yellow label as photo (**45**) on p.97. Rake as (**45**) but with a dark blue frame.*

11. *Britains double-sided concertina leaflet with nine folds 46 x 83 mm, extending to 46 x 448 mm, for inserting into model boxes. This example was found in the box for the tractor in (**15**). The cover read BRITAINS CLOCKWORK SERIES This folder shows only Britains clockwork series. Send for a copy of our fully illustrated Catalogue … BRITAINS LIMITED SUTHERLAND ROAD LONDON, E.17 ENGLAND. It was printed black on light brown paper and not in the very fine detail as in photo (**2**) on p.87. One side was devoted to military items and the other to farm, including the no.127F and 128F Fordson tractors and all their implements. The artwork for no.129F Timber Trailer was identical to that on the box label in photo (**151**) on p.122, which is probably the earliest box for this trailer.*

No. 59F
Tip Lorry

No. 127F
"Fordson" Heavy Duty

No. 128F
"Fordson" Rubber Tyred

No. 129F
Timber Trailer

No. 130F
Farm Trailer

No. 135F
Disc Harrow

135F & 136F
Land Harrow

No. 135F in pairs
Tandem Harrows

No. 137F
Combination Set

No. 138F
4-Furrow Plough

No. 139F
Tractor and Clockwork Unit

BRITAINS CLOCKWORK SERIES

This folder shows only Britains clockwork series

Send for a copy of our fully illustrated Catalogue

United Kingdom 6d.
Dollar Areas 25 cents
Overseas 1/- or equivalent
Dept. 10. Post Free

BRITAINS LIMITED
SUTHERLAND ROAD
LONDON, E.17 ENGLAND.
TRADE *W.Britain* MARK
REGD No.459995

13. *Britains no.40F Farm Horse and Cart in a box as in photo (84) on p.105. Cart with large red slender 16-spoke wheels as in photo (86) on p.105, but painted brown. These wheels were the same as on the Farmer's Gig in photos (79) to (83) on p.104, unlike the smaller, sturdier 12-spoke wheels in photos (84) and (85) which were the same colour as the cart.*

14. *Britains no.40F Farm Horse and Cart, as (13) but green.*

15. *Britains no.128F Fordson Major Tractor in the rare green box which allowed the tractor to be laid on its side, as for no.127F in photo (110) on p.111.*

16. Britains no.9525 Fordson New Performance Super Major Tractor in a sleeve and plinth box as in photo (**116**) on p.113. Tractor as in photo (**123**) on p.115 but with off-white mudguards and red FORDSON SUPER MAJOR transfers. This was the last to have an implement adapter, rather than the model in photo (**123**) as stated in the photo caption.

17. Britains Lilliput no.LV604 Fordson Tractor in a picture box showing a farm scene with a farmhouse, straw stacks, a cow and two sheep. The box identified the model as being made under licence by W.Horton of Middlesbrough. The tractor as photos (**303**) and (**304**) on p.158.

20. *Figure of a walking farmer in a smock, attributed to Charbens in Volume 1 (see photo (**25**) on p.213), but we are not sure. Cream with brown trousers and stick, green base.*

18. *Charbens no.505 Hay Cart and Horse, box as photo (**11**) on p.209 but with a frame around the illustration.*

21. *Charbens no.5 Farm Wagon, box as photos (**88**) and (**89**) on p.231 but covered in pale green paper with blue printing on the label; model as photo (**85**) on p.230.*

19. *Charbens no.513 Tree Wagon and Two Horses, box with small label similar to the Farm Wagon box in photo (**25**) on p.213.*

22. *Charbens no.6 Tractor, as photo (**95**) on p.233 but rare blue colour.*

23. *Nulli Secundus roller and tinplate tractor as photo (5) on p.253, both with E.W.WAGSTAFF LTD FLOREAT SALOPIA crests.*

24. *Nulli Secundus roller with the E.W.WAGSTAFF LTD FLOREAT SALOPIA crest on a red-painted wooden frame, black Bakelite roll and unpainted aluminium drawbar (160 mm).*

25. *Crescent no.1822 Caterpillar Tractor, as photo (57) on p.314 but yellow body with silver grille, red base and green control levers (driver's seat and exhaust broken off). Box with 1822 hand-written on a label on the end*
(photo by Nigel Robertshaw).

26. *Lone Star no.1293 Tractor in the early yellow and red ROADMASTERS packaging (see photo (5) on p.336). Red tractor without labels, silver metal steering wheel*
(photo by Nigel Robertshaw).

Conclusions

Collecting the information

The purpose of this project has been to locate and record as far as possible all British-made farm toy vehicles produced up to 1980. It has taken five years, from 2007 to 2012, and it has involved taking thousands of photographs, seeing many collections and seeking advice from a wide range of collectors and dealers. Everyone to whom we have turned for advice has given unstinting support. Collecting these toys is a fine hobby, and these models represent a very enterprising aspect of British industry.

The book brings together all the information we have found on horse-drawn implements and tractors and their equipment made in Britain between the 1920s and 1980. We have identified 69 manufacturers, which is a much higher figure than we had expected. Their products are illustrated with photographs taken from seventeen different collections. This is a sizeable dataset, but it may not be exhaustive, for there could still be a few small manufacturers to find. We think we have located most model variations; but for a few brands, there was such a wide range of colours that it is impossible to illustrate them all. This is the first time a study of a particular theme like farm vehicles has been undertaken right across the spectrum of British toy manufacturing, and we hope it will inspire others to follow other themes in the same way.

It should be remembered that there was also a small 'backyard' industry, mainly in the 1930s, producing fairly crude lead items without brand names. These rare pieces we have categorised as 'unidentified heavy metal', and they also have a place in the early lead toy-making tradition. For some early lead manufacturers, such as Deltoys and Glyntoys, we do have names, but we know little else about them. Some of the early farm vehicles were quite primitive. W.H. Watson, metal dealers in Brighton in the 1930s, made Sontaw farm carts out of meat tins.

For companies like Maylow we have only the brand name, and nothing else. Documentary searches for some companies have proved particularly difficult, because the industry as a whole is poorly recorded. Manufacturers' papers seldom survive. Even when a famous company like Britains cleared its London offices in 1992 everything was destroyed unless it had a sale value. Details of the quantity of models sold, or to where in the world they were sold, are usually missing.

Two books are of significant importance in tracing the history of particular companies. Marcel R. Van Cleemput's *The Great Book of Corgi 1956–1983* (1989) provides a remarkable record, model by model, of every item produced under the Corgi brand during the heady years of the 1950s and 1960s. Kenneth Brown's *Factory of Dreams – A History of Meccano Ltd, 1901–1979* (2007) was only made possible because a sharp-eyed archivist in 1979 was driving past the Binns Road Meccano factory in Liverpool when it was being cleared and saw contractors loading decades of history into skips. The archivist was not able to save all the papers, but rescued some, mainly from the period after 1964, and they bore the signs of muddy footprints where the disposal bags had split. Another useful book is David Pullen's *Pocket Guide To Britain's Model Tractors 1948-1998* (2003).

Complete runs of sales catalogues for names like Britains, Corgi and Matchbox are not difficult to consult because most, except for some of the early Britains catalogues, are relatively easy to find. For other companies like Crescent and Johillco, just a few catalogues are known, but not many. It is surprising how few companies seem to have produced catalogues. Often for smaller manufacturers the only sources are street and telephone directories. Telephone directories are online, and they can be particularly helpful in providing company addresses and the years when the offices were active. Trade, street and county directories can usually be consulted in public libraries. Advertisements in the trade journals, such as *Games & Toys*, copies of which are held at the British Library Newspaper Library at Colindale, have sometimes provided evidence for particular models, but relevant advertisements are rare.

It should be remembered that toys were mostly produced for play rather than for collecting. The manufacture of models specifically for collectors is a modern phenomenon.

Early examples

While large-scale wooden horse-drawn carts were produced in the nineteenth century, smaller-scale mass-produced lead carts began with a lead tumbrel made by Britains in 1921. Britains followed this with a very fine lead wagon pulled by a pair of horses, then ploughs and a series of other models finishing with a gig by 1928. The next manufacturer on the scene was probably Charbens with a lead horse roller in 1928.

Although by the end of the late 1930s the technology for diecasting zinc alloy toys was well enough developed, the change from lead to zinc alloy took place mainly after the Second World War. Meccano made their long-lived Dinky Massey-Harris tractor from 1948. Almost every small boy in Britain in the 1950s had a diecast Massey-Harris tractor, and with it came a range of implements including a harvest trailer, a manure spreader, a disc harrow, a triple gang mower and a hay rake. These are all still seen at every toy fair. But Britains was the dominant force in the farm toy industry, particularly in the 1960s and 1970s.

When companies like Charbens and Crescent continued to make lead items alongside their diecasting operations, the lead contamination of zinc alloys caused 'metal fatigue' on some early diecast models. Crescent tractors and their implements made in the 1950s are notorious for disintegrating for this reason. That is why the Crescent no.1815 Hayloader, for instance, is so hard to find in an undamaged condition. Lead was withdrawn gradually, but it was not until 1965 that Britains were able to announce that all their toys were lead-free.

For a short time between about 1948 and 1951 a group of companies trading under the brand names Kayron Playthings, A.V.H. Farm Toys and Olson Farminit, all in the Woking area, produced a range of farm machinery made of soldered wire, mainly for horses but also to a limited extent for tractors. While a wire hay rake made sense, a cart made just of wire had limited play value, with the contents falling out! Nevertheless, during that short period they actually had a wider range of farm implements on the market than even Britains managed at the time.

The arrival of plastics

Toy companies in the U.S.A. were making much use of plastics before the Second World War. In Britain, restrictions on the use of metals for the home toy market until 1952 provided a stimulus for the production of plastic farm vehicles, starting with the Airfix Ferguson tractor as a promotion for the real thing in the late 1940s. There were also two plastic promotional tractors by an unknown manufacturer for the David Brown Cropmaster and for the International Farmall M.

Airfix, Chad Valley, Denzil Skinner, Tri-ang, Raphael Lipkin, Lesney and Shackleton all produced large-scale models of contemporary tractors which must have involved the manufacturers of the original machines to some degree. There was inevitably a constant inter-play between toys and the real world of farming throughout the period we have covered.

It is remarkable how much copying there was of designs between toy manufacturers. It was much easier to use an existing model as a pattern than to pay a model maker to start again with a fresh master pattern of a new subject. Sometimes the copying extended to exchange of moulds, particularly with some of the early plastics manufacturers. There were exchanges between Kleeware, Tudor Rose, Thomas, Poplar, Cheerio and Paramount which all had similar products. In 1953 Herald released their first catalogue of injection-moulded plastic figures, and gradually during the 1950s the place of plastic in the making of toys became assured.

In 1932 Tri-ang had 1,500 employees, and in 1935 Meccano had a workforce of 1,600, more than most toy makers in Germany or America. At its height the British toy industry, with brands like Britains, Dinky, Matchbox and Corgi was remarkably successful with strong sales particularly to the U.S.A. and to the British Empire. The late 1950s and 1960s saw the high point of British toy manufacturing, and the Britains

and Chad Valley Fordsons represented a peak of diecasting achievement.

It is remarkable that so many realistic and accurate models were produced as toys. This was partly encouraged by toys being viewed as promotional items, while at the same time almost all the items in the book were meant to be children's playthings. This is in complete contrast with today's toys which are often generic and rather unsatisfactory as models. Children's toys have to be generic to avoid the licensing royalties imposed by manufacturers of the real vehicles. Such royalty payments were not expected in the 1950s and 1960s, when manufacturers of real tractors saw toy look-alikes as free advertising.

This remarkable flowering of accurate and pleasing models in the 1950s and 1960s was made possible by the technology of zinc diecasting and plastic injection moulding, but the high costs of tooling and production still required a mass market for the models as children's toys.

The power of nostalgia and the need to modernise

As the population became more urban, the appeal for farm toys was partly based on nostalgia for the countryside. Farm toys were always popular and continue to be so. Little boys like tractors, whether they live in towns or in the countryside. These toys helped parents to re-engage with the countryside many families had left behind. Indeed, horse-drawn items continued to sell long after they ceased to be seen on the land. As we have argued in the Britains chapter, for many families the countryside still represented the best of Britain. While in reality work on the farms was hard and often backbreaking, memories of working the fields were a strong influence on the toy-buying habits of parents. The Britains tumbrel survived with a plastic horse up until 1984, yet they were seldom seen on the land after the 1950s.

Then there was Wend-al between 1947 and 1956 making their 'unbreakable' aluminium horse-drawn items derived from models by the French company Quiralu. Their hay cutters, hay rakes, ploughs, tumbrels and harrows looked very dated from the start and would have been far more appropriate on pre-war farms.

Meccano showed no interest in horse-drawn farming in their Dinky Toys range, and instead came straight out with their Fordson in 1933 and the Massey-Harris in 1948, followed by a range of implements which found their way into millions of homes all over the world. Not until 1970 did Meccano drop this range when it was looking very tired. Then they brought out a series of David Brown and Leyland tractors, but they never managed to re-capture their early success, probably because they were unable to compete with Britains and Corgi.

Mettoy made a great impact with their Corgi Toys in very attractive packaging from 1959 with a range of tractors and implements whose realism Dinky just could not match. The Corgi items were smaller and more detailed than Britains and they focused on a more limited range of the latest farm machinery. The first combine harvester was made in 1959 by Corgi and was an exceptionally ambitious casting for its day. However, even the Corgi magic had gone by 1971. With the increased use of plastic components and less attention to detail, Corgi lost their appeal. The magic had lasted only some twelve years.

Matchbox stands by itself as an extraordinary achievement. These small high-quality low-price toys cornered a major share of the market, and because they were small and could be stored without taking up space, they were instantly collectable. Unusually, one senses that Lesney saw from the start that collectors could be a key part of the market.

We have looked closely for trends of how full-sized farm vehicles were represented in toy shops, and none clearly emerge. Almost all farm vehicles were depicted at different times. Log wagons were surprisingly popular before and after the Second World War. Britains, at the top end of the price range, were first in the field with their tumbrel, and they remained well ahead of others before the war. But, having been so innovative in the 1920s and1930s, they seem to have sat on their laurels and produced no more farm items until their Fordson tractor came out in 1948, followed by a wide range of implements for the tractor thereafter. In 1961 the Fordson Power Major came out, followed by a series of tractors and implements keeping Britains right up to date. By then the company had indeed become a household word in 'carpet farming'.

It can be seen that some manufacturers tried to have the latest items seen on the land, while others lagged complacently, or deliberately, behind. This was all a part of the shifting pattern of company success and failure.

When the end came

At the end of the 1970s, a deep domestic recession, high interest rates and unfavourable exchange rates brought down most companies which had been familiar names. But also part of the reason for the collapse of the industry was that in the 1970s manufacturers gradually lost sight of the need to produce accurate models and so lost the support of adult enthusiasts buying for themselves and for their children. Britains was better able than most in sticking with accuracy (including, for example, their constant scale) and keeping collectors interested while still producing large-volume quality toys. The position now is quite different, with costs of production and tooling in the Far East making low-volume production viable, so that accurate models can be produced for an adult market. There is, of course, a price premium, but new collectors' models are still quite affordable, even at low volumes.

A very British legacy

The British toy industry started the idea of carpet farming with the Britains Home Farm range in the 1920s, and we dominated the world in diecast model toys for three decades after the war. It is fascinating story and it provides enormous scope for creating collections, large and small.

The flowering of the British industry was over by 1980, and it is now up to collectors like ourselves to collect, study and preserve these important examples of British enterprise. We hope others will be encouraged to take this research still further.

Guide to Harnessed Horses

The purpose of this chapter is to allow easy identification of harnessed horses that have become separated from their original carts or implements. With a few exceptions, we have only included horses which are associated with farm vehicles and therefore also appear elsewhere in the book in the relevant manufacturer's chapter. There are of course many horses that were produced for other types of vehicle, such as delivery carts, stage coaches, royal coaches and 'Wild West' covered wagons. Books such as Joplin, 1993, and Joplin and Dean, 2005, are helpful to identify these. Also, this chapter only includes horses that are approximately 1:32 scale or larger. They are listed alphabetically by manufacturer. For more detailed descriptions of colours etc. and the size of the horses, please refer to the individual manufacturers' chapters.

The interpretation which toy manufacturers put on work horses and their harness varied, and often was not consistent with farming practice. Most people's concept of a carthorse is of a large Shire-type horse with lots of long hair (called feather) around its hooves. However, many manufacturers produced models of clean-legged horses, that is, without feather around their hooves. Some breeds of heavy horse, like the Suffolk Punch, are clean-legged but these were in the minority nationally. In addition, many farm horses were cross-bred with lighter breeds, and as a result could have had clean legs. So maybe clean-legged models reflected many of the horses seen on farms. On the other hand, smaller horses with thinner legs saved metal and were cheaper to produce. After all, major manufacturers like Britains and Johillco did make models of big, feather-legged Shire or 'show' horses, but not in harness and not associated with carts or other implements.

Work-horses had two types of harness – trace harness and cart harness. Trace harness, which comprised a collar and a back strap, was used when horses were simply dragging implements like ploughs or harrows. The back strap was to stop the traces sagging to the ground when slack, so avoiding the risk of a horse stepping over a trace. Trace harness was also used on the front horse when two horses were hitched in tandem to a vehicle like a timber wagon, and the traces were hooked to eyes at the front ends of the shafts. Cart harness, which comprised a collar, a cart saddle (or pad) and breeching, was used when horses were hitched between the shafts of a cart or wagon. The cart saddle carried a chain from one shaft over the back of the horse to the other shaft. With a four-wheeled wagon, this served simply to support the shafts, but with a two-wheeled cart it had to support a proportion of the weight of the cart as well. Breeching was a series of straps over the horse's rump with the thickest strap running around the rear of the horse. This made it possible for the horse either to reverse the vehicle when walked backwards, or to hold the vehicle back when going downhill.

Some toy manufacturers produced models of horses in both types of harness, and generally associated the correct type of harness with the appropriate implement or vehicle. Others, presumably for the sake of economy, or through ignorance, produced models in only one type of harness. This meant, for example, that they might have horses in cart harness pulling ploughs – something which would not have happened in real life.

Harnessed horses usually had leather flaps (called blinkers) on their bridles, at eye-level, to prevent them from being able to see behind them. The eye arrangement of horses and other potential prey animals is such that they have a wide range of vision, so that they can spot predators coming from behind. The argument was that if a horse, when pulling a cart or other implement, caught sight of something coming from behind, it might be inclined to bolt. Casting blinkers would have been difficult, and they would have been fragile and vulnerable to breakage, so they seldom appeared on model horses, however Britains did include them on their later plastic models, shown in photos (**16**) to (**19**).

Further Reading

Joplin, Norman, 1993, *The Great Book of Hollow-Cast Figures*. New Cavendish Books.

Joplin, Norman and Dean, Philip, 2005, *Hollow-Cast Civilian Toy Figures*. Schiffer Publishing Ltd.

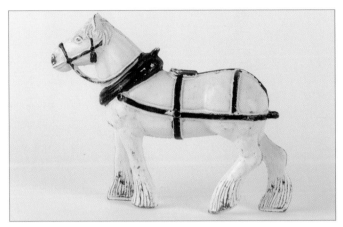

1. Benbros hollow-cast lead horse, post-war, feather-legged, in cart harness, not sold with any cart, marked BENSON ENGLAND.

2. Benbros diecast metal horse, cast in two halves, post-war, clean-legged, in cart harness, unmarked.

3. Britains hollow-cast lead horse for log wagon, pre-war, feather-legged, in cart harness, marked MADE IN ENGLAND COPYRIGHT BRITAINS LTD PROPRIETORS.

4. Britains hollow-cast lead, pre-war, light horse for Farmer's Gig, with loops on pad to take shafts, marked MADE IN ENGLAND COPYRIGHT BRITAINS LTD PROPRIETORS.

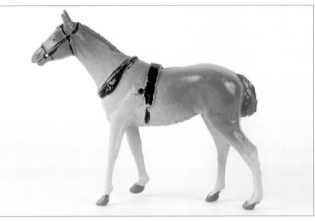

5. Britains hollow-cast lead, post-war, light horse for Farmer's Gig, as (**4**) but with holes in sides to take lugs on shafts, marked MADE IN ENGLAND COPYRIGHT BRITAINS LTD PROPRIETORS.

6. Britains hollow-cast lead, pre-war, light trotting horse for small farm cart, in cart harness, unmarked.

7. Britains hollow-cast lead, pre-war, light walking horse for small farm cart, in cart harness, unmarked

8. Britains hollow-cast lead, post-war, light trotting horse for small farm cart, in cart harness, unmarked.

9. Britains horse as (**8**) but plastic and marked ENGLAND.

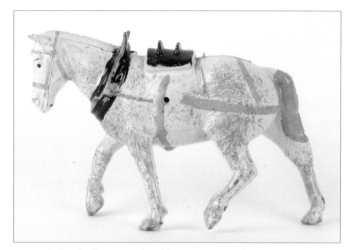

10. *Britains hollow-cast lead horse, pre- and post-war, clean-legged, in long stride and in cart harness, marked* MADE IN ENGLAND BRITAINS LTD PROPRIETORS COPYRIGHT..

11. *Britains horse, as* (**10**) *but plastic and marked* BRITAINS LTD ENGLAND.

12. *Britains hollow-cast lead horse, as* (**10**) *but in short stride.*

13. *Britains horse, as* (**12**) *but plastic and marked* BRITAINS LTD ENGLAND.

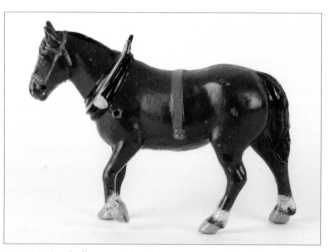

14. *Britains hollow-cast lead horse, pre- and post-war, clean-legged, in short stride and in trace harness, marked* COPYRIGHT BRITAINS PROPRIETORS.

15. *Britains hollow-cast lead horse, pre- and post-war, clean-legged, in long stride and in trace harness, marked* COPYRIGHT BRITAINS PROPRIETORS.

16. *Britains plastic, feather-legged Draught Horse, in trace harness with detachable collar. This appeared briefly in about 1971 with the tumbrel, but was usually sold on its own, marked* BRITAINS LTD, ENGLAND.

17. *Britains plastic, feather-legged Draught Horse, as* (**16**) *but with rider.*

18. *Britains plastic, feather-legged Clydesdale Cart Horse, in cart harness with detachable collar. This first appeared in the 1972 catalogue, alone or pulling the tumbrel. Marked* BRITAINS © 1972 ENGLAND.

19. *Britains plastic, feather-legged Clydesdale Cart Horse, in cart harness with detachable collar. This was a less stocky version of (18) which first appeared alone in the 1973 catalogue and subsequently with the tumbrel in 1983. Marked* BRITAINS LTD © 1972 ENGLAND.

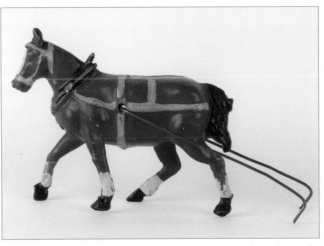

20. *Charbens hollow-cast lead horse, pre-war, clean-legged, without a saddle, shown with the wire traces for the timber wagon, unmarked.*

21. *Charbens hollow-cast lead horse, pre-war, clean-legged, in cart harness, marked* RD, *or* CHARBENS *and* RD.

22. *Charbens diecast metal horse, cast in two halves, post-war, clean-legged, in trace harness, unmarked.*

23. *Charbens diecast metal horse, cast in two halves, post-war, clean-legged, in cart harness, unmarked.*

24. *Charbens hollow-cast lead horse, post-war, feather-legged, in cart harness, unmarked.*

25. *Charbens horse, as (24) but plastic.*

26. *Charbens or Kayron/A.V.H./Olson hollow-cast lead horse, post-war, feather-legged, in partial cart harness, unmarked or with* COPYRIGHT *around the belly band. The unmarked version was probably by Kayron/A.V.H./Olson (see also 35).*

27. *Charbens or Kayron/A.V.H./Olson very large hollow-cast lead horse (length 115 mm), post-war, feather-legged, in partial cart harness, unmarked. The weight of the horse varies considerably – we think the heavier version (around 350g) was by Kayron/A.V.H./Olson, and the lighter version (around 150g) was by Charbens (see also 36).*

28. *Charbens hollow-cast lead, pre- and post-war, small light horse for pony carts, in cart harness, unmarked. There was also a post-war diecast version, unmarked (not illustrated).*

29. *Cherilea plastic horse, post-war, feather-legged in cart harness and with full tail, marked Cherilea MADE IN GT. BRITAIN. There was also a similar hollow-cast metal version, but clean-legged (not illustrated).*

30. *Crescent solid diecast metal horse for the Farmer's Market Wagon, post-war, produced in lead or zinc, in minimal harness, unmarked.*

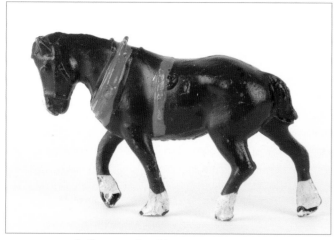

31. *Crescent hollow-cast lead horse, pre- and post-war, feather-legged, in trace harness, marked CRESCENT TOYS MADE IN ENGLAND or unmarked.*

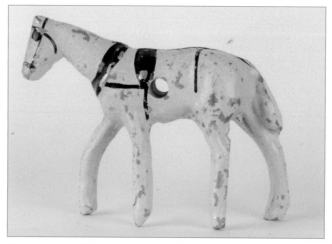

32. *Deltoys solid aluminium small light horse, clean-legged, in cart harness, unmarked. Also produced in solid lead.*

33. *Glyntoys solid lead horse, clean-legged, in cart harness, unmarked.*

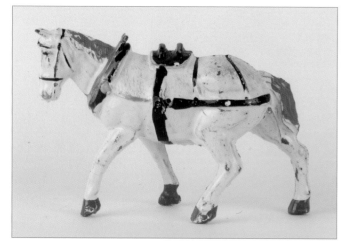

34. *Johillco hollow-cast lead horse, pre- and post-war, clean-legged, in cart harness, marked COPYRIGHT or COPYRIGHT ENGLAND or JOHILLCO ENGLAND.*

35. *Kayron/A.V.H./Olson solid aluminium horse for small carts, post-war, feather-legged, in partial cart harness, unmarked (see also **26**).*

36. *Kayron/A.V.H./Olson very large solid aluminium horse (length 115 mm) for large carts, post-war, feather-legged, in partial cart harness, unmarked (see also **27**).*

37. *Phillip Segal hollow-cast lead horse, post-war, feather-legged, in cart harness, unmarked, not sold with any cart. Phillip Segal was a manufacturer of hollow-cast lead figures in Christchurch, Hampshire, from 1946 to 1951. Another variation of this horse had all four feet flat on the ground*

38. *Pure Rubber Products (see the Benbros chapter in Volume 1) diecast horse, cast in two halves, post-war, feather-legged, in cart harness, unmarked. The photo shows the horse with light blue trim on the harness, which is original, and is a better example than the one pictured in the Benbros chapter (page 41 in Volume 1).*

39. *Taylor & Barrett (pre-war) or F.G. Taylor & Sons (post-war) hollow-cast lead horse for the water cart or brewer's dray, feather-legged, in cart harness, marked COPYRIGHT or ENGLAND FGT & SONS.*

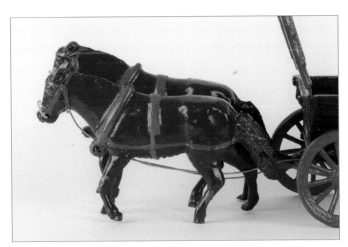

40. *Taylor & Barrett hollow-cast lead horse for the produce wagon, pre-war, clean-legged, in cart harness, marked COPYRIGHT T&B ENGLAND.*

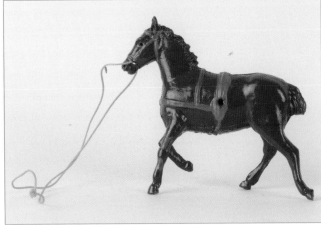

41. *F.G. Taylor & Sons hollow-cast lead, post-war, small light horse in cart harness, marked FGT & SONS ENGLAND. Also issued in plastic (not illustrated).*

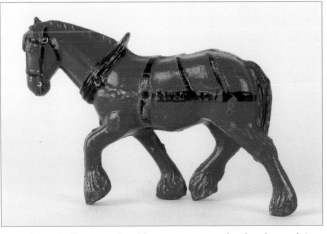

42. *Timpo hollow-cast lead horse, post-war, feather-legged, in cart harness, unmarked or marked TIMPO TOYS on one side and MADE IN ENGLAND on the other along the breeching.*

43. *Timpo hollow-cast lead horse, post-war, clean-legged, in cart harness, unmarked, made in Ireland for the Irish Jaunting Car.*

44. *Timpo hollow-cast lead horse for the timber wagon, post-war, clean-legged, without harness other than a detachable collar, unmarked or with TIMPO ENGLAND underneath. The collar is not shown on this example.*

45. *Unknown maker, hollow-cast lead horse, clean-legged, in trace harness, unmarked.*

46. *Unknown maker, very large hollow-cast lead horse (length 105 mm), feather-legged, in cart harness. Almost as big as the large Charbens horse, this rare item is carefully modelled and a superb example of hollow-casting.*

47. *Wend-al early solid aluminium horse with base, sometimes with a paper label underneath, post-war, clean-legged, in cart harness.*

48. *Wend-al later solid aluminium horse without base, post-war, clean-legged, in cart harness, unmarked.*

49. *Wend-al solid aluminium horse without base, post-war, clean-legged, in cart harness, unmarked, with a full tail, unlike the docked tail in (48).*

General Index

Note: Chapter titles are denoted by bold page numbers in this index.

Index of Models

Note: Page numbers in this index denote the start of the model's description in the text (or an entry in the Addenda).

Model	Manufacturer
Wagons	
- Unspecified	Deltoys, 342, 768
	Glyntoys, 416
	Kayron, A.V.H., Olson, 463
	Paramount, 632
	Taylor & Barrett, 667
	Tri-ang, 710
- Cattle/horse float	Johillco, 442
	Moko Farmette Series, 621
- Farm wagon	Britains, 65, 765
	Charbens, 196, 200, 766
	Crescent, 297
	Hobbies, 427
	Moko Farmette Series, 621
- Hay wagon	Kayron, A.V.H., Olson, 463
	Unknown Manufacturers, 749
- Timber wagon/Log cart (carrier)	
	Benbros, 33
	Benbros T.V. Series & Mighty Midgets, 34
	Britains, 67, 765
	Charbens, 196, 197, 202, 766
	Crescent, 295
	Forest Toys, 412
	Hobbies, 427
	Johillco, 443
	Moko Farmette Series, 621
	Timpo, 686

5. Mechanised Tractor Plant

	Matchbox, 541

6. Mobile Laboratory

	F.G. Taylor & Sons, 669

7. Traction Engines

	Brimtoy, 752
	Matchbox Yesteryear, 542
	Micromodels, 611
	Tri-ang, 710
	Tudor Rose, 735
	Unknown Manufacturers, 748

8. Tractors - generic

Generic Crawlers	
- Cast Metal	Benbros T.V. Series & Mighty Midgets, 35
	Charbens, 197
	Crescent, 296, 767
	Johillco (Tootsietoy copy), 442
- Tinplate	Marx, 517
	Mettoy, 593

Model	Manufacturer
	Tri-ang, 708, 709
	Tri-ang Minic, 711, 712
	Wells-Brimtoy, 752
- Plastic	Kleeware, 495
Generic Tractors	
- Cast Metal	Charbens, 200, 201, 202, 766
	D.C.M.T. Crescent, 331
	Kondor, 496
	Matchbox, 536, 541, 542
	Maylow, 575
	Mears, 580
	Mettoy, 593
	Tri-ang Hi-way Series, 714, 715
	Unidentified Heavy Metal, 744
- Tinplate	Childs & Smith and Wagstaff, 251
	Dinky, 372
	Kayron, A.V.H., Olson, 465, 466
	Marx, 517
	Passall, 639
	Peter Ward, 641
	Primus Engineering, 649
	Tri-ang, 708, 709
	Tri-ang Hi-way Series, 716
	Tri-ang Minic, 711, 712
	Unknown Manufacturers, 748
	Wells-Brimtoy, 752
- Plastic	Charbens, 203
	Cheerio, 248
	Coral, 254
	Fairchild, 408
	Fairylite, 410
	Marx, 518
	Matchbox, 542
	Merit, 583
	Mettoy, 593
	Paramount, 632, 633
	Poplar Plastics, 644, 645
	Rosedale, 735
	T.N. Thomas, 676
	Tri-ang Minic, 713
	Tudor Rose, 733, 734
- Vinyl	Lone Star Eaglet, 331
- Wooden	Chad Valley, 168, 169, 170
	Dragon Toys, 403
	Hercules, 423
	Hitchin Components, 425
	Hobbies, 427
	Kraftoyz, 499

Model	Manufacturer
	Nicoltoys, 630
	S.T., 661
	Taylor Toys, 674
	Tri-ang, 707
	Unknown Manufacturers, 746, 747

9. Tractors – replicas of actual tractors

B.M. Volvo 400	Corgi Juniors, 430
	Husky, 430
Caterpillar	Benbros, 34, 765
	Budgie, 160
	Early Lesney, 502
	Matchbox, 533, 538
Case David Brown 995	Dinky, 371
Cletrac/Blaw Knox	Dinky, 371
	Moko, 619
County Tractors CD50	Denzil Skinner, 351
David Brown	Merit, 582
- Cropmaster	Denzil Skinner, 351
	Promotional Model Tractors, 652
- 990	Authentic Model Books, 28
	Dinky, 371
- Selectamatic	Dinky, 371
- Trackmaster	Shackleton, 654
- 1412	Corgi, 266
Deutz DX110	Britains, 73
Ferguson	Airfix, 17
	Benbros, 32
	Benbros (Pure Rubber Products), 31, 765
	Chad Valley, 172
	Crescent, 297
	Mettoy, 593
	Mills, 613, 614
	Moko, 620
Fiat	Budgie, 160
- 880DT	Britains, 73
Field Marshall	Dinky, 370
Ford	Lone Star (Farm King), 331,767
	Matchbox, 535, 536, 540, 541, 542

10. Tractor-drawn and Tractor-mounted implements

Model	Manufacturer
	Kayron, A.V.H., Olson, 465
	Kleeware, 495
	Master Models, 526
	Maylow, 576
	Mettoy, 591
	Moko, 621
	Paramount, 633
	S.T., 662
- Tedder	Benbros, 33
	Britains, 78
- Triple-gang mower	Dinky, 369
	T.N. Thomas, 677
Miscellaneous	
- Bulldozer blade	Budgie, 160
	Corgi Juniors, 431
	Husky, 430
	Kleeware, 495
	Wells-Brimtoy, 753
- Cargoes/Loads	Britains, 76
	Corgi, 259
- Elevator/conveyor	Britains, 76, 77
	Corgi, 262, 266
	Lone Star, 332
	Micromodels, 611
- Front loader/shovel	Benbros, 33
	Britains, 72, 80
	Budgie, 160
	Cheerio, 248
	Corgi, 258, 260, 263, 265
- Hydraulic scoop	Corgi, 264
- Manure spreader	Britains, 78
	Chad Valley Wee-kin, 170
	Dinky, 369
- Muledozer	Britains, 75
- Post hole digger	Britains, 80
- Road mendng machine	
	Johillco (Tootsietoy copy), 442
- Saw attachment	Corgi, 265
- Threshing machine	Micromodels, 611
- Transplanter	Britains, 79
- Transport box/skip	Britains, 76
	Corgi, 259
- Trenching bucket	Corgi, 265

Model	Manufacturer
Trailers	
- Unspecified	Benbros T.V. Series & Mighty Midgets, 35
	Britains, 74
	Bullock Toys (a trailer but called farm cart), 165
	Chad Valley, 170
	Chad Valley Wee-kin, 170
	Kayron, A.V.H., Olson, 465
	Lone Star, 331
	Dinky, 372
	Hitchin Components, 426
	Johillco (Tootsietoy copy), 441
	Marx, 518
	Merit, 583
	Mettoy, 592, 593
	Mills, 613
	Moko, 621
	Poplar Plastics, 645
	Tri-ang Hi-way Series, 716
	Tudor Rose, 734
	Unknown Manufacturers, 747
- Animal transporter/Beast carrier/Cattle trailer	
	Britains, 78
	Corgi, 262, 263
	Corgi Juniors, 430
	Crescent, 295, 298
	Husky, 430
	Timpo, 687
- Box trailer	Crescent, 295
- Clockwork trailer	Britains, 75
- Dumper/Rear dump	Britains, 77
	Kleeware, 495
	Tri-ang Hi-way Series, 715
- Eight-wheeled trailer	Britains, 80
- Four-wheeled trailer	Charbens, 204
	Crescent, 298
	Dinky, 370
	Kayron, A.V.H., Olson, 466
	Lipkin, 508
	Paramount, 633
	Tri-ang Minic, 712
	Tudor Rose, 733
	Unknown manufacturers, 747
- Grain carrier	Lone Star, 332

Model	Manufacturer
- Halesowen trailer	Dinky, 369
	Moko, 621
- Hay trailer	Lone Star (called a hay cart), 332
	Master Models, 526
	Matchbox, 535, 542
	Mettoy, 594
	Poplar Plastics, 644
	S.T., 661
	Tri-ang Hi-way Series, 715
- High-sided trailer	Britains, 79
	T.N. Thomas, 677
- Horse box	T.N. Thomas, 677
- Platform trailer	Crescent, 295, 767
- Ricklifter trailer	Crescent, 295
- Stake truck	Marx, 517
- Timber wagon/Log carrier	
	Benbros, 33
	Benbros (Pure Rubber Products), 31, 765
	Britains, 73
	Charbens, 201, 766
	Crescent, 296
	Hercules, 423
	Maylow, 575
- Tipper/tipping trailer	Britains, 80
	Corgi, 257, 261, 266
	Corgi Juniors, 430
	Crescent, 294
	Denzil Skinner, 351
	Husky, 430
	Lipkin, 507, 508
	Matchbox, 534, 539, 540, 541
	Mettoy, 591, 592
	Paramount, 633
	S.T., 662
	T.N. Thomas, 677
	Timpo, 687
	Tri-ang, 708, 709
	Tri-ang Pedigree, 714
	Tudor Rose, 735
- Van	Marx, 517
- Vacuum trailer	Britains, 80
- Weeks farm trailer	Dinky, 37